Lecture Notes Electrical Engineering

Volume 50

Yasumichi Hasegawa

Algebraically Approximate and Noisy Realization of Discrete-Time Systems and Digital Images

Springer

Yasumichi Hasegawa
Gifu University
1-1 Yanagido
Gifu, 501-11
Japan
E-mail: yhasega@gifu-u.ac.jp

ISBN 978-3-642-26045-2 e-ISBN 978-3-642-03217-2

DOI 10.1007/978-3-642-03217-2

Softcover reprint of the hardcover 1st edition 2009

Typeset & Coverdesign: Scientific Publishing Services Pvt. Ltd., Chennai, India.

Printed in acid-free paper

9 8 7 6 5 4 3 2 1

springer.com

Preface

This monograph deals with approximation and noise cancellation of dynamical systems which include linear and nonlinear input/output relationships. It also deal with approximation and noise cancellation of two dimensional arrays. It will be of special interest to researchers, engineers and graduate students who have specialized in filtering theory and system theory and digital images. This monograph is composed of two parts. Part I and Part II will deal with approximation and noise cancellation of dynamical systems or digital images respectively. From noiseless or noisy data, reduction will be made. A method which reduces model information or noise was proposed in the reference vol. 376 in LNCIS [Hasegawa, 2008]. Using this method will allow model description to be treated as noise reduction or model reduction without having to bother, for example, with solving many partial differential equations. This monograph will propose a new and easy method which produces the same results as the method treated in the reference. As proof of its advantageous effect, this monograph provides a new law in the sense of numerical experiments. The new and easy method is executed using the algebraic calculations without solving partial differential equations. For our purpose, many actual examples of model information and noise reduction will also be provided.

Using the analysis of state space approach, the model reduction problem may have become a major theme of technology after 1966 for emphasizing efficiency in the fields of control, economy, numerical analysis, and others. Noise reduction problems in the analysis of noisy dynamical systems may have become a major theme of technology after 1974 for emphasizing efficiency in control. However, the subjects of these researches have been mainly concentrated in linear systems.

In common model reduction of linear systems in use today, a singular value decomposition of a Hankel matrix is used to find a reduced order model. However, the existence of the conditions of the reduced order model are derived without evaluation of the resultant model. In the common typical noise reduction of linear systems in use today, the order and parameters of the systems are determined by minimizing information criterion.

Algebraically approximate and noisy realization problems for input/output relations can be roughly stated as follows:

A. The algebraically approximate realization problem.
For any input/output map, find, using only algebraic calculations, one mathematical model such that it is similar to the input/output map and has a lower dimension than the given minimal state space of a dynamical system which has the same behavior to the input/output map.

B. The algebraically noisy realization problem.
For any input/output map which includes noises in output, find, using only algebraic calculations, one mathematical model which has the same input/output map.

Based on these parameters, we have been able to demonstrate that our new method for nonlinear dynamical systems, fully discussed in this monograph, is effective. It is worth remembering that the development of approximate and noisy realization has been strongly stimulated by linear system theory and is well-connected to related mathematics, such as for example, matrix theory or mathematical programming. However, such development of nonlinear dynamical systems has not occurred yet because there have been no suitable mathematical methods for nonlinear systems.

In this monograph, in relation to the approximate quantity of noiseless data as being the noisy part of noisy data, we have identified the approximate realization problem as a noisy realization problem in the sense of how to unify the treatment of these problems. Our method intensively takes a positive attitude toward using computers. We will introduce a new method called the Algebraically Constrained Least Square method, for brevity, Algebraic CLS method. This method can be unified to solve both an approximate and a noisy realization problem. In the reference [Hasegawa, 2008], a CLS method was proposed for solving approximate and noisy realization problems. The method is called the analytically Constrained Least Square method, for brevity, Analytic CLS method. The analytic CLS method demands the solution of partial differential equations. Therefore, the method is too cubersome.
The proposed method seeks to find, using only algebraic calculations, the coefficients of a linear combination without the notion of orthogonal projection, and it can be applied to both approximate and noisy realization problems, i.e., in the sense of a unified manner for both approximate and noisy realization problems, a new method will be proposed that provides effective results. As has already been mentioned, common approximate and noisy realization problems have been mainly discussed via linear systems. On the other hand, there have been few developments regarding nonlinear systems. Our recent monograph, *Realization Theory of Discrete-Time Dynamical Systems* (T. Matsuo and Y. Hasegawa, Lecture Notes in Control and Information

Science, Vol. 296, Springer, 2003), indicated that any input/output map of nonlinear dynamical systems can be characterized by the Hankel matrix or the Input/output matrix. The monograph also demonstrated that obtaining a dynamical system which describes a given input/output map is equal to determining the rank of the matrix of the input/output map and the coefficients of a linear combination of column vectors in the matrix. This new insight leads to the ability of discussing fruitful approximate and noisy realization problems.

Part II deals with approximate and noisy realization problems of digital images, especially two dimensional arrays. In the reference [Hasegawa and Suzuku, 2006], a realization problem of two dimensional arrays was established over a field. In the case of real numbers, we can easily discuss approximate and noisy realization problems of two dimensional arrays in the same manner as in Part I. Therefore, we want to omit the problems over real number fields.

Also in Part II, we discuss approximate and noisy realization problems over a finite field, i.e., the quotient field modulo of the prime number. Because we cannot introduce the norm for a finite field, our problems for two dimensional arrays can be roughly stated as follows:

A. The algebraically approximate realization problem.
For any two dimensional array, find, using only algebraic calculations, one mathematical model such that it is similar to the two dimensional array and has a lower dimension than the given minimal state space of a mathematical model which has the same behavior to the array.

B. The algebraically noisy realization problem.
For any two dimensional array which includes noises in output, find, using only algebraic calculations, one mathematical model which has the same two dimensional array.

The problems will be treated as a sort of non-linear integer programming. We will propose a new method which is suitable for our problems. The method will be called a non-linear integer programming for digital images.

We wish to acknowledge Professor Tsuyoshi Matsuo, who established the foundation for realization theory of continuous and discrete-time dynamical systems, and who taught me much regarding realization theory for discrete-time non-linear systems. He would have been an author of this monograph, but in April sixteen years ago he sadly passed away. We gratefully consider him one of the authors of this manuscript in spirit.

We also wish to thank Professor R. E. Kalman for his suggestions. He stimulated us to research these realization problems directly as well as through his works. We also thank Professor Gary B. White for making the first manuscript into a more readable and elegant one.

On the occasion of becoming 65 Yasumichi Hasegawa

Contents

Part I
Algebraically Approximate and Noisy Realization of Discrete-Time Dynamical Systems

Chapter 1
Introduction

The algebraically approximate and noisy realization problems for discrete-time dynamical systems that we will state here can be stated as the following two problems Ⓐ and Ⓑ. The following notations are used in the problem description. I/O is the set of input/output maps which are partial data of given observed objects. DS is the category of mathematical models with a behavior which is an input/output relationship.

Ⓐ Algebraically approximate realization problem.
For any input/output map $a \in I/O$ without noise, find, using only algebraic calculations, one mathematical model $\sigma \in DS$ such that its behavior is approximately equal to the input/output map a and has a lower dimension than the given minimal state space of a dynamical system which has the behavior a.

Ⓑ Algebraically noisy realization problem.
For any input/output map $a \in I/O$ which includes noises in output, find, using only algebraic calculations, one mathematical model $\sigma \in DS$ which has the same behavior as the given a.

In 1960, R. E. Kalman stated that the realization problem for dynamical systems, i.e., systems with input and output mechanisms, can also be established as the realization theory for linear systems in an algebraic sense. Based on his ideas, we solved a realization problem for a very wide class of discrete-time nonlinear systems [Matsuo and Hasegawa, 2003]. In the monograph, we derived fundamental results of realization theory for nonlinear dynamical systems in an algebraic sense. In particular, by proposing some nonlinear dynamical systems, we could obtain positive results when dynamical systems were characterized by their finite dimensionality through the introduction of a Hankel matrix or input/output matrix suited for these dynamical systems. On the basis of these ideas, we have discussed approximate and noisy realization problems for discrete-time dynamical systems which include any

Y. Hasegawa: Algebra. Approx. & Noisy Reali. of Discrete-Time Sys., LNEE 50, pp. 3–10.
springerlink.com

nonlinear systems in an analytic sense [Hasegawa, 2008]. Developmently, we will propose approximate and noisy realization problems for discrete-time dynamical systems which include any nonlinear systems in an algebraic sense.

As we have said in the reference [Hasegawa, 2008], discrete-time dynamical systems have become ever more important synchronously with the development of computers and the establishment of mathematical programming. Discrete-time linear systems have provided material for many fruitful contributions, as well as for discrete-time nonlinear dynamical systems. R. E. Kalman developed his linear system theory by using algebraic theory. Since then, algebraic theory has provided significant resources for the development of nonlinear dynamical system theory [Matsuo and Hasegawa, 2003] as well. Our algebraic processing methods for approximate and noisy realization of discrete-time nonlinear dynamical systems are the first ones to have been employed using the analytic methods in the reference [Hasegawa, 2008].

Our approach to an approximate problem for nonlinear systems can be stated by using the fact that any input/output relation with causality conditions can be expressed in a sort of Hankel matrix. These relations are discussed in the reference [Matsuo and Hasegawa, 2003]. On this basis, a method regarding an approximate realization problem for discrete-time dynamical systems including nonlinear input/output maps has been proposed, and it can be solved using a Hankel matrix or Input/output matrix. Using the matrix norm for determining the dimension of a given system, we have also proposed a method for determining coefficients for a linear combination of the columns of the Hankel matrix or Input/output matrix by obtaining a minimum value of a rational polynomial in multi variables, i.e., solving the partial differential equations of the polynomial. The method may be called the analytically Constrained Least Square method, for brevity, the analytic CLS method. The approximate realization problems were first conducted on both linear and nonlinear dynamical systems. In this monograph, we will introduce a new method for our approximate realization problems by comparing it with the results of the analytic CLS method. The new method may be called the algebraically Constrained Least Square method, for brevity, the algebraic CLS method.

As a method of modeling for dynamical systems with noise, one typical method is based on AIC (Akaike's Information Criterion). The method has been proposed based on the idea of statistical and probabilistic theory. However, its usage is also restricted to linear systems.

For both linear and nonlinear dynamical systems with noise, we have proposed a noisy realization problem while presenting a similar method to solve the problem of approximate realization, i.e., we have used the matrix norm and the analytically Constrained Least Square method for our purpose [Hasegawa, 2008]. This noisy realization method was first conducted for nonlinear input/output relations with a noisy case. In this monograph, we will also introduce a new method for our noisy realization problems by comparing it with the results of analytically Constrained Least Square method. The new method also may be called the algebraically Constrained Least Square method, for brevity, the algebraic CLS method.

In current approximate realization of discrete-time linear systems, which is different from our method, once a Hankel matrix expressed in z-transformation is given with its singular value decomposition, a reduction model is constructed by proving the existence of the conditions of the reconstructed Hankel matrix from the decomposition. The model reduction method can only be executed without any evaluations which means we don't know how much information has been removed from the original matrix.

The common approximate realization method is put into execution under prior conditions which means that the ideal input/output value is given, signifying that the infinite sequence of the impulse response is already known.

Since this condition never occurs, the method is clearly not practical.

In current noisy realization of discrete-time linear systems which is different from our method, the problem is discussed using knowledge of statistics and probability distribution function in the typical method, AIC. Therefore, this method can only be applied to linear systems.

We wish to stress that in almost all cases, the methods used to treat approximate and noisy realizations need finite data, demonstrating that our situation for our problems is practical.

In this monograph, regarding the approximate quantity of noiseless data as the noisy part of noisy data, we can identify the approximate realization problem as the noisy realization problem in the sense of how to unify the treatment of these problems. Our new method intentionally takes a positive attitude toward using computers because we employ algebraic operations. We will introduce a new method called the algebraically Constrained Least Square method. This method can be unified to solve both an approximate and a noisy realization problem.

The proposed method seeks to find the coefficients of a linear combination without the notion of orthogonal projection which can be applied to both approximate and noisy realization problems. In other words, in the sense of a unified manner for both approximate and noisy realization problems, a new method called the algebraically Constrained Least Square method will be proposed with effective results by comparing it with the analytically Constrained Least Square method.

The general plan of this monograph is to propose the method of how to solve approximate and noisy realization problems in a unified way using an algebraic operation.

In our monograph [Matsuo and Hasegawa, 2003], we proposed the following realization problems A, B and C of nonlinear dynamical systems and solved them by constructing a new and very wide inclusion relation for various nonlinear dynamical systems:

A. The existence and uniqueness in an algebraic sense.
For any input/output map $a \in I/O$, find at least one dynamical system $\sigma \in CD$ such that its behavior is a. Also, prove that any two dynamical systems that have the same behavior a are isomorphic in the sense of the category CD.

B. The finite dimensionality of the dynamical systems.
Clarify when a dynamical system $\sigma \in CD$ is finite dimensional. Because finite dimensional dynamical systems are actually appearing by linear (or nonlinear) circuits or computer programs, it is very important that these conditions become clear.

C. Deriving the dynamical systems from finite data.
Partial realization problems seek to find the minimal dynamical system fit to a given finite input/output's data and to clarify when the minimal dynamical systems are isomorphic.

In our monograph, we introduced General Dynamical Systems, Linear Representation Systems, Affine Dynamical Systems, Pseudo Linear Systems, Almost Linear Systems and So-called Linear Systems.

Their proposed inclusion relations and usual dynamical systems are shown in the figure on the next page, where arrows imply that the above system includes the below system as a subclass.

We will discuss their approximate and noisy realizations except for General Dynamical Systems.

Our realization theory stated here provides a new basis for treating both approximate and noisy realizations of each system. Therefore, after two initial chapters regarding basic matters, this monograph is organized into balanced sections of one chapter for each dynamical system.[1mm]

Each Chapter from 3 to 8 deals with our problem for one dynamical system. The Chapter number and the name of the dynamical system treated in the Chapter are given as follows:

Chapter 3	Linear systems
Chapter 4	So-called linear systems
Chapter 5	Almost linear systems
Chapter 6	Pseudo linear systems
Chapter 7	Affine dynamical systems
Chapter 8	Linear representation systems

Let us preview each chapter in somewhat more detail.

In Chapter 2, we will describe input/output relations and the method used in this monograph. Using the method, we will provide a clear connection between signal and approximated signal and we will provide a clear distinction between signal and noise. The method proposed in this chapter is called the algebraically Constrained Least Square method abbreviated to the algebraic CLS method.

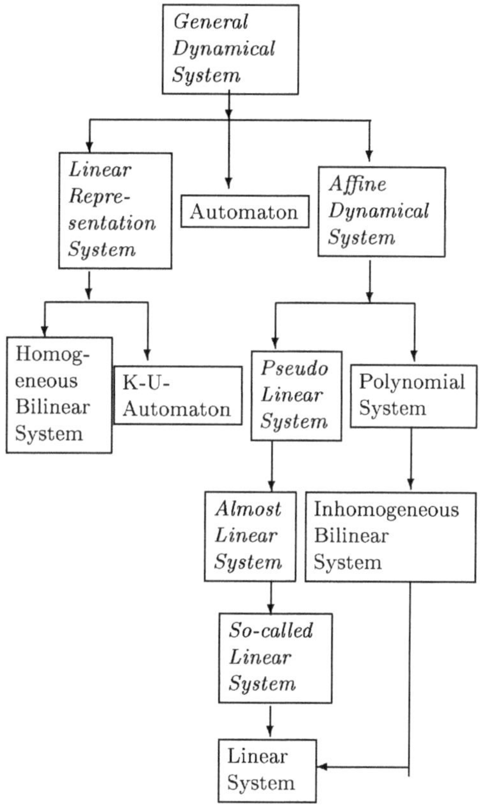

In Chapter 3, we will treat approximate and noisy realization problems for linear systems. Firstly, we will state the facts and establish facts regarding linear systems which are needed for our discussion.

Both problems are solved first by decision of order and next by determination of the parameters of the given linear systems. The decision of order will be performed using singular value decomposition. The determination of parameters will be performed using our algebraically ConstrainedLleast Square method. Singular value decomposition can be used for reduction of information by cutting a part of space for lower eigenvalues. However, normally the degree of information loss is not made clear in proportion to the part of lower eigenvalues. Hence, in this chapter, good approximation and noisy processing will be tested with the ratio of the matrix norm. After selecting a very small ratio, it will be shown that the algebraically Constrained Least Square method can produce a good linear system for a given linear system.

In the noisy case, there is a method which can find good systems, which is called AIC, that is, Akaike's Information Criterion. We will compare our method with the AIC method using some examples. Then we will compare the algebraic CLS method with the analytic CLS method discussed in the reference [Hasegawa, 2008]. We will conclude with the following:

For both approximate and noisy realization problems, we have discovered in our numerical experiments, a new law which says that the linear systems obtained by the algebraic CLS method are the same as the linear systems obtained by the analytic CLS method.

In Chapter 4, we will discuss, algebraically, approximate and noisy realization problems for so-called linear systems which are nonlinear. Such a treatment for problems of nonlinear systems appears for the first time ever in this chapter. Since the characteristic of the systems can be represented by two modified impulse responses, our purpose is to find these two modified impulse responses from the approximated or noisy input/output data. Previously described facts obtained in the monograph [Matsuo and Hasegawa, 2003] and new facts needed in this chapter will be discussed first.

Firstly, we will discuss the algebraically approximate realization problem, and then we will prove that our proposed method is effective by illustrating some examples.

Secondly, discussing the algebraically noisy realization problem, we will ascertain that our method has considerable merit. Then, we will compare our algebraic CLS method with the analytic CLS method discussed in the reference [Hasegawa, 2008]. We will conclude with the following:

For both algebraically approximate and noisy realization problems, we have discovered a new law which says that the so-called linear systems obtained by the algebraic CLS method are the same as the so-called linear systems obtained by the analytic CLS method in the numerical experiments.

In Chapter 5, we will discuss, algebraically, approximate and noisy realization problems for almost linear systems, which are nonlinear systems. Since the characteristic of almost linear systems can be represented by two modified impulse responses, our purpose is to find these two modified impulse responses from the approximated or noisy input/output data of the systems. Previously described facts obtained in the monograph [Matsuo and Hasegawa, 2003] also will be discussed first in this chapter.

Firstly, we will discuss the algebraically approximate realization problem, and we will show that our proposed method is effective by illustrating some examples.

Secondly, discussing the algebraically noisy realization problem, some examples will ascertain that our method has considerable merit.

For both algebraically approximate and noisy realization problems, we have discovered a new law which says that the almost linear systems obtained by the algebraic CLS method are the same as the almost linear systems obtained by the analytic CLS method in the numerical experiments.

In Chapter 6, we will discuss, algebraically, approximate and noisy realization problems for pseudo linear systems, which are nonlinear systems. Since the characteristic of pseudo linear systems can be represented by some modified impulse responses, our purpose is to find these modified impulse responses from the approximate or noisy input/output data of the given pseudo linear systems. Previously described facts obtained in the monograph [Matsuo and

Hasegawa, 2003] and new facts needed in our discussion will be discussed first in this chapter.

Firstly, we will discuss the algebraically approximate realization problem and secondly discuss the algebraically noisy realization problem. Some examples will illustrate that our proposed method is effective.

For both algebraically approximate and noisy realization problems, we have discovered a new law which says that the pseudo linear systems obtained by the algebraic CLS method are the same as the pseudo linear systems obtained by the analytic CLS method in the numerical experiments.

In Chapter 7, we will discuss, algebraically, approximate and noisy realization problems for affine dynamical systems, which are general nonlinear systems and include inhomogeneous bilinear systems as a subclass. Since the characteristic of affine dynamical systems can be represented by an input response map, our purpose is to find this input response map from the approximate or noisy input/output data of the affine dynamical systems. Previously described facts obtained in the monograph [Matsuo and Hasegawa, 2003] and new facts needed in our discussion will be discussed first in this chapter.

We will discuss the algebraically approximate realization problem and then discuss the algebraically noisy realization problem. Affine dynamical systems are general nonlinear systems, nevertheless some examples will illustrate that our proposed method is effective for both cases.

For both algebraically approximate and noisy realization problems, we have discovered a new law which says that the affine dynamical systems obtained by the algebraic CLS method are the same as the affine dynamical systems obtained by the analytic CLS method in the numerical experiments.

In Chapter 8, we will discuss, algebraically, approximate and noisy realization problems for linear representation systems, which are general nonlinear systems and include homogeneous bilinear systems as a subclass. Since the characteristic of linear representation systems can be completely represented by an input response map, our purpose is to find its input response map from the approximate or noisy input/output data of the linear representation systems. Previously described facts obtained in the monograph [Matsuo and Hasegawa, 2003] will be illustrated first in this chapter.

We will discuss the algebraically approximate realization problem and then discuss the algebraically noisy realization problem. The systems are general nonlinear systems, nevertheless some examples will illustrate that our proposed method is effective.

For both algebraically approximate and noisy realization problems, we have discovered a new law which says that the linear representation systems obtained by the algebraic CLS method is the same as the linear representation systems obtained by the analytic CLS method in the numerical experiments.

In this Part, we could obtain the following statement for the algebraically approximate and noisy realization problem of each dynamical system:

Statement
The systems obtained by the analytic CLS method are the same as the system obtained by the analytic CLS method in the numerical experiment.

Since it is considered that the statement cannot be proven using current mathematics, the statement is proposed as a new law.

Notations

$\boldsymbol{R}$: the real number field.
N : the set of non-negative integers.
$F(X, Y)$: the set of all functions from X to Y.
$L(X, Y)$: the set of all linear maps from X to Y.
$L(X)$: the set of all linear maps from X to X.
$\boldsymbol{R}^n$: an n-dimensional coordinate space over the field $\boldsymbol{R}$.
$\boldsymbol{R}^{m \times n}$: the set of all $m \times n$-matrices.
$\mathrm{im}\, f$: the image of a map f.
$\ker f$: the kernel of a map f.
$\ll S \gg$: the smallest linear space which contains a set S.
$\mathrm{Gr}\, T$: the graph of a relation T.
$\mathrm{dom}\, T$: the domain of a relation T.

Acronyms

AIC : an information criterion (equivalently, Akaike's information criterion)
CLS : constrained least square
algebraic CLS : algebraic constrained least square
analytic CLS : analytic constrained least square

Chapter 2
Input/Output Map and Additive Noises

To obtain concrete results, we will consider a case of dynamical systems with an input/output mechanism surrounded by free noise or noise.

2.1 Input Response Maps (Input/Output Maps with Causality)

We will consider a notational method for input/output relations of an object to be observed or to be controlled in a discrete-time case, i.e., a black-box to which any element of the concatenation monoid U^* can be applied and whose output values are in a set of output values, where U^* is the free monoid over the input value's set U. Sometimes, Ω may be used in place of U^*, namely $\Omega = U^*$ always holds. Y is the set of output values. The representation theorems for any input/output map with causality have been given by [Matsuo and Hasegawa 2003]. The theorems can be stated as Lemmas (2.1), (2.5) and (2.8).

Lemma 2.1. *Any input/output relation with causality can be represented as $a \in F(U^*, Y)$. Then, any $a \in F(U^*, Y)$ can be represented as the following equation:*
$\hat{\gamma}(|\omega|) = a(\omega) \in Y$, where $\hat{\gamma}(|\omega|)$ denotes an output value at the time $|\omega|$ for an input ω to have been ended to apply, where $|\omega|$ is the length of the input ω.

Definition 2.2. An element a of $F(U^*, Y)$ is said to be an input response map.

For the convenience of our discussions, we have utilized some kinds of input response maps from [Matsuo and Hasegawa, 2003].

Definition 2.3. If an input response map $a \in F(U^*, Y)$ satisfies the following time-invariant condition, then a is said to be a time-invariant input response map.

Y. Hasegawa: Algebra. Approx. & Noisy Reali. of Discrete-Time Sys., LNEE 50, pp. 11–20.
springerlink.com

Time-invariant condition : $a(\omega_1|\omega) - a(\omega_1) = a(\bar{\omega}_1|\omega) - a(\bar{\omega}_1)$ for any $\omega \in U^*$, and $\omega_1, \bar{\omega}_1 \in U^*$ such that $|\omega_1| = |\bar{\omega}_1|$.

Definition 2.4. For any time-invariant input response map $a \in F(U^*, Y)$, a function $I_a : U \to F(U^*, Y); u \mapsto I_a(u)[; t \mapsto a(u^t) - a(u^{t-1})]$ is said to be a modified impulse response of a, where u^t is given by $u^t(i) = u$ for $i(1 \leq i \leq t)$.

Lemma 2.5. *Representation Theorem*
For any time-invariant input response map $a \in F(U^*, Y)$, there exist uniquely modified impulse responses represented by the following equation. This correspondence is bijective.
$a(\omega) = a(1) + \sum_{j=1}^{|\omega|}\{(I_a(\omega(j)))(|\omega| - j)\}$.

In our case, we consider input/output maps $a \in F(U^*, Y)$ which satisfy the following time-invariant condition and affinity condition. They are said to be time-invariant, affine input response maps, where U is a linear space in this case. We may treat the case where multi-inputs are fed, i.e., $U = \boldsymbol{R}^m$, but conveniently, we will discuss a case where one-input is fed, i.e., $U = \boldsymbol{R}$. And Y is a linear space over the real number field $\boldsymbol{R}$.

Definition 2.6. If an input response map a satisfies the following time-invariant and affinity condition, then a is said to be a time-invariant, affine input response map.
Time-invariant condition:
$a(\omega_1|\omega) - a(\omega_1) = a(\bar{\omega}_1|\omega) - a(\bar{\omega}_1)$
for any ω, ω_1, $\bar{\omega}_1$ such that $|\omega_1| = |\bar{\omega}_1|$.
Affinity condition:
$a : U^* \to Y$ is an affine map, i.e.,
$a(\omega + \bar{\omega}) + a(0^{|\omega|}) = a(\omega) + a(\bar{\omega})$
$a(\lambda\omega) = \lambda a(\omega) + (1 - \lambda)a(0^{|\omega|})$
for any $\omega, \bar{\omega} \in U^*$, $|\omega| = |\bar{\omega}|$ and $\lambda \in K$.

Definition 2.7. For any time-invariant, affine input response map $a \in F(U^*, Y)$, a function $I_a : \{0, 1\} \to F(N, Y); u \mapsto Ia(u)[; t \mapsto a(u^t) - a(u^{t-1})]$ is said to be a modified impulse response of a.

Lemma 2.8. *Representation Theorem*
For any time-invariant, affine input response map $a \in F(U^*, Y)$, there exist uniquely modified impulse responses represented by the following equation. This correspondence is bijective. $a(\omega) = a(1) + \sum_{j=1}^{|\omega|}(\omega(j))(I_a(1)(|\omega| - j + 1)) + (1 - \omega(j))(I_a(0)(|\omega| - j + 1))$ *for any* $\omega \in U^*$.

The problem of approximation and noisy realization for input/output relation with causality are roughly stated as follows:

Problem 2.9. Problem statement for approximate realization
For any given data of the input/output map, find an input response map which is suitable in the sense of approximation, namely, a dynamical system with approximate behavior for the given input response map.

Problem 2.10. Problem statement for noisy realization
For any given data of the input/output map with noises, find an input response map which is suitable in the sense of noisy data, namely, a dynamical system with possibility for the same behavior to the given input response map.

2.2 Analysis for Algebraically Approximate and Noisy Realization

According to our reference [Matsuo and Hasegawa 2003], any input response maps could be combined into a sort of Hankel matrix or Input/output matrix which are respectively suitable for solving our problems.

Here, we will mention the norm of the matrix which is needed to discuss our problems.

First, we will list the facts on singular value decomposition from the reference [R. A. Horn and C. A. Johnson, 1985].

Lemma 2.11. *Let $A \in R^{m\times n}$ with $m \leq n$ and rank $A = k \leq m$. There exists a unitary matrix $X \in R^{m\times m}$, a diagonal matrix $\Lambda \in R^{m\times m}$ with nonnegative diagonal entries $\lambda_1 \geq \lambda_2 \geq \cdots \geq \lambda_k \geq \lambda_{k+1} = \cdots = \lambda_m = 0$, and a matrix $Y \in R^{m\times n}$ with orthogonal rows such that $A = X\Lambda Y^T$. The matrix $\Lambda = diag(\lambda_1, \cdots, \lambda_m)$ is always uniquely determined and $\{\lambda_1^2, \cdots, \lambda_m^2\}$ are eigenvalues of AA^T.*

Lemma 2.12. *Let $A \in R^{m\times n}$ and $A = X\Lambda Y^T$ be the same in lemma (2.11). Let $X \in R^{m\times m}$ and $Y \in R^{m\times n}$ be $X = [\mathbf{x}_1, \mathbf{x}_2, \cdots, \mathbf{x}_n]$ and $Y = [\mathbf{y}_1, \mathbf{y}_2, \cdots, \mathbf{y}_m]$ respectively. Then A can be expressed as follows:*
$A = \lambda_1 \mathbf{x}_1 \mathbf{y}_1^T + \lambda_2 \mathbf{x}_2 \mathbf{y}_2^T + \cdots + \lambda_m \mathbf{x}_m \mathbf{y}_m^T$.
And the following holds:
1) $A\mathbf{y}_j = \lambda_j \mathbf{x}_j$ and $A^T \mathbf{x}_j = \lambda_j \mathbf{y}_j$ hold for $j(j = 1, \cdots, m)$.
2) $A^T A\mathbf{y}_j = \lambda_j^2 \mathbf{y}_j$ and $A\mathbf{y}_j = \lambda_j \mathbf{x}_j$ hold for $j(j = 1, \cdots, m)$.
3) $AA^T \mathbf{x}_j = \lambda_j^2 \mathbf{x}_j$ and $A^T \mathbf{x}_j = \lambda_j \mathbf{y}_j$ hold for $j(j = 1, \cdots, m)$.
4) $\mathbf{x}_i^T \cdot \mathbf{x}_i = \mathbf{y}_j^T \cdot \mathbf{y}_j = 1$ for i $(1 \leq i \leq m,\ 1 \leq j \leq n)$.

Remark: $A = \lambda_1 \mathbf{x}_1 \mathbf{y}_1^T + \lambda_2 \mathbf{x}_2 \mathbf{y}_2^T + \cdots + \lambda_m \mathbf{x}_m \mathbf{y}_m^T$ in lemma (2.12) is called the singular value decomposition.

Let $P_i := \mathbf{x}_i \mathbf{y}_i^T$. Then $P_i^T P_j = P_i P_j^T = \mathbf{0}$ holds for $i \neq j$.

Next, we will discuss the norm of the matrix.

Lemma 2.13. *Let $\| A \|$ be the norm of the matrix $A \in R^{m\times n}$, i.e., $\| A \| = \max_{\|x\|=1} \| Ax \|$. Then $\| A \| = \sqrt{\mu_1(A)}$ holds, where $\mu_1(A)$ is the maximum value of singular values for A.*

[proof] Let a scalar valued function $f(\mathbf{x}, \lambda)$ be $f(\mathbf{x}, \lambda) = \mathbf{x}^T A^T A\mathbf{x} + \lambda[\mathbf{x}^T \mathbf{x} - 1]$ for a Lagrange multiplier $\lambda \in K$. Let the small increment $f(\mathbf{x} + \delta\mathbf{x}, \lambda + \delta\lambda)$

from $f(\mathbf{x}, \lambda)$ be $f(\mathbf{x}+\delta\mathbf{x}, \lambda+\delta\lambda) = (\mathbf{x}+\delta\mathbf{x})^T A^T A(\mathbf{x}+\delta\mathbf{x}) + (\lambda+\delta\lambda)((\mathbf{x}+\delta\mathbf{x})^T(\mathbf{x}+\delta\mathbf{x}) - 1)$. From the equation $f(\mathbf{x}+\delta\mathbf{x}, \lambda+\delta\lambda) - f(\mathbf{x}, \lambda) = 0$, we obtain the equation $A^T A\mathbf{x} = -\lambda\mathbf{x}$. Because of $\mathbf{x}^T A^T A\mathbf{x} = -\lambda\mathbf{x}^T\mathbf{x}$, $\| A\mathbf{x} \|^2 = \mathbf{x}^T A^T A\mathbf{x}$ and $\| \mathbf{x} \|^2 = \mathbf{x}^T\mathbf{x}$, $\| A \| = \max_{\|x\|=1} \| Ax \| = \max_{A^T Ax = -\lambda x} \sqrt{-\lambda}$ holds.

Remark : For the singular values decomposition $A = \lambda_1\mathbf{x}_1\mathbf{y}_1^T + \lambda_2\mathbf{x}_2\mathbf{y}_2^T + \cdots + \lambda_m\mathbf{x}_m\mathbf{y}_m^T$ in lemma (2.12) and $\bar{m} \leq m$, let $B = \lambda_1\mathbf{x}_1\mathbf{y}_1^T + \lambda_2\mathbf{x}_2\mathbf{y}_2^T + \cdots + \lambda_{\bar{m}}\mathbf{x}_{\bar{m}}\mathbf{y}_{\bar{m}}^T$. Then $\| A - B \| = \sqrt{\lambda_{\bar{m}+1}}$ holds.

2.3 Measurement Data with Noise

In this section, we will discuss the case where noises are added to dynamical systems with an input/output mechanism.

For observed values $\gamma(t) \in K^p$ of a time series, a p-dimensional signal $\hat{\gamma}(t) \in K^p$ and an additive noise $\bar{\gamma}(t) \in K^p$ can be considered as the following equation:

$$\gamma(t) = \hat{\gamma}(t) + \bar{\gamma}(t), \text{ where } t \in N.$$

In the sense of a noisy case, the data processing problem can be stated roughly as follows:

For any given data $\{\gamma(t) : t \leq T \text{ for some } T \in N\}$, find the signal $\{\hat{\gamma}(t) : t \leq T\}$ which is the output of a dynamical system.

2.4 Analyses for Approximate and Noisy Data

Let data with noise be a set $\{\mathbf{x}_t \in K^n : t = 1, 2, \cdots, s \in N\}$. Then $\mathbf{x}_t$ is represented by the equation $\mathbf{x}_i = \hat{\mathbf{x}}_i + \bar{\mathbf{x}}_i$, where $\hat{\mathbf{x}}_i$ is exact data and $\bar{\mathbf{x}}_i$ is noise.

Let $\mathbf{x}$ be $\mathbf{x} := [\mathbf{x}_1^T, \mathbf{x}_2^T, \cdots, \mathbf{x}_s^T]^T$, $\hat{\mathbf{x}} := [\hat{\mathbf{x}}_1^T, \hat{\mathbf{x}}_2^T, \cdots, \hat{\mathbf{x}}_s^T]^T$ and $\bar{\mathbf{x}} := [\bar{\mathbf{x}}_1^T, \bar{\mathbf{x}}_2^T, \cdots, \bar{\mathbf{x}}_s^T]^T \in K^{ns\times 1}$ for $\mathbf{x}_i \in K^n$. Then $\mathbf{x} = \hat{\mathbf{x}} + \bar{\mathbf{x}}$ holds.

Let $\mathbf{H}$ be $\mathbf{H} := [\mathbf{x}_1, \mathbf{x}_2, \cdots, \mathbf{x}_n]$, $\mathbf{X} := \mathbf{H^T H}$, $\hat{\mathbf{X}} := \hat{\mathbf{H}}^\mathbf{T}\hat{\mathbf{H}}$ and $\bar{\mathbf{X}} := \bar{\mathbf{H}}^\mathbf{T}\bar{\mathbf{H}}$ for $\hat{\mathbf{H}} := [\hat{\mathbf{x}}_1, \hat{\mathbf{x}}_2, \cdots, \hat{\mathbf{x}}_n]$ and $\bar{\mathbf{H}} := [\bar{\mathbf{x}}_1, \bar{\mathbf{x}}_2, \cdots, \bar{\mathbf{x}}_n]$.

Then $\mathbf{X} = \hat{\mathbf{X}} + \bar{\mathbf{X}}$ holds because of $\hat{\mathbf{x}}_i \perp \bar{\mathbf{x}}_j$.

$$\text{For } \mathbf{Q}^{-1}\mathbf{X}\mathbf{Q} = \begin{bmatrix} \beta_1 & 0 & \cdots & \cdots & 0 \\ 0 & \beta_2 & 0 & \cdots & 0 \\ \vdots & \ddots & \ddots & \vdots & \vdots \\ \vdots & \ddots & \ddots & \beta_{n-1} & 0 \\ 0 & 0 & \cdots & 0 & \beta_n \end{bmatrix} = \begin{bmatrix} \beta_1 & 0 & \cdots & \cdots & 0 \\ 0 & \beta_2 & 0 & \cdots & 0 \\ \vdots & \ddots & \ddots & \vdots & \vdots \\ \vdots & \ddots & \ddots & \beta_{n-1} & 0 \\ 0 & 0 & \cdots & 0 & 0 \end{bmatrix} +$$

$$\begin{bmatrix} 0 & 0 & \cdots & \cdots & 0 \\ 0 & 0 & 0 & \cdots & 0 \\ \vdots & \ddots & \ddots & \vdots & \vdots \\ \vdots & \ddots & \ddots & 0 & 0 \\ 0 & 0 & \cdots & 0 & \beta_n \end{bmatrix}$$, let $\mathbf{Q}^{-1}\hat{\mathbf{X}}\mathbf{Q} = \begin{bmatrix} \beta_1 & 0 & \cdots & \cdots & 0 \\ 0 & \beta_2 & 0 & \cdots & 0 \\ \vdots & \ddots & \ddots & \vdots & \vdots \\ \vdots & \ddots & \ddots & \beta_{n-1} & 0 \\ 0 & 0 & \cdots & 0 & 0 \end{bmatrix}$ and

$\mathbf{Q}^{-1}\bar{\mathbf{X}}\mathbf{Q} = \begin{bmatrix} 0 & 0 & \cdots & \cdots & 0 \\ 0 & 0 & 0 & \cdots & 0 \\ \vdots & \ddots & \ddots & \vdots & \vdots \\ \vdots & \ddots & \ddots & 0 & 0 \\ 0 & 0 & \cdots & 0 & \beta_n \end{bmatrix}$, where $\beta_1 >> \beta_n$, $\mathbf{Q} = [\mathbf{y}_1, \mathbf{y}_2, \cdots, \mathbf{y}_n]$ and $\mathbf{X}\mathbf{y}_i = \beta_i \mathbf{y}_i$.

Proposition 2.14. *Let* $\hat{\mathbf{H}} := [\hat{\mathbf{x}}_1, \hat{\mathbf{x}}_2, \cdots, \hat{\mathbf{x}}_n]$ *and* $\hat{\mathbf{x}}_n = \alpha_1\hat{\mathbf{x}}_1 + \alpha_2\hat{\mathbf{x}}_2 + \cdots + \alpha_{n-1}\hat{\mathbf{x}}_{n-1}$.
Then $[\alpha_1, \alpha_2, \cdots, \alpha_{n-1}]^T = \mathbf{R}_{11}^{-1}\mathbf{R}_{12}$ *holds for* $\mathbf{Q}^{-1} := \begin{bmatrix} \mathbf{R}_{11} & \mathbf{R}_{12} \\ \mathbf{R}_{21} & \mathbf{R}_{22} \end{bmatrix}$, *where* $\mathbf{R}_{11} \in R^{(n-1)\times(n-1)}$, $\mathbf{R}_{12} \in R^{(n-1)\times 1}$, $\mathbf{R}_{21} \in R^{1\times(n-1)}$ *and* $\mathbf{R}_{22} \in R$. *And where,* $\mathbf{H} := [\mathbf{x}_1, \mathbf{x}_2, \cdots, \mathbf{x}_n]$, $\mathbf{X} := \mathbf{H}^\mathbf{T}\mathbf{H} = \hat{\mathbf{X}} + \bar{\mathbf{X}}$, $\hat{\mathbf{X}} := \hat{\mathbf{H}}^\mathbf{T}\hat{\mathbf{H}}$, $\bar{\mathbf{X}} := \bar{\mathbf{H}}^\mathbf{T}\bar{\mathbf{H}}$, $\mathbf{Q}$ *is a matrix composed from eigenvectors of* $\mathbf{X}$,

$$\mathbf{Q}^{-1}\hat{\mathbf{X}}\mathbf{Q} = \begin{bmatrix} \beta_1 & 0 & \cdots & \cdots & 0 \\ 0 & \beta_2 & 0 & \cdots & 0 \\ \vdots & \ddots & \ddots & \vdots & \vdots \\ \vdots & \ddots & \ddots & \beta_{n-1} & 0 \\ 0 & 0 & \cdots & 0 & 0 \end{bmatrix} \text{ and } \mathbf{Q}^{-1}\bar{\mathbf{X}}\mathbf{Q} = \begin{bmatrix} 0 & 0 & \cdots & \cdots & 0 \\ 0 & 0 & 0 & \cdots & 0 \\ \vdots & \ddots & \ddots & \vdots & \vdots \\ \vdots & \ddots & \ddots & 0 & 0 \\ 0 & 0 & \cdots & 0 & \beta_n \end{bmatrix}.$$

[proof] Since $\mathbf{Q}^{-1}\hat{\mathbf{X}} = \mathbf{Q}^{-1}\hat{\mathbf{H}}^T\hat{\mathbf{H}}$

$$= \mathbf{Q}^{-1}\hat{\mathbf{H}}^T[\hat{\mathbf{x}}_1, \hat{\mathbf{x}}_2, \cdots, \hat{\mathbf{x}}_n] = \begin{bmatrix} \beta_1 & 0 & \cdots & \cdots & 0 \\ 0 & \beta_2 & 0 & \cdots & 0 \\ \vdots & \ddots & \ddots & \vdots & \vdots \\ \vdots & \ddots & \ddots & \beta_{n-1} & 0 \\ 0 & 0 & \cdots & 0 & 0 \end{bmatrix} \mathbf{Q}^{-1},$$

$\mathbf{Q}^{-1}\hat{\mathbf{X}}\mathbf{e}_i = \begin{bmatrix} \beta_1 & 0 & \cdots & \cdots & 0 \\ 0 & \beta_2 & 0 & \cdots & 0 \\ \vdots & \ddots & \ddots & \vdots & \vdots \\ \vdots & \ddots & \ddots & \beta_{n-1} & 0 \\ 0 & 0 & \cdots & 0 & 0 \end{bmatrix} \mathbf{Q}^{-1}\mathbf{e}_i$ holds for $\mathbf{e}_i = [0, \cdots, 0, \overset{i}{1}, 0, \cdots, 0]^T$ and $0 \leq i \leq n$.

Hence $\mathbf{Q}^{-1}\hat{\mathbf{H}}^T\hat{\mathbf{x}}_i = \begin{bmatrix} \beta_1 & 0 & \cdots & \cdots & 0 \\ 0 & \beta_2 & 0 & \cdots & 0 \\ \vdots & \ddots & \ddots & \vdots & \vdots \\ \vdots & \ddots & \ddots & \beta_{n-1} & 0 \\ 0 & 0 & \cdots & 0 & 0 \end{bmatrix} \mathbf{Q}^{-1}\mathbf{e}_i$ is obtained.

By using $\hat{\mathbf{x}}_n = \alpha_1\hat{\mathbf{x}}_1 + \alpha_2\hat{\mathbf{x}}_2 + \cdots + \alpha_{n-1}\hat{\mathbf{x}}_{n-1}$, $\mathbf{0} = \mathbf{Q}^{-1}\hat{\mathbf{H}}^T(\alpha_1\hat{\mathbf{x}}_1 + \alpha_2\hat{\mathbf{x}}_2 + \cdots + \alpha_{n-1}\hat{\mathbf{x}}_{n-1} - \hat{\mathbf{x}}_n) = \begin{bmatrix} \beta_1 & 0 & \cdots & \cdots & 0 \\ 0 & \beta_2 & 0 & \cdots & 0 \\ \vdots & \ddots & \ddots & \vdots & \vdots \\ \vdots & \ddots & \ddots & \beta_{n-1} & 0 \\ 0 & 0 & \cdots & 0 & 0 \end{bmatrix} \mathbf{Q}^{-1}[\alpha_1, \alpha_2, \cdots, \alpha_{n-1}, -1]^T$ can be obtained. Hence, $[\mathbf{R}_{11}, \mathbf{R}_{12}][\alpha_1, \alpha_2, \cdots, \alpha_{n-1}, -1]^T = 0$ holds. Therefore, $[\alpha_1, \alpha_2, \cdots, \alpha_{n-1}, -1]^T = \mathbf{R}_{11}^{-1}\mathbf{R}_{12}$.

On the other hand, since $\mathbf{Q}^{-1}\bar{\mathbf{X}}\mathbf{Q} = \mathbf{Q}^{-1}\bar{\mathbf{H}}^{\mathbf{T}}\bar{\mathbf{H}}\mathbf{Q} = \begin{bmatrix} 0 & 0 & \cdots & \cdots & 0 \\ 0 & 0 & 0 & \cdots & 0 \\ \vdots & \ddots & \ddots & \vdots & \vdots \\ \vdots & \ddots & \ddots & 0 & 0 \\ 0 & 0 & \cdots & 0 & \beta_n \end{bmatrix}$ holds, $\mathbf{Q}^{-1}\bar{\mathbf{H}}^{\mathbf{T}}\bar{\mathbf{H}}\mathbf{Q}\mathbf{e}_n = \beta_n\mathbf{e}_n$ is obtained. Hence $\bar{\mathbf{H}}^{\mathbf{T}}\bar{\mathbf{H}}\mathbf{Q}\mathbf{e}_n = \beta_n\mathbf{Q}\mathbf{e}_n$ holds. Therefore, $\bar{\mathbf{H}}^T\bar{\mathbf{H}}\mathbf{y}_n = \beta_n\mathbf{y}_n$ can be obtained. Also, by an equation $\mathbf{H}^T\mathbf{H}\mathbf{Q}\mathbf{e}_n = \mathbf{Q}\begin{bmatrix} \beta_1 & 0 & \cdots & 0 \\ 0 & \beta_2 & \cdots & 0 \\ \vdots & \ddots & \ddots & \vdots \\ 0 & 0 & \cdots & \beta_n \end{bmatrix}\mathbf{e}_n = \beta_n\mathbf{Q}\mathbf{e}_n = \beta_n\mathbf{y}_n$, $\mathbf{H}^T\mathbf{H}\mathbf{y}_n = \beta_n\mathbf{y}_n$ holds. Through the equations $\bar{\mathbf{H}}^T\bar{\mathbf{H}}\mathbf{y}_n = \beta_n\mathbf{y}_n$ and $\mathbf{H}^T\mathbf{H}\mathbf{y}_n = \beta_n\mathbf{y}_n$, $\hat{\mathbf{H}}^T\hat{\mathbf{H}}\mathbf{y}_n = \mathbf{0}$ holds. Since rank $\hat{\mathbf{H}} = n-1$, $\mathbf{y}_n = \beta[\alpha_1, \alpha_2, \cdots, \alpha_{n-1}]^T$ holds for a scalar $\beta \in R$.

Next, we will discuss another decomposition for noisy part and signal part. For $\mathbf{Q}^{-1}\mathbf{X}\mathbf{Q} = \begin{bmatrix} \beta_1 & 0 & \cdots & \cdots & 0 \\ 0 & \beta_2 & 0 & \cdots & 0 \\ \vdots & \ddots & \ddots & \vdots & \vdots \\ \vdots & \ddots & \ddots & \beta_{n-1} & 0 \\ 0 & 0 & \cdots & 0 & \beta_n \end{bmatrix} = \begin{bmatrix} \beta_1-\beta_n & 0 & \cdots & \cdots & 0 \\ 0 & \beta_2-\beta_n & 0 & \cdots & 0 \\ \vdots & \ddots & \ddots & \vdots & \vdots \\ \vdots & \ddots & \ddots & \beta_{n-1}-\beta_n & 0 \\ 0 & 0 & \cdots & 0 & 0 \end{bmatrix} + \begin{bmatrix} \beta_n & 0 & \cdots & \cdots & 0 \\ 0 & \beta_n & 0 & \cdots & 0 \\ \vdots & \ddots & \ddots & \vdots & \vdots \\ \vdots & \ddots & \ddots & \beta_n & 0 \\ 0 & 0 & \cdots & 0 & \beta_n \end{bmatrix}$, let $\mathbf{Q}^{-1}\hat{\mathbf{X}}_{\mathbf{0}}\mathbf{Q} = \begin{bmatrix} \beta_1-\beta_n & 0 & \cdots & \cdots & 0 \\ 0 & \beta_2-\beta_n & 0 & \cdots & 0 \\ \vdots & \ddots & \ddots & \vdots & \vdots \\ \vdots & \ddots & \ddots & \beta_{n-1}-\beta_n & 0 \\ 0 & 0 & \cdots & 0 & 0 \end{bmatrix}$ and

$$\mathbf{Q}^{-1}\bar{\mathbf{X}}_0\mathbf{Q} = \begin{bmatrix} \beta_n & 0 & \cdots & \cdots & 0 \\ 0 & \beta_n & 0 & \cdots & 0 \\ \vdots & \ddots & \ddots & \vdots & \vdots \\ \vdots & \ddots & \ddots & \beta_n & 0 \\ 0 & 0 & \cdots & 0 & \beta_n \end{bmatrix} = \beta_n I,$$

where $\beta_1 >> \beta_n$ and $\mathbf{X} = \hat{\mathbf{X}}_\mathbf{0} + \bar{\mathbf{X}}_\mathbf{0}$, $\mathbf{Q} = [\mathbf{y}_1, \mathbf{y}_2, \cdots, \mathbf{y}_n]$ and $\mathbf{X}\mathbf{y}_i = \beta_i \mathbf{y}_i$.

Proposition 2.15. *Let* $\hat{\mathbf{H}} := [\hat{\mathbf{x}}_{01}, \hat{\mathbf{x}}_{02}, \cdots, \hat{\mathbf{x}}_{0n}]$ *and* $\hat{\mathbf{x}}_{0n} = \alpha_{01}\hat{\mathbf{x}}_{01} + \alpha_{02}\hat{\mathbf{x}}_{02} + \cdots + \alpha_{0n-1}\hat{\mathbf{x}}_{0n-1}$. *Then* $[\alpha_{01}, \alpha_{02}, \cdots, \alpha_{0n-1}]^T = \mathbf{R}_{11}^{-1}\mathbf{R}_{12}$ *holds for* $\mathbf{Q}^{-1} := \begin{bmatrix} \mathbf{R}_{11} & \mathbf{R}_{12} \\ \mathbf{R}_{21} & \mathbf{R}_{22} \end{bmatrix}$, *where* $\beta_1 >> \beta_n$ *and* $\mathbf{X} := \mathbf{H}^\mathbf{T}\mathbf{H} = \hat{\mathbf{X}}_\mathbf{0} + \bar{\mathbf{X}}_\mathbf{0}$, $\mathbf{Q} = [\mathbf{y}_1, \mathbf{y}_2, \cdots, \mathbf{y}_n$ *and* $\mathbf{X}\mathbf{y}_i = \beta_i \mathbf{y}_i$.
And $\bar{\mathbf{x}}_i^T \bar{\mathbf{x}}_j = \beta_n \delta_{i,j}$ *holds, where* $\delta_{i,j} = 1$ *for* $i = j$ *and* $\delta_{i,j} = 0$ *for* $i \neq j$.

$$\mathbf{Q}^{-1}\hat{\mathbf{X}}_0\mathbf{Q} = \begin{bmatrix} \beta_1 - \beta_n & 0 & \cdots & \cdots & 0 \\ 0 & \beta_2 - \beta_n & 0 & \cdots & 0 \\ \vdots & \ddots & \ddots & \vdots & \vdots \\ \vdots & \ddots & \ddots & \beta_{n-1} - \beta_n & 0 \\ 0 & 0 & \cdots & 0 & 0 \end{bmatrix} \text{ and } \mathbf{Q}^{-1}\bar{\mathbf{X}}\mathbf{Q} = \begin{bmatrix} 0 & 0 & \cdots & \cdots & 0 \\ 0 & 0 & 0 & \cdots & 0 \\ \vdots & \ddots & \ddots & \vdots & \vdots \\ \vdots & \ddots & \ddots & 0 & 0 \\ 0 & 0 & \cdots & 0 & \beta_n \end{bmatrix}.$$

[proof] This Proposition can be proved in the same manner as Proposition 2.15.

2.5 Algebraically Constrained Least Square Method

Let $\hat{\mathbf{x}}_i \in K^n$, $\mathbf{x}_i = \hat{\mathbf{x}}_i + \bar{\mathbf{x}}_i$ and $\mathbf{x} = [\mathbf{x}_1^T, \mathbf{x}_2^T, \cdots, \mathbf{x}_s^T]^T$. We will cite the following three lemmas from the reference [Hasegawa, 2008].

Lemma 2.16. *If* $\bar{\mathbf{x}}^T\bar{\mathbf{x}}$ *is minimum under the condition* $S\hat{\mathbf{x}} = 0$ *for a proper full rank matrix* S,
then $\bar{\mathbf{x}}$ *is given by the equation* $\bar{\mathbf{x}} = S^T[SS^T]^{-1}S\mathbf{x}$.

[proof] Let a scalar valued function $f(\bar{\mathbf{x}}, \lambda)$ be
$f(\bar{\mathbf{x}}, \lambda) = \bar{\mathbf{x}}^T\bar{\mathbf{x}} + \lambda^T S[\mathbf{x} - \bar{\mathbf{x}}] + [\mathbf{x} - \bar{\mathbf{x}}]^T S^T \lambda$ for a Lagrange multiplier matrix $\lambda \in K^{sn\times 1}$.

Let the small increment $f(\bar{\mathbf{x}}+\delta\bar{\mathbf{x}}, \lambda+\delta\lambda)$ from $f(\bar{\mathbf{x}}, \lambda)$ be $f(\bar{\mathbf{x}}+\delta\bar{\mathbf{x}}, \lambda+\delta\lambda) = [\bar{\mathbf{x}}+\delta\bar{\mathbf{x}}]^T[\bar{\mathbf{x}}+\delta\bar{\mathbf{x}}] + [\lambda+\delta\lambda]^T S[\mathbf{x}-\bar{\mathbf{x}}-\delta\bar{\mathbf{x}}] + [\mathbf{x}-\bar{\mathbf{x}}-\delta\bar{\mathbf{x}}]^T S^T[\lambda+\delta\lambda]$. From the equation $f(\bar{\mathbf{x}}+\delta\bar{\mathbf{x}}, \lambda+\delta\lambda) - f(\bar{\mathbf{x}}, \lambda) = [\bar{\mathbf{x}}^T - \lambda^T S]\delta\bar{\mathbf{x}} + \delta\bar{\mathbf{x}}[\bar{\mathbf{x}} - S^T\lambda] = 0$, we obtain the equation $\bar{\mathbf{x}} = S^T\lambda$. Because of $\lambda = [SS^T]^{-1}S\bar{\mathbf{x}}$, $\bar{\mathbf{x}} = S^T[SS^T]^{-1}S\bar{\mathbf{x}}$ holds.

Next, we will discuss concretely a least square of noise $\bar{\mathbf{x}}_i$ for $1 \leq i \leq s$.

Let a matrix $A \in K^{q\times n}$ with a full rank satisfy an equation $A\hat{\mathbf{x}}_i = 0$ for any $1 \leq i \leq s$.

Lemma 2.17. *Under the constraint* $A\hat{\mathbf{x}}_i = 0$ *for* i $(1 \le i \le s)$, *let* $\sum_{i=1}^{s} \bar{\mathbf{x}}_i . \bar{\mathbf{x}}_i$ *take a minimum value, where* $\bar{\mathbf{x}}_i . \bar{\mathbf{x}}_i$ *denotes the inner product of the vectors* $\bar{\mathbf{x}}_i$ *and* $\bar{\mathbf{x}}_i$. *Then* $\bar{\mathbf{x}}_\mathbf{i} = A^T[AA^T]^{-1}A\mathbf{x}_i$ *holds for* i $(1 \le i \le s)$.

[proof] Let $\mathbf{x} \in K^{sn\times 1}$ and $S \in K^{qs\times ns}$ be $\mathbf{x} := [\mathbf{x}_1^T, \mathbf{x}_2^T, \cdots, \mathbf{x}_s^T]^T$ and
$S = \begin{bmatrix} A^T & 0 & \cdots & 0 \\ 0 & A^T & \cdots & 0 \\ \vdots & \ddots & \ddots & \vdots \\ 0 & 0 & \cdots & A^T \end{bmatrix}$, where $\mathbf{x} = \hat{\mathbf{x}} + \bar{\mathbf{x}}$, $\hat{\mathbf{x}} := [\hat{\mathbf{x}}_1^T, \hat{\mathbf{x}}_2^T, \cdots, \hat{\mathbf{x}}_s^T]^T$, $\bar{\mathbf{x}} :=$ $[\bar{\mathbf{x}}_1^T, \bar{\mathbf{x}}_2^T, \cdots, \bar{\mathbf{x}}_s^T]^T$.

Let a scalar function $f(\bar{\mathbf{x}}, \lambda)$ be $f(\bar{\mathbf{x}}, \lambda) = \bar{\mathbf{x}} . \bar{\mathbf{x}} + \lambda^T S[\mathbf{x} - \bar{\mathbf{x}}] + [\mathbf{x} - \bar{\mathbf{x}}]^T S^T \lambda$ for a Lagrange multiplier vector $\lambda \in K^{qs\times 1}$.

Let the small increment $f(\bar{\mathbf{x}}+\delta\bar{\mathbf{x}}, \lambda+\delta\lambda)$ from $f(\bar{\mathbf{x}}, \lambda)$ be $f(\bar{\mathbf{x}}+\delta\bar{\mathbf{x}}, \lambda+\delta\lambda) = [\bar{\mathbf{x}} + \delta\bar{\mathbf{x}}] . [\bar{\mathbf{x}} + \delta\bar{\mathbf{x}}] + [\lambda^T + \delta\lambda^T]S[\mathbf{x} - \bar{\mathbf{x}} - \delta\bar{\mathbf{x}}] + [\mathbf{x} - \bar{\mathbf{x}} - \delta\bar{\mathbf{x}}]^T S^T[\lambda + \delta\lambda]$. From the equation $f(\bar{\mathbf{x}} + \delta\bar{\mathbf{x}}, \lambda + \delta\lambda) - f(\bar{\mathbf{x}}, \lambda) = 0$ and Lemma (2.14), we obtain the equation $\bar{\mathbf{x}} = S^T[SS^T]^{-1}S\mathbf{x}$.

Therefore, $\bar{\mathbf{x}}_\mathbf{i} = A^T[AA^T]^{-1}A\mathbf{x}_i$ holds for i $(1 \le i \le s)$.

Lemma 2.18. *Under the constraint* $A\hat{\mathbf{x}}_i = 0$, $\bar{\mathbf{x}}_i = [\mathbf{0}^T, \bar{\mathbf{x}}_{i,2}^T] \in K^n$, $\mathbf{0}^T \in K^{n_1}$ *and* $\bar{\mathbf{x}}_{i,2}^T \in K^{n_2}$ *for* i $(1 \le i \le s)$, *let* $\sum_{i=1}^{s} \bar{\mathbf{x}}_i . \bar{\mathbf{x}}_i$ *take minimum value, where* $\bar{\mathbf{x}}_i . \bar{\mathbf{x}}_i$ *denotes the inner product of the vectors* $\bar{\mathbf{x}}_i$ *and* $\bar{\mathbf{x}}_i$. *Then* $\begin{bmatrix} \lambda_2^T \\ \bar{\mathbf{x}}_\mathbf{i} \end{bmatrix} = A^T[AA^T]^{-1}A\left(\hat{\mathbf{x}}_i + \begin{bmatrix} \lambda_2^T \\ \bar{\mathbf{x}}_\mathbf{i} \end{bmatrix}\right)$ *holds for* i $(1 \le i \le s)$, *where* $\lambda_2^T =$ $\mathbf{a}_1[\mathbf{a}_2^T\mathbf{a}_2]^{-1}\mathbf{a}_2^T\bar{\mathbf{x}}_{i,2}$ *for* $A := [\mathbf{a}_1^T\mathbf{a}_2^T]$.

[proof] Let $\mathbf{x} \in K^{sn\times 1}$, S_1 and $S_2 \in K^{qs\times ns}$ be $\mathbf{x} := [\mathbf{x}_1^T, \mathbf{x}_2^T, \cdots, \mathbf{x}_s^T]^T$
$S_1 = \begin{bmatrix} A^T & 0 & \cdots & 0 \\ 0 & A^T & \cdots & 0 \\ \vdots & \ddots & \ddots & \vdots \\ 0 & 0 & \cdots & A^T \end{bmatrix}$, $S_2 = \begin{bmatrix} B^T & 0 & \cdots & 0 \\ 0 & B^T & \cdots & 0 \\ \vdots & \ddots & \ddots & \vdots \\ 0 & 0 & \cdots & B^T \end{bmatrix}$, where $\mathbf{x} = \hat{\mathbf{x}} + \bar{\mathbf{x}}$, $\hat{\mathbf{x}} :=$ $[\hat{\mathbf{x}}_1^T, \hat{\mathbf{x}}_2^T, \cdots, \hat{\mathbf{x}}_s^T]^T$, $\bar{\mathbf{x}} := [\bar{\mathbf{x}}_1^T, \bar{\mathbf{x}}_2^T, \cdots, \bar{\mathbf{x}}_s^T]^T$ and $B^T = [1, \cdots, 1, 0, \cdots, 0]$.
Let a scalar function $f(\bar{\mathbf{x}}, \boldsymbol{\lambda}_1, \boldsymbol{\lambda}_2)$ be $f(\bar{\mathbf{x}}, \boldsymbol{\lambda}_1, \boldsymbol{\lambda}_2) = \bar{\mathbf{x}} . \bar{\mathbf{x}} + \lambda_1^T S_1[\mathbf{x} - \bar{\mathbf{x}}] + \lambda_2^T S_2[\mathbf{x} - \bar{\mathbf{x}}] + [\mathbf{x} - \bar{\mathbf{x}}]^T S_1^T \lambda_1 + [\mathbf{x} - \bar{\mathbf{x}}]^T S_2^T \lambda_2$ for a Lagrange multiplier vector λ_1, $\lambda_2 \in K^{qs\times 1}$.

Let the small increment $f(\bar{\mathbf{x}} + \delta\bar{\mathbf{x}}, \boldsymbol{\lambda}_1 + \delta\boldsymbol{\lambda}_1, \boldsymbol{\lambda}_2 + \delta\boldsymbol{\lambda}_2)$ from $f(\bar{\mathbf{x}}, \boldsymbol{\lambda}_1, \boldsymbol{\lambda}_2)$ be $f(\bar{\mathbf{x}}+\delta\bar{\mathbf{x}}, \boldsymbol{\lambda}_1+\delta\boldsymbol{\lambda}_1, \boldsymbol{\lambda}_2+\delta\boldsymbol{\lambda}_2) = [\bar{\mathbf{x}}+\delta\bar{\mathbf{x}}] . [\bar{\mathbf{x}}+\delta\bar{\mathbf{x}}] + [\boldsymbol{\lambda}_1^T + \delta\boldsymbol{\lambda}_1^T]S_1[\mathbf{x}-\bar{\mathbf{x}}-\delta\bar{\mathbf{x}}] + [\boldsymbol{\lambda}_2^T + \delta\boldsymbol{\lambda}_2^T]S_2[\mathbf{x}-\bar{\mathbf{x}}-\delta\bar{\mathbf{x}}] + [\mathbf{x}-\bar{\mathbf{x}}-\delta\bar{\mathbf{x}}]^T S_1^T[\boldsymbol{\lambda}_1+\delta\boldsymbol{\lambda}_1] + [\mathbf{x}-\bar{\mathbf{x}}-\delta\bar{\mathbf{x}}]^T S_2^T[\boldsymbol{\lambda}_2+\delta\boldsymbol{\lambda}_2]$. From the equation $f(\bar{\mathbf{x}}+\delta\bar{\mathbf{x}}, \boldsymbol{\lambda}_1+\delta\boldsymbol{\lambda}_1, \boldsymbol{\lambda}_2+\delta\boldsymbol{\lambda}_2) - f(\bar{\mathbf{x}}, \boldsymbol{\lambda}_1, \boldsymbol{\lambda}_2) = 0$, we obtain the equation $\bar{\mathbf{x}} = S_1^T\boldsymbol{\lambda}_1 + S_2^T\boldsymbol{\lambda}_2$.

Hence, $\bar{\mathbf{x}}_\mathbf{i} = A\boldsymbol{\lambda}_1^T - [1, \cdots, 1, 0, \cdots, 0]^T\boldsymbol{\lambda}_2^T = \begin{bmatrix} \mathbf{a}_1\boldsymbol{\lambda}_1^T - \boldsymbol{\lambda}_2 \\ \mathbf{a}_2\boldsymbol{\lambda}_1^T \end{bmatrix}$.

Therefore, $\mathbf{a}_1\boldsymbol{\lambda}_1^T = \boldsymbol{\lambda}_2$ and $\mathbf{x}_i = \begin{bmatrix} \mathbf{0} \\ \bar{\mathbf{x}}_{i,2} \end{bmatrix} = \begin{bmatrix} \mathbf{0} \\ \mathbf{a}_2\boldsymbol{\lambda}_1^T \end{bmatrix}$ hold.

From the equations $\mathbf{a}_1\boldsymbol{\lambda}_1^T = \boldsymbol{\lambda}_2$ and $\bar{\mathbf{x}}_{i,2} = \mathbf{a}_2\boldsymbol{\lambda}_1^T$, $\boldsymbol{\lambda}_2 = \mathbf{a}_1[\mathbf{a}_2^T\mathbf{a}_2]^{-1}\mathbf{a}_2^T\bar{\mathbf{x}}_{i,2}$ holds.

By the relation $\begin{bmatrix} \boldsymbol{\lambda}_2^T \\ \bar{\mathbf{x}}_{i,2} \end{bmatrix} = \begin{bmatrix} \mathbf{a}_1\boldsymbol{\lambda}_1^T \\ \mathbf{a}_2\boldsymbol{\lambda}_1^T \end{bmatrix} = A\boldsymbol{\lambda}_1^T$, $A^T \begin{bmatrix} \boldsymbol{\lambda}_2^T \\ \bar{\mathbf{x}}_{i,2} \end{bmatrix} = A^T A\boldsymbol{\lambda}_1^T$ and $\boldsymbol{\lambda}_1^T = [A^T A]^{-1} A^T \begin{bmatrix} \boldsymbol{\lambda}_2^T \\ \bar{\mathbf{x}}_{i,2} \end{bmatrix}$ can be obtained. Using the relation $A^T\hat{\mathbf{x}}_i = 0$, we obtain $\begin{bmatrix} \boldsymbol{\lambda}_2^T \\ \bar{\mathbf{x}}_\mathbf{i} \end{bmatrix} = A^T[AA^T]^{-1}A\left(\hat{\mathbf{x}}_i + \begin{bmatrix} \boldsymbol{\lambda}_2^T \\ \bar{\mathbf{x}}_\mathbf{i} \end{bmatrix}\right)$, which holds for i $(1 \le i \le s)$ with the relation $\boldsymbol{\lambda}_2^T = \mathbf{a}_1[\mathbf{a}_2^T\mathbf{a}_2]^{-1}\mathbf{a}_2^T\bar{\mathbf{x}}_{i,2}$.

2.5.1 Algebraic CLS Method and Analytic CLS Method

Definition 2.19. Based on Proposition 2.14 or 2.15, determine the coefficients of a given dynamical system and determine the error vector $\bar{\mathbf{x}}$ by the equation of Lemma 2.17. The method to obtain such a dynamical system is called the algebraic CLS method, i.e., algebraically constrained least square method.

Definition 2.20. Based on Lemma 2.17, determine the coefficients of a given dynamical system such that the inner product of the error vector $\bar{\mathbf{x}}$ takes a minimum value and determine the error vector $\bar{\mathbf{x}}$ by the equation of Lemma 2.17. The method to obtain this dynamical system is called the analytic CLS method, i.e., analytically constrained least square method.

2.6 Historical Notes and Concluding Remarks

In the field of model reduction of discrete-time systems, singular values decomposition and polynomial equations are used as effective methods [Glover, 1981]. On the other hand, in the field of modeling under a noisy environment, various methods such as AIC (Akaike's Information Criterion) have been proposed from the viewpoint of probabilistic sense.

In the monograph [Hasegawa, 2008], it was shown for the first time that approximate realization (model reduction) and noisy realization with solving partial differential equations were proposed for linear and nonlinear dynamical systems with an input and output mechanism. Note that the usual methods for approximation and noisy realization are limited to only linear systems. Of course, our methods in this monograph will be applied not only to linear systems but also to nonlinear systems in a unified manner by using algebraic operations.

It is noteworthy that our methods are quite different from usual methods and unified for any input/output relations with causality condition.

Also note that our methods are geared only toward the linear combination of vectors. Furthermore, it is also noteworthy that we have shown that any input/output relations with causality condition can be expressed in a Hankel matrix or input/output matrix which can serve as a linear operator.

The relation between the algebraic CLS method proposed in this monograph and the analytic CLS method discussed in the reference [Hasegawa, 2008] becomes clear.

Chapter 3
Algebraically Approximate and Noisy Realization of Linear Systems

Let the set Y of output's values be a linear space over the real number field $\boldsymbol{R}$. It is well known that Linear System Theory was established in the algebraic sense [Kalman, 1969]. The main theorem says that for any causal linear input/output map, there exist at least two canonical (controllable and observable) Linear Systems which realize (faithfully describe) it and any two canonical Linear Systems with the same behavior are isomorphic.

Details of finite dimensional Linear Systems were investigated. The criterion for the canonical finite dimensional Linear Systems and various standard canonical Linear Systems were given.

Their partial realization was also discussed according to the above results. We will state an algorithm to obtain a canonical partial realization from a given partial input/output map.

Based on fundamentally established results, an approximate realization problem and a noisy realization problem were discussed through solving partial differential equations of rational polynomial in multi-variables [Hasegawa, 2008].

In this chapter, approximate realization problem and noisy realization problems will be discussed by executing only algebraic operations and comparing with the results of the reference [Hasegawa, 2008].

3.1 Basic Facts Regarding Linear Systems

We will summarize fundamentally established facts, which are needed for the approximate and noisy realization problems.

Definition 3.1. Linear Systems

(1) A system represented by the following equations is written as a collection $\sigma = ((X, F), g, h)$ and it is said to be a linear system:

Y. Hasegawa: Algebra. Approx. & Noisy Reali. of Discrete-Time Sys., LNEE 50, pp. 21–51.
springerlink.com

$$\begin{cases} x(t+1) = Fx(t) + g\omega(t+1) \\ x(0) \quad\;\; = 0 \\ \hat{\gamma}(t) \quad\;\; = hx(t) \end{cases}$$

for any $t \in N$, $x(t) \in X$, $\gamma(t) \in Y$, where X is a linear space over the field $\boldsymbol{R}$, F is a linear operator on X, $g \in X$ and $h : X \to Y$ is a linear operator.
(2) The input response map $a_\sigma : U^* \to Y; \omega \mapsto h[\sum_{j=1}^{|\omega|} F^{|\omega|-j} g\omega(j)]$ is said to be the behavior of σ. For an input response map $a \in F(U^*, Y)$, σ which satisfies $a_\sigma = a$ is called a realization of a.
(3) For the linear system σ, $I_\sigma(i) = hF^i g$ is said to be an impulse response of σ. Note that there is a one-to-one correspondence between the behavior of σ and the impulse response of σ.
(4) A linear system σ is said to be reachable if the reachable set $\{\sum_{j=1}^{|\omega|} F^{|\omega|-j} g\omega(j); \omega \in U^*\}$ is equal to X.
(5) A linear system σ is called observable if $hF^i x_1 = hF^i x_2$ for any $i \in N$ implies $x_1 = x_2$.
(6) A linear system σ is called canonical if σ is reachable and observable.

Remark 1: It is meant for σ to be a faithful model for the input response map a that σ realizes a.

Remark 2: A canonical linear system $\sigma = ((X, F), g, h)$ is a system that has the most reduced state space X among systems that have the behavior a_σ.

Remark 3: The linear system $\sigma = ((X, F), g, h)$ obtained by the following common linear system equation and a transformation is a canonical linear system with the same behavior.

$$\begin{cases} x(t+1) = Fx(t) + g\omega(t+1), \\ x(0) \quad\;\; = 0, \\ \hat{\gamma}(t) \quad\;\; = hx(t), \end{cases} \quad \begin{cases} \underline{x}(t+1) = A\underline{x}(t) + \mathbf{b}\omega(t) \\ \underline{x}(0) \quad\;\; = 0 \\ \hat{\gamma}(t) \quad\;\; = c\underline{x}(t) \end{cases}$$

The transformation is given as follows:

$$x(t) = \begin{bmatrix} \underline{x}(t) \\ \omega(t) \end{bmatrix}, \; F = \begin{bmatrix} A & \mathbf{b} \\ 0 & 0 \end{bmatrix}, \; g = \begin{bmatrix} 0 \\ 1 \end{bmatrix}, \; h = \begin{bmatrix} c \; 0 \end{bmatrix}.$$

Example 3.2. Let $\boldsymbol{R}[z]$ be a set of polynomials in one variable z over the field $\boldsymbol{R}$. The variable $z : \boldsymbol{R}[z] \to \boldsymbol{R}[z]; \lambda \mapsto z\lambda$ is a linear operator. Let $a \in F(U^*, Y)$ be regarded as a linear operator : $\boldsymbol{R}[z] \to Y; z^i \mapsto a(i)$. Then $\sigma_I = ((\boldsymbol{R}[z], z), 1, a)$ is a linear system which is a realization of a. The impulse response I_{σ_I} of the linear system σ_I is given by $I_{\sigma_I}(i) = a(0|\cdots|0|1)$, where i is the length of an input $0|\cdots|0|1$.

Remark: For $a \in F(U^*, Y)$, an operator $\tilde{a} : \boldsymbol{R}[z] \to Y; \alpha_i z^i \mapsto \alpha_i a(i)$ is regarded as a linear operator $: N \to Y; i \mapsto a(i)$. This correspondence is one to one.

Example 3.3. Let $i \in N$, $a \in F(U^*, Y)$, $S_l : F(N, Y) \to F(N, Y); \gamma \mapsto S_l\gamma\ [;t \mapsto \gamma(t+1)]$ and let $0 : F(N, Y) \to Y; \gamma \mapsto \gamma(0)$ be a linear operator. Then $\sigma_F = ((F(N, Y), S_l), a, 0)$ is a linear system which is a realization of $a \in F(U^*, Y)$.

The impulse response I_{σ_F} of the linear system σ_F is given by $I_{\sigma_F}(i) = a(0|\cdots|0|1)$, where i is the length of an input $0|\cdots|0|1$.

Theorem 3.4. *For an input response map $a \in F(U^*, Y)$, the following two linear systems are both canonical realizations of a:*

1) $((\boldsymbol{R}[z]/_{\equiv a}, \dot{z}), [1], \dot{a})$, where $\boldsymbol{R}[z]/_{\equiv a}$ is a quotient space defined by an equivalence relation $\lambda_1 = \sum_i \lambda_1(i)z^i \equiv \lambda_2 = \sum_i \lambda_2(i)z^i \iff a(\lambda_1) = a(\lambda_2)$, $[1]$ is defined as a map $\boldsymbol{R}[z] \to \boldsymbol{R}[z]/_{\equiv a}; 1 \mapsto [1]$. $\dot{a}$ is defined by $\dot{a}([\lambda]) = a(\lambda)$ for any $\lambda \in \boldsymbol{R}[z]$.
2) $((\ll \{S_l^i a : i \in N\} \gg, S_l), a, 0)$, where $\ll S \gg$ is the linear hull generated by the set S.

Definition 3.5. Let $\sigma_1 = ((X_1, F_1), g_1, h_1)$ and $\sigma_2 = ((X_2, F_2), g_2, h_2)$ be linear systems. Then a linear operator $T : X_1 \to X_2$ is said to be a linear system morphism $T : \sigma_1 \to \sigma_2$ if T satisfies $TF_1 = F_2T$, $Tg_1 = g_2$ and $h_1 = h_2T$. If $T : X_1 \to X_2$ is bijective, then $T : \sigma_1 \to \sigma_2$ is said to be an isomorphism.

Corollary 3.6. *Let T be a linear system morphism $T : \sigma_1 \to \sigma_2$. Then $a_{\sigma_1} = a_{\sigma_2}$ holds.*

[proof] The definitions of the behavior and linear system morphism lead to this corollary.

Theorem 3.7. *Realization Theorem of linear systems*

Existence: For any input response map $a \in F(U^*, Y)$, there exist at least two canonical linear systems which realize a.
Uniqueness: Let σ_1 and σ_2 be any two canonical linear systems that realize $a \in F(U^*, Y)$. Then there exists an isomorphism $T : \sigma_1 \to \sigma_2$.

3.2 Finite Dimensional Linear Systems

In this section, a canonical form of finite-dimensional linear systems will be treated based on the realization theorem (3.7). Many results of linear systems have already been illustrated in a reference [Kalman, 1969]. In the following sections, these results have been summarized for this monograph to be self-contained.

At first, the conditions when a finite dimensional linear system is canonical is presented.

Secondly, the canonical form which is suitable for approximate and noisy realization problems is defined. We introduce a standard system as a representative in their equivalence classes.

Thirdly, a criterion for the behavior of finite dimensional linear systems, that is, the rank condition of an infinite Hankel matrix is presented.

Finally, a procedure to obtain the reachable standard system which realizes a given input response map is presented.

There is a fact about finite dimensional linear spaces that an n-dimensional linear space over the field $\boldsymbol{R}$ is isomorphic to $\boldsymbol{R}^n$ and $L(\boldsymbol{R}^n, \boldsymbol{R}^m)$ is isomorphic to $\boldsymbol{R}^{m\times n}$ (See Halmos [1958]). Therefore, without loss of generality, we can consider a n-dimensional linear system as $\sigma = ((\boldsymbol{R}^n, F), g, h)$, where $F \in \boldsymbol{R}^{n\times n}$, $g \in \boldsymbol{R}^n$ and $h \in \boldsymbol{R}^{p\times n}$.

Lemma 3.8. *A linear system $\sigma = ((\boldsymbol{R}^n, F), g, h)$ is canonical if and only if the following conditions 1) and 2) hold:*

1) rank $[g, Fg, \cdots, F^{n-1}g] = n$.
2) rank $[h^T, (hF)^T, \cdots, (hF^{n-1})^T] = n$.

Definition 3.9. A canonical linear system $\sigma_s = ((\boldsymbol{R}^n, F_s), \mathbf{e}_1, h_s)$ is said to be a reachable standard system if $\mathbf{e}_i = F_s^{i-1}\mathbf{e}_1$ and $F_s^n\mathbf{e}_1 = \sum_{i=1}^n \alpha_i F_s^{i-1}\mathbf{e}_1$ hold. Such F_s is presented as follows:

$$F_s = \begin{bmatrix} 0 & \cdots & \cdots & 0 & \alpha_1 \\ 1 & \ddots & & \vdots & \alpha_2 \\ \vdots & \ddots & \ddots & \vdots & \vdots \\ \vdots & \ddots & \ddots & 0 & \vdots \\ 0 & \cdots & 0 & 1 & \alpha_n \end{bmatrix}.$$

Lemma 3.10. *Lemma for equivalence classes*

For any finite dimensional canonical linear system, there exists a uniquely determined isomorphic reachable standard system.

Definition 3.11. For any input response map $a \in F(U^*, Y)$, the corresponding linear input/output map $A : \boldsymbol{R}[z] \to F(N, Y)$ satisfies $A(z^i)(j) = a(0|\cdots|0|1) = I_a(i+j)$ for $i,\ j \in N$ and the length of an input $0|\cdots|0|1$ is $i+j$.

Hence, A is represented by the following infinite matrix $\hat{H}_a$. This $\hat{H}_a$ is said to be a Hankel matrix of a.

$$\hat{H}_a = \begin{array}{c} \\ \\ \\ j \end{array} \left(\begin{array}{ccc} & & i \\ & & \vdots \\ & & \vdots \\ \cdots & \cdots & I_a(i+j) \\ & & \end{array} \right)$$

Note that for the linear input/output map $A : \boldsymbol{R}[z] \to F(N, Y)$, there exists a unique function $I_a : N \to Y$ such that $I_a(i+j) = A(z^i)(j)$ holds.

It is also noted that the column vectors of $\hat{H}_a$ denote $S_l^i I_a$.

Theorem 3.12. *Theorem for existence criterion*

For an input response map $a \in F(U^*, Y)$, the following conditions are equivalent:

1) The input response map $a \in F(U^*, Y)$ has the behavior of a n-dimensional canonical linear system.
2) There exist n-linearly independent vectors and no more than n-linearly independent vectors in a set $\{S_l^i a; i \leq n \text{ for } i \in N\}$.
3) The rank of the Hankel matrix $\hat{H}_a$ of a is n.

Theorem 3.13. *Theorem for a realization procedure*

Let $a \in F(U^*, Y)$ be an input response which satisfies the condition of Theorem (3.12). Then the reachable standard system $\sigma_s = ((\boldsymbol{R}^n, F_s), \mathbf{e_1}, h_s)$ which realizes the input response map a is obtained by the following procedure:

1) Select the linearly independent vectors $\{S_l^i a; 0 \leq i \leq n-1\}$ from the set $\{S_l^i a; i \in N\}$.
2) Let the state be $\mathbf{e}_1$, where $\mathbf{e}_1 = [1, 0, \cdots, 0]^T \in \boldsymbol{R}^n$.
3) Let the output map h_s be $h_s = [a(1), a(1), a(0|1), \cdots, a(0|0|\cdots|0|1)]$.
4) Let F_s be the same as in the reachable standard system defined in Definition (3.9) for $S_l^n a = \sum_{i=1}^n \alpha_i S_l^{i-1} a$.

3.3 Partial Realization Theory of Linear Systems

In this section, we consider a partial realization problem of linear systems. Let $\underline{a}$ be an $\underline{N}$-sized input response map($\in F(U_{\underline{N}}^*, Y)$), where $\underline{N} \in N$ and $U_{\underline{N}}^* := \{\omega \in U^*; |\omega| \leq \underline{N}\}$. The $\underline{a}$ is said to be a partial input response map. A finite dimensional linear system $\sigma = ((X, F), g, h)$ is called a partial realization of $\underline{a}$ if $hF^{|\omega|-1}g = \underline{a}(0|\cdots|0|1)$ holds for any $\omega \in U_{\underline{N}}^*$, $|\omega| = |0|\cdots|0|1|$.

A partial realization problem of linear systems is roughly stated as follows: < For any given $\underline{a} \in F(U_{\underline{N}}^*, Y)$, find a partial realization σ of $\underline{a}$ such that the dimension of state space X of σ is minimum. Then the σ is said to be a minimal partial realization of $\underline{a}$. Moreover, show an algorithm to obtain the minimal partial realization.>

For a partial input response map $\underline{a} \in F(U^*_{\underline{N}}, Y)$, the following matrix $\hat{H}_{\underline{a}\ (p,\bar{p})}$ is said to be a finite-sized Hankel matrix of $\underline{a}$.

$$\hat{H}_{\underline{a}\ (p,\bar{p})} = \begin{pmatrix} & & \bar{i} \\ & & \vdots \\ & & \vdots \\ i \quad \cdots & \cdots & I_a(i+\bar{i}) \end{pmatrix},$$

where $i \leq p$ and $\bar{i} \leq \bar{p}$.

Note that the column vectors of $\hat{H}_{\underline{a}\ (p,\bar{p})}$ is represented by $\underline{S}_l^i I_{\underline{a}}$.

When we actually treat the approximate and noisy realization problem, we will use a notation $H_{\underline{a}\ (n_1, \underline{N}-n_1)}$ expressed as follows:

$$H_{\underline{a}\ (n_1, \underline{N}-n_1)} = [I_{\underline{a}}, \cdots, S_l^{n_1-1} I_{\underline{a}}].$$

Proposition 3.14. *Let the rank of a finite-sized Hankel matrix* $\hat{H}_{\underline{a}\ (p,\bar{p})}$ *be* n. *Then a minimal partial realization* $\sigma_{\underline{a}} = ((\boldsymbol{R}^n, F_s), \mathbf{e}_1, h_s)$ *of the impulse response* $I_{\underline{a}}$ *is obtained by the following algorithm:*

1) Let F_s *be the same as* F_s *in Definition (3.9) for* $\underline{S}_l^n I_{\underline{a}} = \sum_{i=1}^n \alpha_i \underline{S}_l^{i-1} I_{\underline{a}}$.
2) Let $\mathbf{e}_1$ *be* $\mathbf{e}_1 = [1, 0, \cdots, 0]^T$.
3) Let h_s *be* $h_s = [I_{\underline{a}}(1), I_{\underline{a}}(2), \cdots, I_{\underline{a}}(n)]$.

[proof] It is obvious from the definition of behavior of the system.

3.4 Algebraically Approximate Realization of Linear Systems

In this section, we discuss an algebraically approximate realization problem for linear systems which is stated as follows:

<For any given finite-length impulse response, find, using only algebraic calculation, a linear system which approximates it.>

In order to make our discussion simple, we assume that the set Y of output is the set $\boldsymbol{R}$ of real numbers, namely, 1-output.

Theorem 3.15. *Algebraic algorithm for approximate realization*

Let $\underline{a}$ *be a considered object which is a linear system. Then an approximate realization* $\sigma = ((\boldsymbol{R}^n, F_s), g, h_s)$ *of* $\underline{a}$ *based on the algebraic CLS method is given by the following algorithm:*

1) Based on the ratio of the square root of eigenvalues for a matrix $H_{\underline{a}\ (p,\bar{p})}H^T_{\underline{a}\ (p,\bar{p})}$*, determine the value* n *of rank for the matrix* $H_{\underline{a}\ (p,\bar{p})}$*, where* $n \le p$*.*
Namely, determine the value n *of rank for the matrix* $H_{\underline{a}\ (p,\bar{p})}$ *such that the ratio of the square root of eigenvalues for the covariance matrix becomes very small. The small ratio means the nearness of approximation degree.*

2) We use the algebraic CLS method as follows:
① Based on Proposition (2.14), determine coefficients $\{\alpha_i : 1 \le i \le n\}$*. The* Q *in Proposition (2.14) can be considered as the matrix composed from the eigenvectors of* $H_{\underline{a}\ (n+1,L)}H^T_{\underline{a}\ (n+1,L)}$*.*
Let a matrix $A \in \boldsymbol{R}^{1\times(n+1)}$ *be* $A = [\alpha_1, \alpha_2, \cdots, \alpha_n, -1]$*.*
② Determine the error vectors $\{\underline{S_l^i}\bar{I}_{\underline{a}} \in \boldsymbol{R}^{L\times 1} : 0 \le i \le n\}$ *by using the equation* $[\bar{I}_{\underline{a}}, \underline{S_l}\bar{I}_{\underline{a}}, \cdots, \underline{S_l^n}\bar{I}_{\underline{a}}]^T := A^T[AA^T]^{-1}AH^T_{\underline{a}\ (n+1,L)}$ *and* $H_{\underline{a}\ (n+1,L)} := [I_{\underline{a}}, \cdots, S_l^{n-1}I_{\underline{a}}, S_l^n I_{\underline{a}}]$*.*
③ Let $F \in \boldsymbol{R}^{n\times n}$ *be given as below. Let* g_0 *be* $g_0 = \mathbf{e}_1$*, where* $\mathbf{e}_1 = [1, 0, \cdots, 0]^T \in \boldsymbol{R}^n$*.*
④ Let h_s *be* $h_s = [I_{\underline{a}}(1) - \bar{I}_{\underline{a}}(1), I_{\underline{a}}(1) - \bar{I}_{\underline{a}}(1), \cdots, I_{\underline{a}}(0^{n_1-2}|1) - \bar{I}_{\underline{a}}(0^{n_1-2}|1)]$*.*

$$F_0 = \begin{bmatrix} 0 & \cdots & 0 & \alpha_1 \\ 1 & \ddots & & \alpha_2 \\ \vdots & \ddots & 0 & \vdots \\ 0 & & 1 & \alpha_n \end{bmatrix}.$$

[proof] By 1), the approximate part of the data can be excluded in the sense of the norm of Hankel matrix $H_{\underline{a}\ (p,\bar{p})}$. Note that Q in Proposition (2.14) corresponds to the matrix composed of eigenvectors of $H_{\underline{a}\ (p,\bar{p})}$. The matrix A in 2) corresponds to the matrix A in Lemma (2.17). Hence, if we determine the coefficients $\{\alpha_i : 1 \le i \le n\}$ by Proposition (2.14), we can obtain the approximation error of the given linear system by using Lemma (2.17) in the sense of a linear combination.

Therefore, we obtain the approximate Hankel matrices $\hat{H}_{\underline{a}\ (n_1+1,\bar{p})}(n_1+1, 0)$. Finally, we apply Proposition (3.14) to the $\hat{H}_{\underline{a}\ (n_1+1,\bar{p})}(n_1+1, 0)$.

Example 3.16. Let a signal be the impulse response of the following 3-dimensional linear system: $\sigma = ((\boldsymbol{R}^3, F), g, h)$, where $F = \begin{bmatrix} 0 & 0 & -0.55 \\ 1 & 0 & -0.04 \\ 0 & 1 & 1.1 \end{bmatrix}$, $h = [10, 2, -5], g = [1, 0, 0]^T$.
Then the algebraically approximate realization problem is solved as follows:

covariance matrix	eigenvalues			
	1	2	3	4
$H^T_{\underline{a}\ (2,50)} H_{\underline{a}\ (2,50)}$	5023	458		
$H^T_{\underline{a}\ (3,50)} H_{\underline{a}\ (3,50)}$	6523	1649	0.63	
$H^T_{\underline{a}\ (4,50)} H_{\underline{a}\ (4,50)}$	7265	3596	0.65	0
covariance matrix	square root of eigenvalues			
$H^T_{\underline{a}\ (3,50)} H_{\underline{a}\ (3,50)}$	80.8	40.6	0.8	
$H^T_{\underline{a}\ (4,50)} H_{\underline{a}\ (4,50)}$	85.2	60	0.8	0

1) Since the ratio $\frac{0.8}{85.2} = 0.01$ obtained by the square root of $H^T_{\underline{a}\ (4,50)} H_{\underline{a}\ (4,50)}$ is small, the approximate linear system obtained by the algebraic CLS method may be a good model for the original system.
2) After determining the number n of dimensions which is 2, we execute the algebraic algorithm for approximate realization.

In this connection, the approximate linear system obtained by the algebraic Constrained Least Square (algebraic CLS) method is a two-dimensional linear system $\sigma_1 = ((\boldsymbol{R}^2, F_1),\ g_1,\ h_1)$, where $F_1 = \begin{bmatrix} 0 & -0.97 \\ 1 & 1.66 \end{bmatrix}$, $h_1 = [9.7,\ 2.5]$, $g_1 = [1,\ 0]^T$.

On the other hand, the linear system $\sigma_2 = ((\boldsymbol{R}^3,\ F_2),\ \mathbf{e}_1,\ h_2)$ obtained by the algebraic CLS method is expressed as follows:

$$F_2 = \begin{bmatrix} 0 & 0 & -0.55 \\ 1 & 0 & -0.04 \\ 0 & 1 & 1.1 \end{bmatrix},\ h_2 = [10,\ 2,\ -5].$$

In this case, the system σ_2 completely represents the original system.

We can show that the algorithm for approximate realization given by the analytic CLS method in the reference [Hasegawa, 2008] produces the same systems as the above ones in the sense of the numerical calculation.

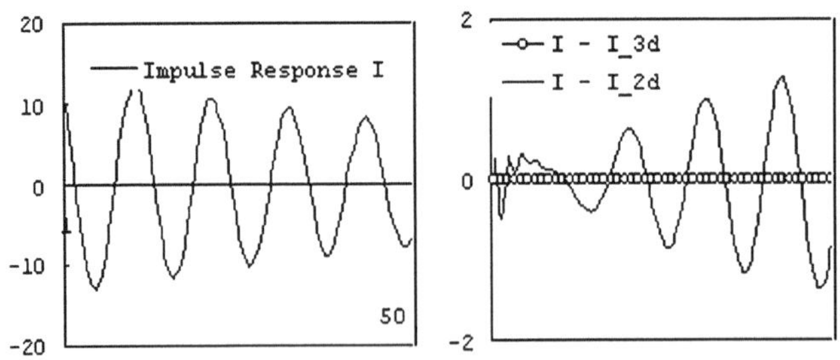

Fig. 3.1 The left is the original impulse response, the right is the difference between the original one and impulse responses approximated by a two or three-dimensional linear system in Example (3.16)

The following table is a comparison with the mean values of the square root for the sum of the square of the original signal, the obtained signal by algebraic CLS method and the error to the original signal ratio in terms of the selection of the state space dimension. This table indicates that the 2-dimensional linear system reconstructs the original signal with a 8 % error to signal ratio, and the 3-dimensional linear system completely reconstructs the original system.

Just as we thought, the following table and Fig. 3.1 indicate that the 2-dimensional linear system is a somewhat good approximation to the original 3-dimensional linear system.

dimen-sion	ratio of matrices	mean values of square root for sum of signal	signal by CLS	error	cosine ① and ②	error ratio
		①	②	③	$\cos\theta$	③/①
2	0.01	1.06	1.0	0.09	0.998	0.08
3	0	1.06	1.06	0	1	0

Example 3.17. Let a signal be the impulse response of the following 4-dimensional linear system: $\sigma = ((\boldsymbol{R}^4, F), g, h)$, where $F = \begin{bmatrix} 0\,0\,0 & 0.6 \\ 1\,0\,0 & 0.3 \\ 0\,1\,0 & -0.2 \\ 0\,0\,1 & 0.1 \end{bmatrix}$, $h = [15,\ 7,\ -10,\ 1], g = [1,\ 0,\ 0,\ 0]^T$.

Then the algebraically approximate realization problem is solved as follows:

covariance matrix	eigenvalues				
	1	2	3	4	5
$H^T_{\underline{a}\ (3,50)} H_{\underline{a}\ (3,50)}$	1776	862	138		
$H^T_{\underline{a}\ (4,50)} H_{\underline{a}\ (4,50)}$	1805	1451	238	2	
$H^T_{\underline{a}\ (5,50)} H_{\underline{a}\ (5,50)}$	2474	1483	255	2	0
covariance matrix	square root of eigenvalues				
$H^T_{\underline{a}\ (4,50)} H_{\underline{a}\ (4,50)}$	42.5	38.1	15.4	1.4	
$H^T_{\underline{a}\ (5,50)} H_{\underline{a}\ (5,50)}$	49.7	38.5	16	1.4	0

1) Since the ratio $\frac{1.4}{49.7} = 0.03$ obtained by the square root of $H^T_{\underline{a}\ (5,50)} H_{\underline{a}\ (5,50)}$ is small, the approximate linear system obtained by the algebraic CLS method may be somewhat good.

2) After determining the number n of dimensions which is 3, we execute the algebraic algorithm for approximate realization.

The approximate linear system obtained by the algebraic CLS method is a 3-dimensional linear system $\sigma_1 = ((\boldsymbol{R}^3, F_1),\ g_1,\ h_1)$, where

$$F_1 = \begin{bmatrix} 0\ 0 & 0.86 \\ 1\ 0 & -0.75 \\ 0\ 1 & 0.81 \end{bmatrix},\ h_1 = [15.4,\ 6.63,\ -9.6],\ g_1 = [1,\ 0,\ 0]^T.$$

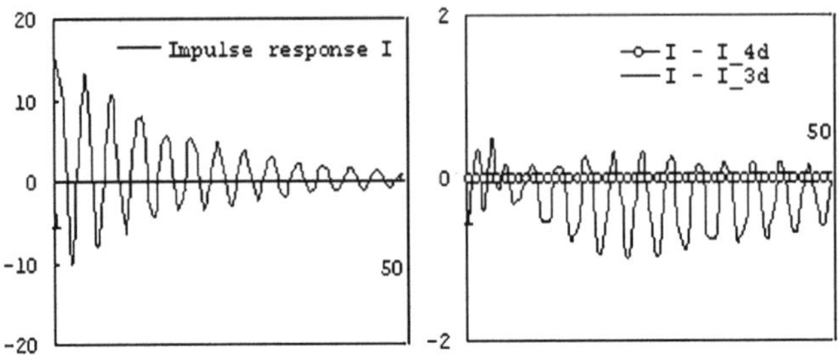

Fig. 3.2 The left is the original impulse response, the right is the difference between the original one and the impulse responses of 3 or 4-dimensional linear systems obtained by the algebraic CLS method in Example (3.17)

For reference, a 4-dimensional linear system $\sigma_2 = ((\boldsymbol{R}^4,\ F_2),\ \mathbf{e}_1,\ h_2)$ obtained by the algebraic CLS method is expressed as follows:

$$F_2 = \begin{bmatrix} 0\,0\,0 & 0.6 \\ 1\,0\,0 & 0.3 \\ 0\,1\,0 & -0.2 \\ 0\,0\,1 & 0.1 \end{bmatrix},\ h_2 = [15,\ 7,\ -10,\ 1].$$

In this case, the system σ_2 completely represents the original system.

We can show that the algorithm for approximate realization given by the analytic CLS method in the reference [Hasegawa, 2008] produces the same systems as the above ones in the sense of the numerical calculation.

For reference, in the following table, we list the mean values of the sum of the square for the original signal, the obtained signal and the error to signal ratio.

This table indicates that the 3-dimensional linear system reconstructs the original signal with a 10 % error to signal ratio, and the 4-dimensional linear system completely reconstructs the original system.

Just as we thought, the following table and Fig. 3.2 indicate that the 3-dimensional linear system is not such a good approximation to the original 4-dimensional linear system.

dimen-sion	ratio of matrices	mean values of square root for sum of signal	signal by CLS	error	cosine ① and ②	error ratio
		①	②	③	$\cos\theta$	③/①
3	0.03	0.661	0.66	0.066	0.995	0.1
4	0	0.661	0.661	0	1	0

Example 3.18. Let a signal be the impulse response of the following 4-dimensional linear system: $\sigma = ((\boldsymbol{R}^4, F), g, h)$, where $F = \begin{bmatrix} 0\,0\,0 & 0.6 \\ 1\,0\,0 & 0.55 \\ 0\,1\,0 & -0.05 \\ 0\,0\,1 & 0.2 \end{bmatrix}$,

$h = [15,\ 9,\ -5,\ 1], g = [1,\ 0,\ 0,\ 0]^T$.

Then the algebraically approximate realization problem is solved as follows:

covariance matrix	eigenvalues				
	1	2	3	4	5
$H^T_{\underline{a}\ (3,50)} H_{\underline{a}\ (3,50)}$	1052	535	188		
$H^T_{\underline{a}\ (4,50)} H_{\underline{a}\ (4,50)}$	1228	551	422	0.3	
$H^T_{\underline{a}\ (5,50)} H_{\underline{a}\ (5,50)}$	1630	575	422	0.36	0
covariance matrix	square root of eigenvalues				
$H^T_{\underline{a}\ (4,50)} H_{\underline{a}\ (4,50)}$	35	23.5	20.5	0.5	
$H^T_{\underline{a}\ (5,50)} H_{\underline{a}\ (5,50)}$	40.4	24	20.5	0.6	0

1) Since the ratio $\frac{0.6}{40.4} = 0.01$ obtained by the square root of $H^T_{\underline{a}\ (5,50)} H_{\underline{a}\ (5,50)}$ is somewhat small, the approximate linear system obtained by the algebraic CLS method may be good.
2) After determining the number n of dimensions which is 3, we execute the algebraic algorithm for the approximate realization.

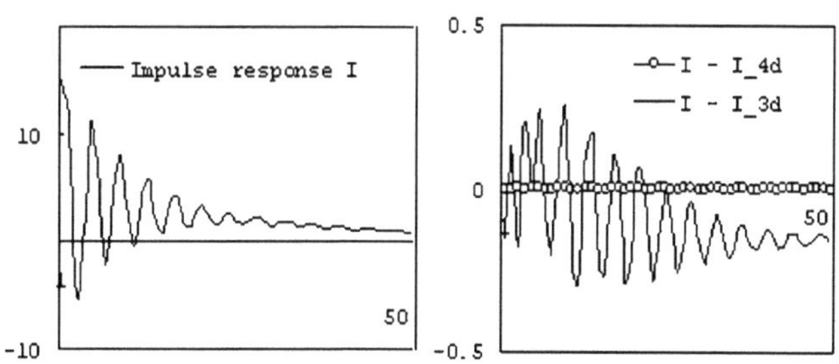

Fig. 3.3 The left is the original impulse response, the right is the difference between the original one and impulse responses of the 3 or 4-dimensional linear systems obtained by the algebraic CLS method in Example (3.18)

The approximate linear system obtained by the algebraic CLS method is a 3-dimensional linear system $\sigma_1 = ((\boldsymbol{R}^3, F_1),\ g_1,\ h_1)$, where

$$F_1 = \begin{bmatrix} 0\,0 & 0.75 \\ 1\,0 & -0.65 \\ 0\,1 & 0.85 \end{bmatrix},\ h_1 = [15.2,\ 9.9,\ -4.83],\ g_1 = [1,\ 0,\ 0]^T.$$

For reference, a 4-dimensional linear system $\sigma_2 = ((\boldsymbol{R}^4,\ F_2),\ \mathbf{e}_1,\ h_2)$ obtained by the algebraic CLS method is expressed as follows:

$$F_2 = \begin{bmatrix} 0 & 0 & 0 & 0.5 \\ 1 & 0 & 0 & 0.3 \\ 0 & 1 & 0 & -0.1 \\ 0 & 0 & 1 & 0.2 \end{bmatrix},\ h_2 = [15,\ 10,\ -5,\ 1].$$

This system completely reconstructs the original system.

We can show that the algorithm for approximate realization given by the analytic CLS method in the reference [Hasegawa, 2008] produces the same systems as the above ones in the sense of the numerical calculation.

For reference, in the following, we list the mean values of the sum of the square for the original signal, the obtained signal and the error to signal ratio.

This table indicates that the 3-dimensional linear system reconstructs the original signal with a 4 % error to signal ratio, and the 4-dimensional linear system completely reconstructs the original system.

The following table and Fig. 3.3 indicate that the 3-dimensional linear system is a somewhat good approximation to the original 4-dimensional linear system.

dimen-sion	ratio of matrices	mean values of square root for sum of			cosine	error
		signal	signal by CLS	error	① and ②	ratio
		①	②	③	$\cos\theta$	③/①
3	0.01	0.556	0.56	0.02	0.999	0.04
4	0	0.556	0.556	0	1	0

Example 3.19. Let a signal be the impulse response of the following 5-dimensional linear system: $\sigma = ((\boldsymbol{R}^5, F), g, h)$, where $F = \begin{bmatrix} 0 & 0 & 0 & 0 & 0.9 \\ 1 & 0 & 0 & 0 & -0.0384 \\ 0 & 1 & 0 & 0 & -0.112 \\ 0 & 0 & 1 & 0 & 0.52 \\ 0 & 0 & 0 & 1 & -0.4 \end{bmatrix}$,

$h = [8,\ 4,\ -2,\ -1,\ 1],\ g = [1,\ 0,\ 0,\ 0,\ 0]^T.$

Then the algebraically approximate realization problem is solved as follows:

covariance matrix	eigenvalues					
	1	2	3	4	5	6
$H^T_{\underline{a}\ (4,50)} H_{\underline{a}\ (4,50)}$	15179	2697	232	160		
$H^T_{\underline{a}\ (5,50)} H_{\underline{a}\ (5,50)}$	17816	5221	299	161	107	
$H^T_{\underline{a}\ (6,50)} H_{\underline{a}\ (6,50)}$	19129	9396	340	163	109	0
covariance matrix	square root of eigenvalues					
$H^T_{\underline{a}\ (5,50)} H_{\underline{a}\ (5,50)}$	133	72.3	17.3	12.7	10.3	
$H^T_{\underline{a}\ (6,50)} H_{\underline{a}\ (6,50)}$	138	97	18.4	13	10.4	0

1) Since the ratio $\frac{10.4}{138} = 0.075$ obtained by the square root of $H^T_{\underline{a}\ (6,50)} H_{\underline{a}\ (6,50)}$ is not so small, the approximate linear system obtained by the algebraic CLS method may be not good.
2) After determining the number n of dimensions which is 4, we execute the algebraic algorithm for the approximate realization.

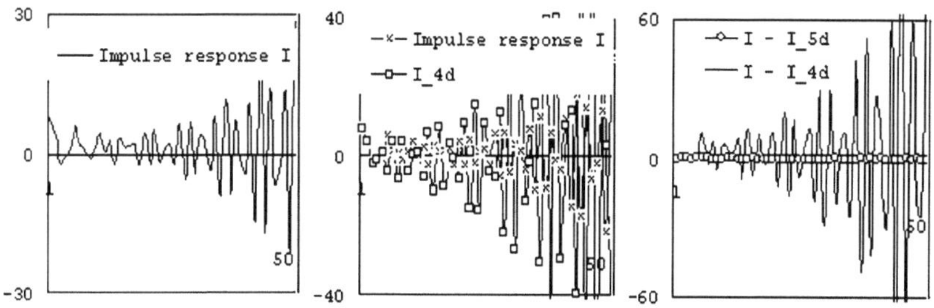

Fig. 3.4 The left is the original impulse response, the middle is the original one and the impulse responses of 4 -dimensional linear system obtained by the algebraic CLS method, and the right is the difference between the original one and impulse responses of the 4 or 5-dimensional linear systems obtained by the algebraic CLS method in Example (3.19)

The approximate linear system obtained by the algebraic CLS method is a 4-dimensional linear system $\sigma_1 = ((\boldsymbol{R}^4, F_1),\ g_1,\ h_1)$, where

$$F_1 = \begin{bmatrix} 0\ 0\ 0 & -0.197 \\ 1\ 0\ 0 & 1.002 \\ 0\ 1\ 0 & 0.961 \\ 0\ 0\ 1 & -0.792 \end{bmatrix},\ h_1 = [8.02,\ 3.91,\ -2.08,\ -0.93],\ g_1 = [1,\ 0,\ 0,\ ,0]^T.$$

For reference, a 5-dimensional linear system $\sigma_2 = ((\boldsymbol{R}^5,\ F_2),\ \mathbf{e}_1,\ h_2)$ obtained by the algebraic CLS method is expressed as follows:

$$F_2 = \begin{bmatrix} 0\ 0\ 0\ 0 & 0.9 \\ 1\ 0\ 0\ 0 & -0.0384 \\ 0\ 1\ 0\ 0 & -0.112 \\ 0\ 0\ 1\ 0 & 0.52 \\ 0\ 0\ 0\ 1 & -0.4 \end{bmatrix},\ h_2 = [8,\ 4,\ -2,\ -1,\ 1].$$

This system completely reconstructs the original 5-dimensional linear system. We can show that the algorithm for approximate realization given by the analytic CLS method in the reference [Hasegawa, 2008] produces the same systems as the above ones in the sense of the numerical calculation.

For reference, in the following table, we list the mean values of the sum of the square for the original signal, the obtained signal and the error to signal ratio.

This table indicates that the 4-dimensional linear system reconstructs the original signal with a 370 % error to signal ratio, and the 5-dimensional linear system completely reconstructs the original system.

The following table and Fig. 3.4 indicate that the 4-dimensional linear system is not a good approximation to the original 5-dimensional linear system regardless of its small ratio in cutting the number of dimensions. This result means our expectations have not been met. Why this occurs may be caused by the divergence of the impulse response.

dimen-sion	ratio of matrices	mean values of square root for sum of			cosine	error
		signal	signal by CLS	error	① and ②	ratio
		①	②	③	$\cos\theta$	③/①
4	0.075	1.19	3.68	4.43	-0.53	3.7
5	0	1.19	1.19	0	1	0

Example 3.20. Let a signal be the impulse response of the following 6-dimensional linear system: $\sigma=((\boldsymbol{R}^6, F), g, h)$, where $F=\begin{bmatrix} 0\,0\,0\,0\,0 & 0.1 \\ 1\,0\,0\,0\,0 & -0.0384 \\ 0\,1\,0\,0\,0 & 0.0272 \\ 0\,0\,1\,0\,0 & 0.164 \\ 0\,0\,0\,1\,0 & 0.2 \\ 0\,0\,0\,0\,1 & -0.5 \end{bmatrix}$,

$h = [10,\ 2,\ -5,\ -1,\ 3,\ -2]$, $g = [1,\ 0,\ 0,\ 0,\ 0,\ 0]^T$.

Then the algebraically approximate realization problem is solved as follows:

covariance matrix	eigenvalues					
	1	2	3	4	5	6	7
$H^T_{\underline{a}\,(4,50)} H_{\underline{a}\,(4,50)}$	174	59	28	3.2			
$H^T_{\underline{a}\,(5,50)} H_{\underline{a}\,(5,50)}$	177	61	40	5.8	1.1		
$H^T_{\underline{a}\,(6,50)} H_{\underline{a}\,(6,50)}$	178	64	44	5.8	1.3	0.4	
$H^T_{\underline{a}\,(7,50)} H_{\underline{a}\,(7,50)}$	182	65	46	5.9	1.5	0.4	0
covariance matrix	square root of eigenvalues						
$H^T_{\underline{a}\,(4,50)} H_{\underline{a}\,(4,50)}$	13.2	7.7	5.3	1.8			
$H^T_{\underline{a}\,(5,50)} H_{\underline{a}\,(5,50)}$	13.3	7.8	6.3	2.4	1.05		
$H^T_{\underline{a}\,(6,50)} H_{\underline{a}\,(6,50)}$	13.3	8	6.6	2.4	1.1	0.6	
$H^T_{\underline{a}\,(7,50)} H_{\underline{a}\,(7,50)}$	13.5	8.1	6.8	2.4	1.2	0.6	0

1) Since the ratio $\frac{1.8}{13.2} = 0.14$ obtained by the square root of $H^T_{\underline{a}\,(4,50)} H_{\underline{a}\,(4,50)}$ is not small, the approximate linear system obtained by the algebraic CLS method may not be good.
2) After determining the number n of dimensions which is 3, we execute the algebraically approximate realization algorithm.

The approximate linear system obtained by the algebraic CLS method is a 3-dimensional linear system $\sigma_1 = ((\boldsymbol{R}^3, F_1),\ g_1,\ h_1)$, where

$$F_1 = \begin{bmatrix} 0 & 0 & -0.46 \\ 1 & 0 & -0.64 \\ 0 & 1 & -1.18 \end{bmatrix},\ h_1 = [10.15,\ 2.21,\ -4.6],\ g_1 = [1,\ 0,\ 0]^T.$$

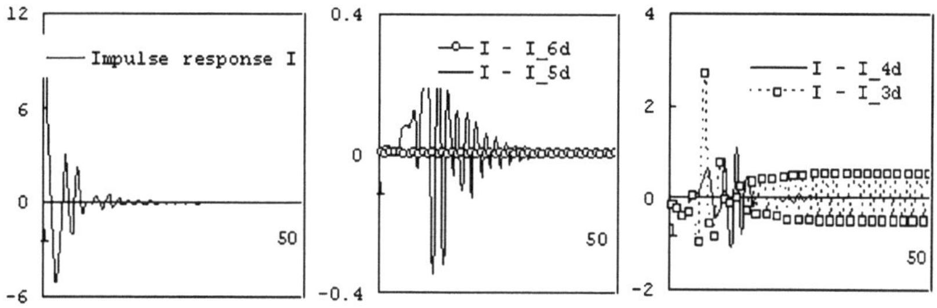

Fig. 3.5 The left is the original impulse response, the middle is the difference between the original one and impulse responses of the 5 or 6-dimensional linear systems obtained by the algebraic CLS method, and the right is the difference between the original one and impulse responses of the 3 or 4-dimensional linear systems obtained by the algebraic CLS method in Example (3.20)

3) Since the ratio $\frac{1.05}{13.3} = 0.08$ obtained by the square root of $H^T_{\underline{a}\ (5,50)} H_{\underline{a}\ (5,50)}$ is not so small, the approximate linear system obtained by the algebraic CLS method may not be so good.
4) After determining the number n of dimensions which is 4, we execute the approximate realization algorithm by the algebraic CLS method.

The approximate linear system obtained by the algebraic CLS method is a 4-dimensional linear system $\sigma_2 = ((\boldsymbol{R}^4, F_2),\ g_2,\ h_2)$, where

$$F_2 = \begin{bmatrix} 0 & 0 & 0 & -0.155 \\ 1 & 0 & 0 & -0.264 \\ 0 & 1 & 0 & -0.735 \\ 0 & 0 & 1 & -1.31 \end{bmatrix},\ h_2 = [10,\ 1.99,\ -5.02,\ -1.04],\ g_2 = [1,\ 0,\ 0,\ ,0]^T.$$

5) Since the ratio $\frac{0.6}{13.3} = 0.045$ obtained by the square root of $H^T_{\underline{a}\ (6,50)} H_{\underline{a}\ (6,50)}$ is somewhat small, the approximate linear system obtained by the CLS method may be good.
6) After determining the number n of dimensions which is 5, we execute the algebraic algorithm for the approximate realization.

The linear system $\sigma_3 = ((\boldsymbol{R}^5,\ F_3),\ \mathbf{e}_1,\ h_3)$ obtained by the algebraic CLS method is expressed as follows:

$$F_3 = \begin{bmatrix} 0\,0\,0\,0 & -0.139 \\ 1\,0\,0\,0 & -0.265 \\ 0\,1\,0\,0 & -0.267 \\ 0\,0\,1\,0 & -0.11 \\ 0\,0\,0\,1 & -0.55 \end{bmatrix}, \; h_3 = [10,\ 2,\ -5,\ -1,\ 3].$$

For reference, a 6-dimensional linear system $\sigma_4 = ((\boldsymbol{R}^6,\ F_4),\ \mathbf{e}_1,\ h_4)$ obtained by the algebraic CLS method is expressed as follows:

$$F_4 = \begin{bmatrix} 0\,0\,0\,0\,0 & 0.1 \\ 1\,0\,0\,0\,0 & -0.0384 \\ 0\,1\,0\,0\,0 & 0.0272 \\ 0\,0\,1\,0\,0 & 0.164 \\ 0\,0\,0\,1\,0 & 0.2 \\ 0\,0\,0\,0\,1 & -0.5 \end{bmatrix}, \; h_4 = [10,\ 2,\ -5,\ -1,\ 3,\ -2].$$

We note that this system σ_4 completely reconstructs the original 6-dimensional linear system σ.

We can show that the algorithm for approximate realization given by the analytic CLS method in the reference [Hasegawa, 2008] produces the same systems as the above ones in the sense of the numerical calculation.

For reference, in the following table, we list the mean values of the sum of the square for the original signal, the obtained signal and the error to signal ratio.

This table indicates that the 5-dimensional linear system reconstructs the original signal with an 8 % error to signal ratio, and the 6-dimensional linear system almost reconstructs the original system.

Fig. 3.5 and this table indicate that the 5-dimensional linear system is a somewhat good approximation to the original 6-dimensional linear system.

dimen-sion	ratio of matrices	mean values of square root for sum of signal ①	signal by CLS ②	error ③	cosine ① and ② $\cos\theta$	error ratio ③/①
3	0.14	0.244	0.246	0.087	0.94	0.36
4	0.08	0.244	0.244	0.044	0.98	0.18
5	0.045	0.244	0.243	0.02	0.997	0.08
6	0	0.244	0.244	0	1	0

3.5 Algebraically Noisy Realization of Linear Systems

In this section, we discuss the algebraically noisy realization problem of linear systems. Firstly, we must refer to the information criterion method AIC in noisy cases of linear systems which is more commonly used.

We will compare our algorithm by the algebraic CLS method with the AIC method.

In order to make our discussion simple, we assume that the set Y of outout is the set $\boldsymbol{R}$ of real numbers, namely 1-output.

***AIC* criterion 3.21**
The information criterion for linear systems is given by the following equation:

$AIC = (-2)\log(\text{maximum likelihood}) + 2\times(\text{number of unknown parameters})$.

The $AIC(\underline{N},\ n)$ of n-dimensional linear systems with the data number $\underline{N}$ is concretely expressed by $AIC(\underline{N},\ n) = \underline{N}\log((1/\underline{N}) * (\sum_{i=1}^{\underline{N}}(da(i) - \hat{da}(i))^2)) + 2 * 2 * n$, where $\{da(i) : i \leq \underline{N}\}$ are noisy original data obtained by experiments and $\{\hat{da}(i) : i \leq \underline{N}\}$ are cleaned-up signals.

A situation for algebraically noisy realization problem 3.22
Let the observed object be a linear system and noise be added to output. Then we will obtain the data $\{\gamma(t) = \hat{\gamma}(t) + \bar{\gamma}(t) : 0 \leq t \leq \underline{N}\}$ for some integer $\underline{N} \in N$, where $\hat{\gamma}(t)$ is the exact signal which comes from the observed linear system and $\bar{\gamma}(t)$ is the noise added at observation.

For a given $\{\hat{\gamma}(|\omega|) + \bar{\gamma}(|\omega|) : \omega \in U^*\}$, σ which satisfies $\{a_\sigma(\omega) \approx \hat{\gamma}(|\omega|) : \omega \in U^*\}$ is called a noisy realization of a.

We can propose the following algebraically noisy realization problem:

For a given $\{\hat{\gamma}(|\omega|)+\bar{\gamma}(|\omega|) : \omega \in U^*\}$, find, using only algebraic calculations, a linear system σ which satisfies $a_\sigma(\omega) \approx \hat{\gamma}(|\omega|)$ for any $\omega \in U^*$.

Problem statement of algebraically noisy realization for linear systems 3.23
Let $H_{\underline{a}\ (p,\bar{p})}$ be the measured finite-sized Hankel matrix. Then find the cleaned-up signal Hankel matrix $\hat{H}_{\underline{a}\ (p,\bar{p})}$ such that $H_{\underline{a}\ (p,\bar{p})} = \hat{H}_{\underline{a}\ (p,\bar{p})} + \bar{H}_{\underline{a}\ (p,\bar{p})}$ holds.

Namely, find, using only algebraic calculations, a minimal dimensional linear system $\sigma = ((\boldsymbol{R}^n, F), g, h))$ which realizes $\hat{H}_{\underline{a}\ (p,\bar{p})}$.

Theorem 3.24. *Algebraic algorithm for noisy realization*

Let $\underline{a}$ be a considered object which is a linear system. Then a noisy realization $\sigma = ((\boldsymbol{R}^n, F_s), g, h_s)$ of $\underline{a}$ is given by the following algorithm:

1) *Based on the square root of eigenvalues for a matrix $H_{\underline{a}\ (p,\bar{p})}H^T_{\underline{a}\ (p,\bar{p})}$, determine the value n of rank for the matrix $H_{\underline{a}\ (p,\bar{p})}$, where $n \leq p$. Namely, determine the value n of rank for the matrix $H_{\underline{a}\ (p,\bar{p})}$ such that a set of the square roots of eigenvalues for the covariance matrix composed of relatively small and equally-sized numbers is excluded, where the signal part effected by the set may be the noisy part.*

2) We use the algebraic CLS method as follows:
① *Based on Proposition (2.14), determine coefficients* $\{\alpha_i : 1 \leq i \leq n\}$*.*
The Q *in Proposition (2.14) can be considered as the matrix composed from the eigenvectors of* $H_{\underline{a}\ (n+1,L)} H^T_{\underline{a}\ (n+1,L)}$.
Let a matrix $A \in \boldsymbol{R}^{1\times(n+1)}$ *be* $A = [\alpha_1, \alpha_2, \cdots, \alpha_n, -1]$.
② *Determine the noisy vectors* $\{\underline{S}_l^i \bar{I}_{\underline{a}} \in \boldsymbol{R}^{L\times 1} : 0 \leq i \leq n\}$ *by using the equation* $[\bar{I}_{\underline{a}}, \underline{S}_l \bar{I}_{\underline{a}}, \cdots, \underline{S}_l^n \bar{I}_{\underline{a}}]^T := A^T[AA^T]^{-1} A H^T_{\underline{a}\ (n+1,L)}$
and $H_{\underline{a}\ (n+1,L)} := [I_{\underline{a}}, \cdots, S_l^{n-1} I_{\underline{a}}, S_l^n I_{\underline{a}}]$.
③ *Let* $F \in \boldsymbol{R}^{n\times n}$ *be given as below. Let* g *be* $g = \mathbf{e}_1$*,*
where $\mathbf{e}_1 = [1, 0, \cdots, 0]^T \in \boldsymbol{R}^n$.
④ *Let* h_s *be* $h_s = [I_{\underline{a}}(1) - \bar{I}_{\underline{a}}(1), I_{\underline{a}}(1) - \bar{I}_{\underline{a}}(1), \cdots, I_{\underline{a}}(0^{n-2}|1) - \bar{I}_{\underline{a}}(0^{n-2}|1)]$.

$$F_0 = \begin{bmatrix} 0 & \cdots & 0 & \alpha_1 \\ 1 & \ddots & & \alpha_2 \\ \vdots & \ddots & 0 & \vdots \\ 0 & & 1 & \alpha_n \end{bmatrix}.$$

[proof] By 1), the noisy part in the data can be excluded in the sense of the number of dimensions. The matrix A in 2) corresponds to the matrix A in Lemma (2.17). If we determine the coefficients $\{\alpha_i : 1 \leq i \leq n\}$, we can obtain the noise part of the finite Hankel matrices $H_{\underline{a}\ (n+1,\bar{p})}$ by using Lemma (2.17).

Therefore, we obtain the cleaned-up Hankel matrices $\hat{H}_{\underline{a}\ (n,\bar{p})}$. Finally, we apply Proposition (3.15) to the $\hat{H}_{\underline{a}\ (n+1,\bar{p})}$.

Remark 1: A determination method of the degree n in the linear system $\sigma = ((\boldsymbol{R}^n, F_s), g, h_s)$ is found in the Principal Component Method. The method is very popular.
Remark 2: Let S and N be the norm of a signal and a noise. Then the selected ratio of matrices in the algorithm may be considered as $\frac{N}{S+N}$.
Remark 3: This algebraically noisy realization method is very new.
Remark 4: For a noisy case, the AIC method is famous for determining linear systems including dimensions of the state spaces.

Definition 3.25. The algebraic algorithm for noisy realization (3.24) is called an algebraic Constrained Least Square method, abbreviated, the algebraic CLS method.

We show examples for the algebraic CLS method.

We can show that the analytic approximate realization algorithm given in the reference [Hasegawa, 2008] produces the same systems as the ones obtained by the algebraic CLS method in the sense of the numerical calculation. In addition, we compare the method with the common method for noise processing which is called AIC.

Example 3.26. Let a signal be the impulse response of the following 3-dimensional linear system: $\sigma = ((\boldsymbol{R}^3, F), \mathbf{e}_1,\ h)$,

where $F = \begin{bmatrix} 0 & 0 & 0.9 \\ 1 & 0 & 0.3 \\ 0 & 1 & -0.41 \end{bmatrix}$, $h = [10,\ 5,\ -5]$.

Let an added noise be given in Fig. 3.6.

Then the algebraically noisy realization problem is solved as follows:

covariance matrix	eigenvalues					
	1	2	3	4	5	6
$H^T_{\underline{a}\ (4,50)} H_{\underline{a}\ (4,50)}$	2412	2150	321	10.3		
$H^T_{\underline{a}\ (5,50)} H_{\underline{a}\ (5,50)}$	3239	2203	435	10.4	9.8	
$H^T_{\underline{a}\ (6,50)} H_{\underline{a}\ (6,50)}$	3830	2553	491	10.8	10	7.9
covariance matrix	square root of eigenvalues					
$H^T_{\underline{a}\ (4,50)} H_{\underline{a}\ (4,50)}$	49	46.4	18	3.2		
$H^T_{\underline{a}\ (5,50)} H_{\underline{a}\ (5,50)}$	57	47	20.8	3.3	3.1	
$H^T_{\underline{a}\ (6,50)} H_{\underline{a}\ (6,50)}$	61.9	50.5	22.2	3.3	3.2	2.8

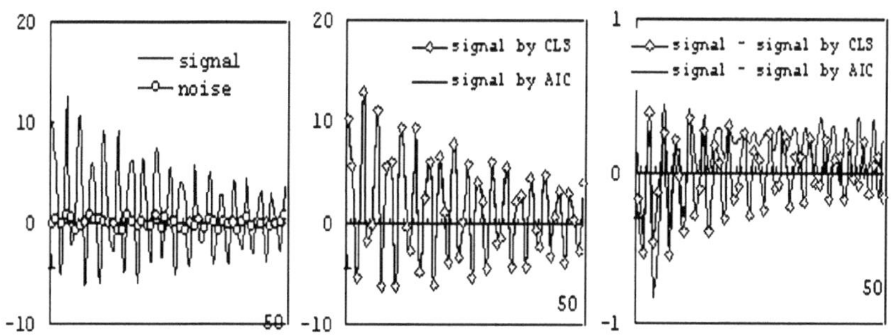

Fig. 3.6 The left is the exact signal of a 3-dimensional linear system and noise, the middle is signals by the algebraic CLS and AIC methods, and the right is the difference between the original signal and the signal obtained by the algebraic CLS or AIC methods in Example (3.26)

1) Since a set $\{3.3,\ 3.2,\ 2.8\}$ is composed of relatively small and equally-sized numbers in the square root of $H^T_{\underline{a}\ (6,50)} H_{\underline{a}\ (6,50)}$, the noisy realization of a linear system obtained by the algebraic CLS method may be good for 3-dimensional space.
2) After determining the number n of dimensions which is 3, we will continue the noisy realization algorithm by the algebraic CLS method.

Therefore, the modified impulse response $I(0)$ of a linear system obtained by the algebraic CLS method is realized by a 3-dimensional linear system $\sigma_c = ((\boldsymbol{R}^3,\ F_c),\ \mathbf{e}_1,\ h_c)$.

It is expressed as follows:

$$F_c = \begin{bmatrix} 0 & 0 & 0.9 \\ 1 & 0 & 0.3 \\ 0 & 1 & -0.4 \end{bmatrix}, \ h_c = [10.2, \ 5.54, \ -5.4].$$

We can show that the algorithm for noisy realization given by the analytic CLS method in the reference [Hasegawa, 2008] produces the same systems as the above one in the sense of the numerical calculation.

Moreover, the linear system obtained by the AIC method is a 5-dimensional linear system $\sigma_a = ((\boldsymbol{R}^5, \ F_a), \ \mathbf{e}_1, \ h_a)$ given as follows:

$$F_a = \begin{bmatrix} 0 & 0 & 0 & 0 & 0.46 \\ 1 & 0 & 0 & 0 & -0.23 \\ 0 & 1 & 0 & 0 & 0.54 \\ 0 & 0 & 1 & 0 & -0.06 \\ 0 & 0 & 0 & 1 & 0.02 \end{bmatrix}, \ h_a = [10, \ 5.4, \ -5.3, \ 13.3, \ -1.5].$$

In this example, the original signal is considered as the impulse response of a 3-dimensional linear system and the desirable impulse response is obtained by the two methods, that is, the algebraic CLS and AIC methods.

For reference, in the following table, we list the mean values of the sum of the square for the original signal, the obtained signal and the error to signal ratio.

This table indicates that a 3-dimensional linear system reconstructs the original signal with a 4 % error to signal ratio and 0.05 noise to signal ratio, please refer to Remark 2 in Theorem 3.24 for the noise to signal ratio.

The model obtained by the algebraic CLS method is a 3-dimensional linear system which has the same number of dimensions as the number of the original system. The model obtained by the AIC method is a 5-dimensional linear system.

Nevertheless, Fig. 3.6 indicates that the 5-dimensional linear system obtained by AIC is the system with the same error as in the algebraic CLS method.

dimen-sion	ratio of matrices	mean values of square root for sum of			cosine	error
		signal	signal by CLS	error	① and ②	ratio
		①	②	③	$\cos\theta$	③/①
3	0.05	0.71	0.74	0.03	1.000	0.04

Example 3.27. Let a signal be the impulse response of the following 3-dimensional linear system: $\sigma = ((\boldsymbol{R}^3, F), \mathbf{e}_1, \ h)$,

where $F = \begin{bmatrix} 0 & 0 & -0.7 \\ 1 & 0 & 0.5 \\ 0 & 1 & 0.8 \end{bmatrix}, \ h = [12, \ -8.5, \ 1].$

Let an added noise be given in Fig. 3.7.

Then the algebraically noisy realization problem is solved as follows:

covariance matrix	eigenvalues					
	1	2	3	4	5	6
$H^T_{\underline{a}\ (4,50)} H_{\underline{a}\ (4,50)}$	799	742	154	4.5		
$H^T_{\underline{a}\ (5,50)} H_{\underline{a}\ (5,50)}$	831	760	298	5.2	4.1	
$H^T_{\underline{a}\ (6,50)} H_{\underline{a}\ (6,50)}$	932	806	339	5.7	4.9	2.9
covariance matrix	square root of eigenvalues					
$H^T_{\underline{a}\ (4,50)} H_{\underline{a}\ (4,50)}$	28.3	27.2	12.4	2.1		
$H^T_{\underline{a}\ (5,50)} H_{\underline{a}\ (5,50)}$	28.8	27.6	17.3	2.3	2.1	
$H^T_{\underline{a}\ (6,50)} H_{\underline{a}\ (6,50)}$	30.5	28.4	18.4	2.4	2.2	1.7

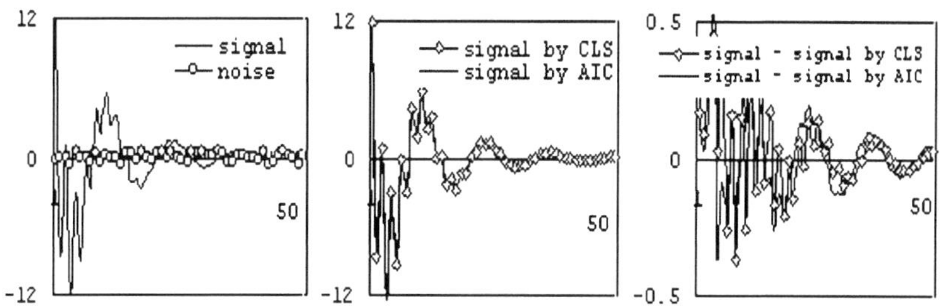

Fig. 3.7 The left is the signal of a 3-dimensional linear system and noise, the middle is signals of a 3-dimensional linear system obtained by the algebraic CLS method and a 5-dimensional linear system obtained by the AIC method, and the right is the difference between the exact signal and signal obtained by the algebraic CLS or AIC in Example (3.27)

1) Since a set $\{2.4,\ 2.2,\ 1.7\}$ is composed of relatively small and equally-sized numbers in the square root of $H^T_{\underline{a}\ (6,50)} H_{\underline{a}\ (6,50)}$, the noisy realization of a linear system obtained by the algebraic CLS method may be good for 3-dimensional space.
2) After determining the number n of dimensions which is 3, we will continue the noisy realization algorithm by the CLS method.

Therefore, the modified impulse response $I(0)$ of a linear system obtained by the algebraic CLS method is realized by a 3-dimensional linear system.

Therefore, the linear system obtained by the algebraic CLS method is a 3-dimensional linear system and the system is given by $\sigma_c = ((\boldsymbol{R}^3,\ F_c),\ \mathbf{e}_1,\ h_c)$, where

$$F_c = \begin{bmatrix} 0 & 0 & -0.71 \\ 1 & 0 & 0.53 \\ 0 & 1 & 0.77 \end{bmatrix},\ h_c = [11.8,\ -8.59,\ 0.76].$$

We can show that the algorithm for noisy realization given by the analytic CLS method in the reference [Hasegawa, 2008] produces the same systems as the above one in the sense of the numerical calculation.

Moreover, the linear system obtained by the AIC method is a 5-dimensional linear system $\sigma_a = ((\boldsymbol{R}^5, F_a), \mathbf{e}_1, h_a)$ given as follows:

$$F_a = \begin{bmatrix} 0 & 0 & 0 & 0 & -0.47 \\ 1 & 0 & 0 & 0 & 0.13 \\ 0 & 1 & 0 & 0 & -0.01 \\ 0 & 0 & 1 & 0 & 0.08 \\ 0 & 0 & 0 & 1 & 0.481 \end{bmatrix}, \; h_a = [11.7, \; -8.5, \; 0.85, \; -12.4, \; -2.7].$$

In this example, the original signal is considered as the impulse response of a 3-dimensional linear system and the desirable impulse response is obtained by the two methods, that is, the algebraic CLS and AIC methods.

For reference, in the following table, we list the mean values of the sum of the square for the original signal, the obtained signal and the error to signal ratio. This table indicates that the 3-dimensional linear system reconstructs the original signal with a 4 % error to signal ratio and 0.08 noise to signal ratio, please refer to Remark 2 in Theorem 3.24 for the noise to signal ratio.

The 3-dimensional linear system obtained by the algebraic CLS method has the same number of dimensions as the number of the original system. The model obtained by the AIC method is a 5-dimensional linear system.

Nevertheless, Fig. 3.7 indicates that the model obtained by the algebraic CLS method causes the same degree of error as the model obtained by AIC.

dimen-sion	ratio of matrices	mean values of square root for sum of			cosine	error
		signal	signal by CLS	error	① and ②	ratio
		①	②	③	$\cos\theta$	③/①
3	0.08	0.471	0.479	0.02	0.999	0.04

Example 3.28. Let a signal be the impulse response of the following 4-dimensional linear system: $\sigma = ((\boldsymbol{R}^4, F), \mathbf{e}_1, \; h)$,

$$\text{where } F = \begin{bmatrix} 0 & 0 & 0 & 0.6 \\ 1 & 0 & 0 & 0.4 \\ 0 & 1 & 0 & -0.3 \\ 0 & 0 & 1 & 0.1 \end{bmatrix}, \; h = [9, \; 15, \; -5, \; 10].$$

Let an added noise be given in Fig. 3.8.

Then the algebraically noisy realization problem is solved as follows:

covariance matrix	eigenvalues						
	1	2	3	4	5	6	7
$H^T_{\underline{a}\,(5,50)} H_{\underline{a}\,(5,50)}$	4057	2307	1924	165	10.1		
$H^T_{\underline{a}\,(6,50)} H_{\underline{a}\,(6,50)}$	4084	3509	2014	166	12.5	7.9	
$H^T_{\underline{a}\,(7,50)} H_{\underline{a}\,(7,50)}$	4403	3545	2951	166.2	13.8	10.1	7
covariance matrix	square root of eigenvalues						
$H^T_{\underline{a}\,(5,50)} H_{\underline{a}\,(5,50)}$	63.7	48	43.9	12.8	3.2		
$H^T_{\underline{a}\,(6,50)} H_{\underline{a}\,(6,50)}$	63.9	59.2	44.9	12.9	3.5	2.8	
$H^T_{\underline{a}\,(7,50)} H_{\underline{a}\,(7,50)}$	66.4	59.5	54.3	12.9	3.7	3.2	2.6

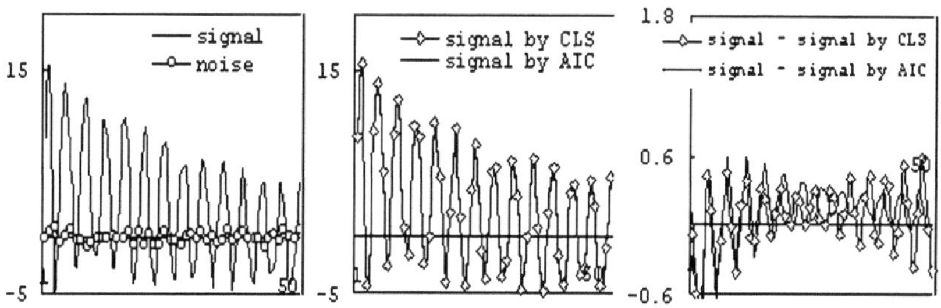

Fig. 3.8 The left is the exact signal of a 4-dimensional linear system and noise, the middle is the signals of a 4-dimensional linear system obtained by the algebraic CLS method and the signal of a 4-dimensional linear system obtained by the AIC method, and the right is the difference between the original signal and the signal by the algebraic CLS and AIC methods in Example (3.28)

1) Since a set $\{3.7,\ 3.2,\ 2.6\}$ is composed of relatively small and equally-sized numbers in the square root of $H^T_{\underline{a}\,(7,50)} H_{\underline{a}\,(7,50)}$, the algebraically noisy realization of a linear system may be good for a 4-dimensional space.
2) After determining the number n of dimensions which is 4, we will continue the noisy realization algorithm by the algebraic CLS method.

Therefore, the modified impulse response $I(0)$ of a linear system obtained by the algebraic CLS method is realized by a 4-dimensional linear system.

Therefore, the linear system obtained by the algebraic CLS method is a 4-dimensional linear system. The system is given by $\sigma_c = ((\boldsymbol{R}^4,\ F_c),\ \mathbf{e}_1,\ h_c)$,

where $F_c = \begin{bmatrix} 0\ 0\ 0\ 0.61 \\ 1\ 0\ 0\ \ 0.4 \\ 0\ 1\ 0\ {-0.3} \\ 0\ 0\ 1\ 0.08 \end{bmatrix}$, $h_c = [9.1,\ 15.6,\ -4.3,\ 9.6]$.

We can show that the algorithm for noisy realization given by the analytic CLS method in the reference [Hasegawa, 2008] produces the same systems as the above one in the sense of the numerical calculation.

Moreover, the linear system obtained by the AIC method is a 4-dimensional linear system $\sigma_a = ((\boldsymbol{R}^4,\ F_a),\ \mathbf{e}_1,\ h_a)$ given as follows:

$$F_a = \begin{bmatrix} 0 & 0 & 0 & 0.6 \\ 1 & 0 & 0 & 0.39 \\ 0 & 1 & 0 & -0.3 \\ 0 & 0 & 1 & 0.08 \end{bmatrix},\ h_a = [8.9,\ 15.5,\ -4.2,\ 9.6].$$

In this example, the original signal is considered as the impulse response of a 4-dimensional linear system and the desirable impulse response is obtained by the two methods, that is, the algebraic CLS and AIC methods.

For reference, in the following table, we list the mean values of the sum of the square for the original signal, the obtained signal and the error to signal ratio.

This table indicates that the 4-dimensional linear system obtained by the algebraic CLS method reconstructs the original signal with a 4 % error to signal ratio and 0.06 noise to signal ratio, please refer to Remark 2 in Theorem 3.24 for the noise to signal ratio.

The 4-dimensional linear system obtained by the algebraic CLS method has the same number of dimensions as the number of the original system. The AIC method also produces a 4-dimensional linear system.

Also, Fig. 3.9 indicates that the model obtained by the algebraic CLS method causes the same degree of error as the model obtained by AIC.

dimen-sion	ratio of matrices	mean values of square root for sum of			cosine	error
		signal	signal by CLS	error	① and ②	ratio
		①	②	③	$\cos\theta$	③/①
4	0.06	0.87	0.88	0.04	0.999	0.04

Example 3.29. Let a signal be the impulse response of the following 5-dimensional linear system: $\sigma = ((\boldsymbol{R}^5, F), \mathbf{e}_1,\ h)$,

$$\text{where } F = \begin{bmatrix} 0 & 0 & 0 & 0 & 0.5 \\ 1 & 0 & 0 & 0 & -0.3 \\ 0 & 1 & 0 & 0 & -0.2 \\ 0 & 0 & 0 & 1 & 0.5 \\ 0 & 0 & 0 & 1 & -0.3 \end{bmatrix},\ h = [11,\ -5,\ 3,\ -1,\ 1].$$

Let an added noise be given in Fig. 3.9.

Then the algebraically noisy realization problem is solved as follows:

covariance matrix	eigenvalues							
	1	2	3	4	5	6	7	8
$H^T_{\underline{a}\ (6,80)} H_{\underline{a}\ (6,80)}$	3721	2662	44.6	26.01	21.5	13.1		
$H^T_{\underline{a}\ (7,80)} H_{\underline{a}\ (7,80)}$	3746	3670	50.2	26.2	21.5	14.6	11.8	
$H^T_{\underline{a}\ (8,80)} H_{\underline{a}\ (8,80)}$	4757	3687	50.2	27.4	22.2	15.2	13.6	10.6
covariance matrix	square root of eigenvalues							
$H^T_{\underline{a}\ (6,80)} H_{\underline{a}\ (6,80)}$	61	51.6	6.7	5.1	4.6	3.6		
$H^T_{\underline{a}\ (7,80)} H_{\underline{a}\ (7,80)}$	61.2	60.6	7.1	5.1	4.6	3.8	3.4	
$H^T_{\underline{a}\ (8,80)} H_{\underline{a}\ (8,80)}$	69	60.7	7.1	5.2	4.7	3.9	3.7	3.3

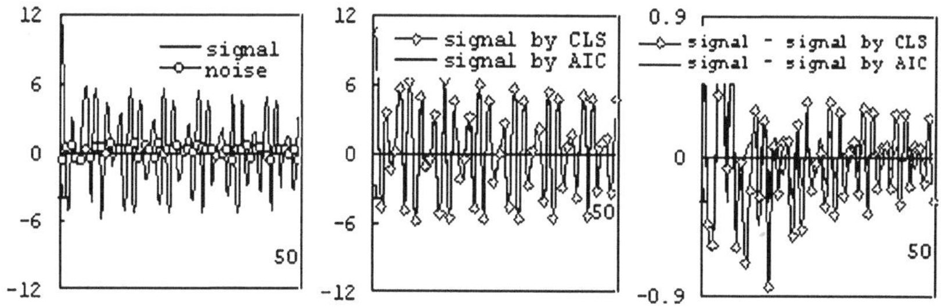

Fig. 3.9 The left is the exact signal of a 5-dimensional linear system and noise, the middle is signals of a 5-dimensional linear system obtained by the algebraic CLS method and a 8-dimensional linear system obtained by the AIC method, and the right is the difference between the signal by the CLS or AIC methods in Example (3.29)

1) Since a set {3.9, 3.7, 3.3} is composed of relatively small and equally-sized numbers in the square root of $H^T_{\underline{a}\ (8,80)} H_{\underline{a}\ (8,80)}$, the algebraically noisy realization of a linear system may be good for a 5-dimensional space.
2) After determining the number n of dimensions which is 5, we will continue the noisy realization algorithm by the algebraic CLS method.

Therefore, the linear system obtained by the algebraic CLS method is a 5-dimensional linear system. The system is given by $\sigma_c = ((\boldsymbol{R}^5,\ F_c),\ \mathbf{e}_1,\ h_c)$,

$$\text{where } F_c = \begin{bmatrix} 0\ 0\ 0\ 0 & 0.62 \\ 1\ 0\ 0\ 0 & -0.05 \\ 0\ 1\ 0\ 0 & -0.11 \\ 0\ 0\ 1\ 0 & 0.37 \\ 0\ 0\ 0\ 1 & -0.39 \end{bmatrix},\ h_c = [10.4,\ -4.6,\ 3.6,\ -1.4,\ 0.22].$$

We can show that the algorithm for noisy realization given by the analytic CLS method in the reference [Hasegawa, 2008] produces the same systems as the above one in the sense of the numerical calculation.

Moreover, the linear system obtained by the AIC method is a 8-dimensional linear system $\sigma_a = ((\boldsymbol{R}^8,\ F_a),\ \mathbf{e}_1,\ h_a)$ given as follows:

$$F_a = \begin{bmatrix} 0\ 0\ 0\ 0\ 0\ 0\ 0 & -0.16 \\ 1\ 0\ 0\ 0\ 0\ 0\ 0 & 0.32 \\ 0\ 1\ 0\ 0\ 0\ 0\ 0 & -0.03 \\ 0\ 0\ 1\ 0\ 0\ 0\ 0 & 0.25 \\ 0\ 0\ 0\ 1\ 0\ 0\ 0 & 0.006 \\ 0\ 0\ 0\ 0\ 1\ 0\ 0 & 0.06 \\ 0\ 0\ 0\ 0\ 0\ 1\ 0 & 0.15 \\ 0\ 0\ 0\ 0\ 0\ 0\ 1 & -0.29 \end{bmatrix},\ h_a = [10.3, -4.6, 3.6, -1.5, 0.3, 5.9, -4.9, 6.1].$$

In this example, the original signal is considered as the impulse response of a 5-dimensional linear system and the desirable impulse response is obtained by two methods, that is, the algebraic CLS and AIC methods.

For reference, in the following table, we list the mean values of the sum of the square for the original signal, the obtained signal and the error to signal ratio.

This table indicates that the 5-dimensional linear system obtained by the algebraic CLS method reconstructs the original signal with a 9 % error to signal ratio and 0.06 noise to signal ratio, please refer to Remark 2 in Theorem 3.24 for the noise to signal ratio.

The 5-dimensional linear system obtained by the algebraic CLS method has the same number of dimensions as the number of the original system.

The AIC method produces a 8-dimensional linear system.

Nevertheless, Fig. 3.9 indicates that the model obtained by the algebraic CLS method causes the same degree of error as the model obtained by AIC.

dimen-	ratio of	mean values of square root for sum of			cosine	error
sion	matrices	signal	signal by CLS	error	① and ②	ratio
		①	②	③	$\cos\theta$	③/①
5	0.06	0.588	0.616	0.05	0.997	0.09

Example 3.30. Let a signal be the impulse response of the following 6-dimensional linear system: $\sigma = ((\boldsymbol{R}^6, F), \mathbf{e}_1,\ h)$,

$$\text{where } F = \begin{bmatrix} 0\,0\,0\,0\,0 & 0.1 \\ 1\,0\,0\,0\,0 & -0.4 \\ 0\,1\,0\,0\,0 & -0.3 \\ 0\,0\,0\,1\,0 & 0.2 \\ 0\,0\,0\,1\,0 & 0.5 \\ 0\,0\,0\,0\,1 & -0.5 \end{bmatrix},\ h = [10,\ 2,\ -5,\ -1,\ 3,\ -2].$$

Let an added noise be given in Fig. 3.10.

Then the algebraically noisy realization problem is solved as follows:

covariance matrix	eigenvalues							
	1	2	3	4	5	6	7	8
$H^T_{\underline{a}\,(6,20)} H_{\underline{a}\,(6,20)}$	199.5	140.3	97.3	23.7	21	11		
$H^T_{\underline{a}\,(7,20)} H_{\underline{a}\,(7,20)}$	233	143	97.6	26.2	23.4	13.5	2.6	
$H^T_{\underline{a}\,(8,20)} H_{\underline{a}\,(8,20)}$	242	151	98.8	27.5	23.9	18.7	3.6	1.8
covariance matrix	square root of eigenvalues							
$H^T_{\underline{a}\,(6,20)} H_{\underline{a}\,(6,20)}$	14.1	11.8	9.9	4.9	4.6	3.3		
$H^T_{\underline{a}\,(7,20)} H_{\underline{a}\,(7,20)}$	15.2	11.9	9.9	5.1	4.8	3.7	1.6	
$H^T_{\underline{a}\,(8,20)} H_{\underline{a}\,(8,20)}$	15.6	12.3	9.9	5.2	4.9	4.3	1.9	1.3

1) Since a set $\{1.9,\ 1.3\}$ is composed of relatively small and equally-sized numbers in the square root of $H^T_{\underline{a}\,(8,20)} H_{\underline{a}\,(8,20)}$, the algebraically noisy realization of a linear system may be good for a 6-dimensional space.

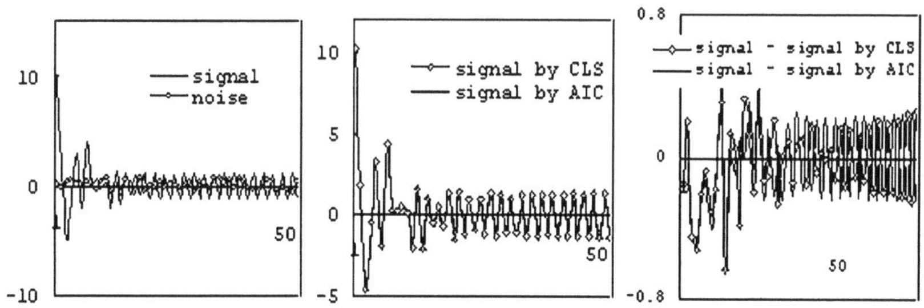

Fig. 3.10 The left is the exact signal of a 6-dimensional linear system and noise, the middle is signals of a 6-dimensional linear system obtained by the algebraic CLS method and a 8-dimensional linear system obtained by the AIC method, and the right is the difference between the exact signal and the signal by the algebraic CLS or AIC methods in Example (3.30)

2) After determining the number n of dimensions which is 6, we will continue the noisy realization algorithm by the algebraic CLS method.

Therefore, the linear system obtained by the algebraic CLS method is a 6-dimensional linear system. The system is given by $\sigma_c = ((\boldsymbol{R}^6,\ F_c),\ \mathbf{e}_1,\ h_c)$,

$$\text{where } F_c = \begin{bmatrix} 0\,0\,0\,0\,0 & 0.14 \\ 1\,0\,0\,0\,0 & -0.44 \\ 0\,1\,0\,0\,0 & -0.26 \\ 0\,0\,1\,0\,0 & 0.34 \\ 0\,0\,0\,1\,0 & 0.53 \\ 0\,0\,0\,0\,1 & -0.51 \end{bmatrix},\ h_c = [10.2,\ 1.8,\ -4.6,\ -0.47,\ 3.2,\ -1.92].$$

We can show that the algorithm for noisy realization given by the analytic CLS method in the reference [Hasegawa, 2008] produces the same systems as the above one in the sense of the numerical calculation.

Moreover, the linear system obtained by the AIC method is a 8-dimensional linear system $\sigma_a = ((\boldsymbol{R}^8,\ F_a),\ \mathbf{e}_1,\ h_a)$ given as follows:

$$F_a = \begin{bmatrix} 0\,0\,0\,0\,0\,0\,0 & 0.04 \\ 1\,0\,0\,0\,0\,0\,0 & -0.25 \\ 0\,1\,0\,0\,0\,0\,0 & -0.16 \\ 0\,0\,1\,0\,0\,0\,0 & -0.25 \\ 0\,0\,0\,1\,0\,0\,0 & 0.15 \\ 0\,0\,0\,0\,1\,0\,0 & 0.1 \\ 0\,0\,0\,0\,0\,1\,0 & -0.16 \\ 0\,0\,0\,0\,0\,0\,1 & -0.73 \end{bmatrix},\ h_a = [10.1, 1.84, -4.5, -0.5, 3.2, -1.87, 4.4, 0.33].$$

In this example, the original signal is considered as the impulse response of a 6-dimensional linear system and the desirable impulse response is obtained by the two methods, that is, the algebraic CLS and AIC methods.

For reference, in the following table, we list the mean values of the sum of the square for the original signal, the obtained signal and the error to signal ratio.

This table indicates that the 6-dimensional linear system obtained by the algebraic CLS method reconstructs the original signal with a 10 % error to signal ratio and 0.12 noise to signal ratio, please refer to Remark 2 in Theorem 3.24 for the noise to signal ratio.

The 6-dimensional linear system obtained by the algebraic CLS method has the same number of dimensions as the number of the original system.

The AIC method produces a 8-dimensional linear system.

Nevertheless, Fig. 3.10 indicates that the model obtained by the algebraic CLS method causes the same degree of error as the model obtained by the AIC.

dimen-sion	ratio of matrices	mean values of square root for sum of			cosine	error
		signal	signal by CLS	error	① and ②	ratio
		①	②	③	$\cos\theta$	③/①
6	0.12	0.289	0.298	0.03	0.995	0.10

3.5.1 *Comparative Table of the Algebraic CLS and AIC Method*

We have proposed an algebraically noisy realization problem to clean up noise from actual observed data. To solve the problem, up to now, we could only use the analytic CLS method and the AIC method. Here we have introduced a new method which is the algebraic CLS method. However, it is well-known that the analytic method is troublesome and that the AIC method is only applied to linear systems.

Here, we list the difference between the algebraic CLS and AIC methods through our examples.

The error in the next table means the mean value of the square root of the following value: $(1/V) * \sum_{i=1}^{V}(\text{signal}(i) - \text{obtained signal}(i))^2$.

No.	26	27	28	29	30
dim.	3	3	4	5	6
CLS	3	3	4	5	6
error	0.03	0.02	0.04	0.05	0.03
AIC	5	5	4	8	8
error	0.038	0.02	0.04	0.034	0.036

'No.' denotes the number of examples in this chapter.
'dim.' denotes the number of dimensions of the original systems.
Numbers in the upper stand denote the number of dimensions of the obtained ones.
Numbers in the lower stand denote the root mean square error.

3.6 Historical Notes and Concluding Remarks

Algebraically approximate realization and noisy realization problems of linear systems were studied with the notion of the ratio of Hankel norm and the algebraic CLS method. The ratio of Hankel norm is used for determining the dimension of a state space and the algebraic CLS method is used for determining the parameters of linear systems.

In our treatment of its approximate and noisy realization problems, there may be an advantage in using singular value decomposition and algebraically Constrained Least Square (CLS). In the reference [Kalman, 1997], Kalman pointed out that the identification problem from noisy data should be treated without any prejudice, hence, should be approached in a statistical sense, not in a probabilistic sense. Here, we only insist that the signal and the noise are not correlated. Allowing for this, we could discuss algebraically approximate and noisy realization problems for linear systems with a unified method.

Since our determination method of the dimensions for linear systems is directly executed without any restrictions, our method is very useful and convenient for both approximate realization, equivalency, model reduction, and noisy realization problems.

However, we cannot fully apply the approximate realization algorithm to impulse responses which increase in numerical value, which can be seen in Example 3.19. For the noisy realization problem, we cannot fully apply the noisy realization algorithm to impulse responses which have rapid damping or values near zero, which can be seen in Examples 3.30.

In multivariable analysis which is a traditional method for analysis in economics, biology, psychology and others, it is known that the factor number is determined by the number of eigenvalues of the covariance matrix which are greater than one. Our determination for the dimension of linear systems is based on the ratio of the Hankel norm. This direction is presented by showing examples.

For normal noisy realization, we can easily perform AIC as a typical example. It is known that this method has been proposed with the notion of both ideas of statistical and probabilistic view points. Therefore, the AIC method is considered as a very technical idea. Since we only stress statistical idea for our treatment of noisy data, we can easily connect the idea of the Hankel norm and Constrained Least Square method.

In order to show that our method for approximate and noisy realization is effective, we provided several examples. Based on the result of these examples, we demonstrated that the ratio of the square root of singular values imply the degree of approximation. For our noisy realization problems, we have demonstrated that we can determine the dimension of linear systems when the set of equally-sized numbers of the square root of singular values can be found.

Roughly speaking, our several examples for the algebraically approximate realization problem suggest that the smaller the ratio of the matrix norm is, the smaller the error to signal ratio is. The change of ratio ranges from one to eight percent for the error to signal ratio per 0.01 ratio of the matrix norm.

The several examples suggest that three unique features can be expressed as follows:

(1) : The ratio of the matrix norm determines the degree of the crossed angle between directions of the approximated signal and the original signal.
(2) : We could propose a new law which says that linear systems obtained by the algebraic CLS method are the same as ones obtained by the analytic CLS method proposed in the reference [Hasegawa, 2008]. The law is said to be a law of a constrained least square.
(3) : The algebraic CLS method determines the coefficients of linearly dependent vectors such that the error between the approximate signal and the original signal has a minimum value in the sense of a square norm while conserving the crossed angle.

Intuitively, our several examples for the algebraically noisy realization problem show that the smaller the ratio of the matrices is, the smaller the error to signal ratio is. The change ratio ranges from one to two percent by the error to signal ratio per 0.01 ratio of the matrix norm.

The several examples suggest that three unique features can be expressed as follows:

(1) : The ratio of matrices determines the degree of the crossed angle between directions of the obtained signal and the original signal.
(2) : We could propose a new law which says that linear systems obtained by the algebraic CLS method are the same as ones obtained by the analytic CLS method proposed in the reference [Hasegawa, 2008].
(3) : The algebraic CLS method determines the coefficients of linearly dependent vectors such that the error between the obtained signal and original signal has a minimum value in the sense of a square norm while conserving the crossed angle.

In subsection (3.5.1), we compared our algebraic CLS method with the AIC method and we demonstrated that the algebraic CLS method is more useful and easier than the AIC method in the sense of noisy realization because our algebraic CLS method especially results in less dimensional state space than the AIC method.

We want to state that our algebraic approximate and noisy realizations were developed by our algebraic realization procedure for obtaining the reachable standard system from a given input response map and a partial realization algorithm.

The analtic CLS method for determing n variables is reduced to the minimization of the following rational polynomial $p(x_1, x_2, \cdots, x_n)$ in n variables:

$$p(x_1, x_2, \cdots, x_n) = \frac{\sum_{c_1, c_2, \cdots, c_n} \alpha(c_1, c_2, \cdots, c_n) \times x_1^{c_1} x_2^{c_2} \cdots x_n^{c_n}}{(1 + x_1^2 + x_2^2 + \cdots + x_n^2)^2},$$

where $\sum_{c_1, c_2, \cdots, c_n}$ means all summations of any combination with the conditions $0 \le c_i \le 4$ and $c_1 + c_2 + \cdots + c_n \le 4$ for any $1 \le i \le n$ and $\alpha(c_1, c_2, \cdots, c_n) \in \boldsymbol{R}$.

Therefore, our new Law shows that approximate and noisy problems can be solved using only algebraic calculation, namely, without treating partial differential equations.

Chapter 4
Algebraically Approximate and Noisy Realization of So-called Linear Systems

Let the set Y of output's values be a linear space over the real number field $\boldsymbol{R}$.

Almost linear systems were introduced in the monograph [Matsuo and Hasegawa, 2003], and it was also shown that the systems contain so-called linear systems as a sub-class, where so-called linear systems are linear systems with a non-zero initial state.

It is well known that a common method to obtain so-called linear systems is solved through two problems.

One is the realization problem to obtain linear systems with a zero initial state and the other is the state estimation problem for systems with a non-zero initial state. Based upon the prejudice that so-called linear systems are completely the same as linear systems, so-called linear systems were treated separately.

In the monograph, it was also shown that so-called linear systems can be obtained from input/output data from a single experiment.

In this chapter, based on the results regarding so-called linear systems, we will discuss algebraically approximate and noisy realization of the systems. For our discussion, we will present for the first time a concrete and easy method to discuss algebraically approximate and noisy realization problems from partial data, equivalently, i.e., data obtained in finite real time. Hence, this new method is very useful and practical.

Note that because of the system's nonlinearity, these problems were discussed by using the analytic CLS method the first time in the reference [Hasegawa, 2008].

In order to be self-contained, we will list the main results needed for our discussion from our monograph.

In order to solve our problems, we will use singular value decomposition and the algebraically Constrained Least Square method, which is abbreviated to the algebraic CLS method which has been discussed in Chapter 2. The singular value decomposition is used to determine the dimension of so-called

Y. Hasegawa: Algebra. Approx. & Noisy Reali. of Discrete-Time Sys., LNEE 50, pp. 53–85.
springerlink.com

linear systems and the algebraic CLS method is used to determine parameters of a so-called linear system.

At first, we will discuss algebraically approximate realization problems and give many examples to ascertain the effectiveness of our algorithm. Next, we will discuss algebraically noisy realization problems and give several examples to ascertain the effectiveness of our algebraically noisy realization algorithm. Both an approximate realization problem and a noisy realization problem will be discussed through executing only algebraic operations in comparison with the analytic CLS method in the reference [Hasegawa, 2008].

4.1 Basic Facts about So-called Linear Systems

Definition 4.1. So-called Linear Systems

1) A system given by the following system equation is said to be a so-called linear system $\sigma = ((X,F),x^0,g,h)$. This system is a linear system with a non-zero initial state.

$$\begin{cases} x(t+1) = Fx(t) + g\omega(t+1) \\ x(0) \quad\;\; = x^0 \\ \gamma(t) \quad\;\;\; = hx(t) \end{cases},$$

where $F \in L(X)$, $\omega(t+1) \in U$, $g,\ x^0 \in X$. In addition, h is a linear operator : $X \to Y$ for any $t \in N$, $\gamma(t) \in Y$.

2) The input response map $a_\sigma : U^* \to Y; \omega \mapsto h(\sum_{j=1}^{|\omega|} F^{|\omega|-j}(Fx^0 + g\omega(j))$ is said to be the behavior of σ.

3) For the so-called linear system σ and any $i \geq 1$, $I_\sigma(1)(i) := a_\sigma(0^i|1) - a_\sigma(0^i) = hF^i(g^0+g)$ and $I_\sigma(0)(i) := a_\sigma(0^{i+1}) - a_\sigma(0^i) = hF^ig^0$ are said to be modified impulse responses of σ, where $0^0 := 1,\ g^0 := Fx^0 - x^0$.

Note that there is a one-to-one correspondence between the behavior of σ and the modified impulse responses $I_\sigma(0)$ and $I_\sigma(1) \in F(N,Y)$ of σ by the relations $a_\sigma(\omega) = (\sum_{j=1}^{|\omega|}(I_\sigma(0)(|\omega|-j+1) + I_\sigma(1)(|\omega|-j+1) \times \omega(j))$.

4) A so-called linear system σ is said to be reachable if the reachable set $\{\sum_{j=1}^{|\omega|} F^{|\omega|-j}(g^0+g\omega(j)); \omega \in U^*\}$ is equal to X and the system σ is called to be observable if $hF^ix_1 = hF^ix_2$ for any $i \in N$ implies $x_1 = x_2$, where $g^0 := Fx^0 - x^0$.

5) A so-called linear system σ is called canonical if σ is reachable and observable.

Remark 1: It is meant for σ to be a faithful model for the input response map a that σ realizes a.

Remark 2: Notice that a canonical so-called linear system $\sigma = ((X,F),x^0,g,h)$ is a system that has the most reduced state space X among systems that have the behavior a_σ.

Proposition 4.2. *For any so-called linear system* $\tilde{\sigma} = ((X, F), x^0, g, h)$, *there exists an almost linear system* $\sigma = ((X, F), g^0, g, h, h^0)$ *with the same input/output relation which satisfies* $g^0 = Fx^0 - x^0$ *and* $h^0 = hx^0$.

Remark: For details of almost linear systems, see Definition (5.1) in Chapter 5.

Lemma 4.3. *Let* $\sigma = ((X, F), x^0, g, h)$ *be a canonical (controllable and observable) so-called linear system, then the almost linear system* σ *obtained by Proposition (4.2) is intrinsically canonical.*

Conversely, let $\sigma = ((X, F), g^0, g, h, h^0)$ *be an intrinsically canonical almost linear system, then so-called linear system* $\tilde{\sigma}$ *obtained by Proposition (4.2) is canonical.*

Example 4.4. Let $F(N, Y) := \{$ any function $f : N \to Y\}$. Let $S_l\gamma(t) = \gamma(t+1)$ for any $\gamma \in F(N, Y)$ and $t \in N$, then $S_l \in L(F(N, Y))$. Let a map $\chi^0 \in F(N, Y)$ be $(\chi^0)(t) := a(\omega|0) - a(\omega)$ and $\chi \in F(N, Y)$ be $\chi(t) := a(\omega|1) - a(\omega)$ for any $t \in N$, a time-invariant, affine input response map $a \in F(U^*, Y)$ and $\omega \in U^*$ such that $|\omega| = t$. Moreover, let a linear map 0 be $F(N, Y) \to Y; \gamma \mapsto \gamma(0)$. Then a collection $((F(N, Y), S_l), \chi^0, \chi, 0, a(1))$ is an observable almost linear system that realizes a.

Theorem 4.5. *The following almost linear system is the canonical realizations of any time-invariant, affine input response map* $a \in F(U^*, Y)$.
$((\ll S_l^N(\chi(U)) \gg, S_l), \chi^0, \chi, 0, a(1))$,
where $\ll S_l^N(\chi(U)) \gg$ *is the smallest linear space that contains* $S_l^N(\chi(U)) := \{S_l^i(\chi^0 + \chi \times u); u \in \boldsymbol{R}, i \in N,\ S_l^i(\chi^0 + \chi u)(t) = (\chi(u)(t+1) = a(\omega|u) - a(\omega), \omega \in U^*\}$.

Proposition 4.6. *Let* $\sigma = ((\ll S_l^N(\chi(U)) \gg, S_l), \chi^0, \chi, 0, a(1))$ *be the intrinsically canonical almost linear system which is given in Theorem (4.5).*

The so-called linear system $((\ll S_l^N(\chi(U)) \gg, S_l), x^0, \chi, 0))$ *is given by* σ *if and only if there exists a* $x^0 \in \ll S_l^N(\chi(U)) \gg$ *such that* $\chi^0 = S_l x^0 - x^0$ *and* $a(1) = 0x^0$.

Definition 4.7. Let $\sigma_1 = ((X_1, F_1), g_1^0, g_1, h_1, h^0)$ and $\sigma_2 = ((X_2, F_2), g_2^0, g_2, h_2, h^0)$ be almost linear systems. Then a linear operator $T : X_1 \to X_2$ is said to be an almost linear system morphism $T : \sigma_1 \to \sigma_2$ if T satisfies $TF_1 = F_2T$, $Tg_1^0 = g_2^0$, $Tg_1 = g_2$ and $h_1 = h_2T$.
If $T : X_1 \to X_2$ is bijective, then $T : \sigma_1 \to \sigma_2$ is said to be an isomorphism.

Corollary 4.8. *Let* T *be an almost linear system morphism* $T : \sigma_1 \to \sigma_2$. *Then* $a_{\sigma_1} = a_{\sigma_2}$ *holds.*

4.2 Finite Dimensional So-called Linear Systems

We will state facts regarding finite-dimensional so-called linear systems in this section. Since many results of so-called linear systems have been shown in a monograph [Matsuo and Hasegawa, 2003], the main results are cited from the monograph.

Firstly, we introduce conditions in which a finite dimensional so-called linear system is canonical.

Secondly, we introduce a canonical form which is suitable for approximate and noisy realization problems.

Namely, we introduce a standard system as a representative in their equivalence classes.

Thirdly, we introduce a criterion for the behavior of finite dimensional so-called linear systems, i.e., a rank condition of infinite Input/output matrix.

Lastly, we introduce a procedure to obtain a real-time standard system which realizes a given input response map.

There is a fact regarding finite dimensional linear spaces that an n-dimensional linear space over the field $\boldsymbol{R}$ is isomorphic to $\boldsymbol{R}^n$ and $L(\boldsymbol{R}^n, \boldsymbol{R}^m)$ is isomorphic to $\boldsymbol{R}^{m\times n}$ (See Halmos [1958]). Therefore, without loss of generality, we can consider a n-dimensional linear system as $\sigma = ((\boldsymbol{R}^n, F), g, h)$, where $F \in \boldsymbol{R}^{n\times n}$, $g \in \boldsymbol{R}^n$ and $h \in \boldsymbol{R}^{p\times n}$.

Definition 4.9. For any time-invariant, affine input response map $a \in F(U^*, Y)$, the corresponding linear input/output map $A : (A(N \times \{0,1\}), S_r) \to (F(N,Y), S_l)$ satisfies $A(\mathbf{e}_{(\mathbf{s},\mathbf{u})})(t) = a(u^{s+t+1}) - a(u^{s+t})$ for any $u \in \{0,1\}$.
Therefore, A is represented by the next infinite matrix $(I/O)_a$.
This $(I/O)_a$ is said to be an Input/output matrix of a.
For the $(A(N \times \{0,1\}), S_r)$, see Example (5.2) in chapter 5.

$$
(I/O)_a = \begin{array}{c} \\ \\ \\ t \end{array}
\begin{pmatrix}
 & & (s,u) \\
 & & \vdots \\
 & & \vdots \\
\cdots & \cdots & a(u^{s+t+1}) - a(u^{s+t})
\end{pmatrix}
$$

Note that for the linear input/output map $A : A(N \times \{0,1\}) \to F(N,Y)$, there exists a unique function $I_a : \{0,1\} \to F(N,Y)$ such that $I_a(u)(i+j) = A(\mathbf{e}_{(\mathbf{i},\mathbf{u})})(j) = a(u^{i+j+1}) - a(u^{i+j})$ holds for $u \in \{0,1\}$.

Also note that column vectors of $(I/O)_a$ denote $S_l^i I_a(u)$.

Theorem 4.10. *Theorem for existence criterion*

For a time-invariant, affine input response map $a \in F(U^*, Y)$, the following conditions are equivalent:

1) The input response map $a \in F(U^*, Y)$ has the behavior of a n-dimensional canonical almost linear system.

2) There exist n linearly independent vectors and no more than n linearly independent vectors in a set $\{S_l^i I_a(u) \in \ll S_l^N(\chi(U)) \gg; i \leq n$ *for* $i \in N, u \in \{0,1\}\}$.

3) The rank of the Input/output matrix $(I/O)_a$ *of a is n.*

Definition 4.11. Let $\sigma_r = ((\boldsymbol{R}^n, F_r), g_r^0, g_r, h_r, h_r^0)$ be a canonical almost linear system. The σ_r which satisfies the following conditions is called a real time standard system.

1) $g_r^0 = \mathbf{e}_1$, $F_r^{i-1} g_r^0 = \mathbf{e}_i$, $1 \leq i \leq n_1$ and $F_r^{n_1} g_r^0 = \sum_{i=1}^{n_1} \alpha_{1i} F_r^{i-1} g_r^0$, $\alpha_{1i} \in \boldsymbol{R}$ hold.
2) $g_r^0 + g_r = \mathbf{e}_{n_1+1}$, $F_r^{i-1}(g_r^0 + g_r) = \mathbf{e}_{n_1+i}$, $1 \leq i \leq n_2$ and $F_r^{n_2}(g_r^0 + g_r) = \sum_{i=1}^{n_1} \alpha_{2i} F_r^{i-1} g_r^0 + \sum_{i=n_1+1}^{n_1+n_2} \alpha_{2i} F_r^{i-1} g_r$, α_{1i}, $\alpha_{2i} \in \boldsymbol{R}$ hold.
3) $n = n_1 + n_2$ holds.
4) F_r is given as follows:

$$F_r = \begin{bmatrix}
0 & \cdots & 0 & \alpha_{11} & 0 & \cdots & \cdots & 0 & \alpha_{21} \\
1 & \ddots & & \alpha_{12} & 0 & \cdots & & 0 & \alpha_{22} \\
\vdots & \ddots & 0 & \vdots & \vdots & & & \vdots & \vdots \\
0 & & 1 & \alpha_{1n_1} & \vdots & & & \vdots & \alpha_{2n_1} \\
0 & 0 & \cdots & 0 & 0 & \cdots & \cdots & 0 & \alpha_{2n_1+1} \\
0 & 0 & \cdots & \vdots & 1 & \ddots & & \vdots & \alpha_{2n_1+2} \\
0 & 0 & \cdots & \vdots & 0 & \ddots & \ddots & \vdots & \vdots \\
0 & \ddots & \cdots & \vdots & \vdots & \ddots & 1 & 0 & \alpha_{2n-1} \\
0 & 0 & \cdots & 0 & 0 & \cdots & 0 & 1 & \alpha_{2n}
\end{bmatrix}.$$

For the real time standard system $\sigma_r = ((\boldsymbol{R}^n, F_r), g_r^0, g_r, h_r, h_r^0)$, its modified impulse responses $I(0)(i) := h_r F_r^i g_r^0$ and $I(1)(i) := h_r F_r^i (g_r^0 + g_r)$ may be written as $I(0)\_(n_1, n_2)$ and $I(1)\_(n_1, n_2)$ respectively.

Theorem 4.12. *Representation Theorem for equivalence classes*

For any finite dimensional canonical almost linear system, there exists a uniquely determined isomorphic real time standard system.

[proof] Let $\sigma = ((\boldsymbol{R}^n, F), g^0, g, h, h^0)$ be any finite dimensional canonical almost linear system. We select the set of linearly n independent vectors $\{g^0, Fg^0, F^2 g^0, \cdots, F^{n_1-1} g^0, g^0+g, F(g^0+g), F^2(g^0+g), \cdots, F^{n_2-1}(g^0+g); n = n_1+n_2\}$ among $\{F^i g^0, F^j g; 1 \leq i \leq n, 1 \leq j \leq n\}$ in the order of a set $\{g^0, Fg^0, F^2 g^0, \cdots, F^{n_1-1} g^0, g^0+g, F(g^0+g), F^2(g^0+g), \cdots, F^{n_2-1}(g^0+g)\}$. Then we introduce a linear operator $T : \boldsymbol{R}^n \to \boldsymbol{R}^n$ by setting $TF^{i-1} g^0 = \mathbf{e}_i$ for $i (1 \leq i \leq n_1)$ and $TF^{j-1}(g^0 + g) = \mathbf{e}_{n_1+j}$ for j $(1 \leq j \leq n_2)$, then T is a regular matrix. Let $F_r := TFT^{-1}$. Then $F_r \in \boldsymbol{R}^{n \times n}$. Since T is a regular matrix, $T : \boldsymbol{R}^n \to \boldsymbol{R}^n$ preserves linear independence and dependence. Also, T satisfies the equations $F_r T = TF, Tg^0 = g_r^0$ and $Tg = g_r$

by the construction of T. Let $h_r = hT^{-1}$. Then T is an almost linear system morphism : $\sigma = ((K^n, F), g^0, g, h, h^0) \to \sigma_r = ((K^n, F_r), g_r^0, g_r, h_r, h^0)$. T is bijective and σ_r is the only real time standard system by the selection of T. By Corollary (4.9), the behaviors of σ and σ_r are the same.

Moreover, we can show that its uniqueness comes from the selection of $\{F^i g^0,\ F^j(g^0+g)\ 1 \le n_1,\ n_2 \le n,\ n = n_1 + n_2\}$.

Theorem 4.13. *Theorem for a realization procedure*

Let a time-invariant, affine input response map $a \in F(U^*, Y)$ satisfy rank $(I/O)_a = n$. Then the real time standard system $\sigma_r = ((\boldsymbol{R}^n, F_r), g_r^0,\ g_r,\ h_r,\ a(1))$ which realizes a is obtained by the following procedure:

1) Select n_1 independent vectors on the vectors $\{S_l^s I_a(0) : 0 \le s \le n\}$. Next select n_2 independent vectors in $\{S_l^s I_a(1) : 0 \le s \le n\}$.
2) Let the state space be $\boldsymbol{R}^n$. And let g_r^0 and g_r be as follows: $g_r^0 = \mathbf{e}_1$, $g_r^0 + g_r = \mathbf{e}_{n_1+1}$. Moreover, $n = n_1 + n_2$ and $\mathbf{e}_i = [0, \cdots, 0, \overset{i}{1}, 0, \cdots, 0]^T$ hold.
3) $F_r \in \boldsymbol{R}^{n \times n}$ is given in Definition (4.11), where $S_l^{n_1} I_a(0) = \sum_{i=1}^{n_1} \alpha_{1i} S_l^{i-1} I_a(0)$,
$S_l^{n_2} I_a(1) = \sum_{i=1}^{n_1} \alpha_{2i} S_l^{i-1} I_a(0) + \sum_{j=1}^{n_2} \alpha_{2n_1+j} S_l^{j-1} I_a(1)$.
4) Let h_r be $h_r = [a(0) - a(1), a(0^2) - a(0), \cdots, a(0^{n_1}) - a(0^{n_1-1}),$
$a(1) - a(1), a(0|1) - a(0), \cdots, a(0^{n_1-1}|1) - a(0^{n_1-1})]$.
5) Let h^0 be $h^0 = a(1)D$

[proof] Since a time-invariant, affine input response map $a \in F(U^*, Y)$ satisfies rank $(I/O)_a = n$, the system $((\ll S_l^N(\chi(U)) \gg, S_l), \chi^0, \chi, 0, a(1))$ which realizes a is a canonical n-dimensional almost linear system by theorem (4.5). Select the linearly independent vectors $\{S_l^{i-1}\chi^0; 1 \le i \le n_1\}$ and select the linearly independent vectors $\{S_l^{j-1}(\chi^0 + \chi); 1 \le j \le n_2\}$ from the linearly independent vectors $\{S_l^i(\chi^0 + \chi u); u \in \{0,1\}, i \in N, 1 \le i \le n\}$. Let a linear map $T : \ll S_l^N(\chi(U)) \gg \to \boldsymbol{R}^n$ be $TS_l^{i-1}\chi^0 = \mathbf{e}_i$, $1 \le i \le n_1$ and $TS_l^{j-1}(\chi^0 + \chi) = \mathbf{e}_{n_1+j}$, $1 \le j \le n_2$. Then, by step 2), $T\chi^0 = g_r^0$ and $T\chi = g_r$ hold and by step 3), $h_r T = 0$ holds. By step 4), $F_r T = TS_l$ holds. Consequently, T is bijective and an almost linear system morphism : $((\ll R(\chi) \gg, S_l), \chi^0, \chi, 0, a(1)) \to \sigma_r = ((\boldsymbol{R}^n, F_r), g_r^0, g_r, h_r, a(1))$. By Corollary (4.9), the behavior of σ_s is a. It follows from the choice of $\{S_l^{i-1}\chi^0,\ S_l^{j-1}\chi; 1 \le i \le n_1,\ 1 \le j \le n_2\}$ and the determination of map T that σ_r is the real time standard system.

4.3 Partial Realization of So-called Linear Systems

Here we consider a partial realization problem by multi-experiment. Let $\underline{a}$ be an $\underline{N}$ sized time-invariant, affine input response map ($\underline{a} \in F(U_{\underline{N}}^*, Y)$), where $\underline{N} \in N$ and $U^*_{\underline{N}} := \{\omega \in U^*; |\omega| \leq \underline{N}\}$. The $\underline{a}$ is said to be a partial time-invariant, affine input response map.

A finite dimensional so-called linear system $\sigma = ((K^n, F), x^0, g, h)$ is said to be a partial realization of $\underline{a}$ if $h(\sum_{j=1}^{|\omega|} F^{|\omega|-j}(Fx^0 + g\omega(j))) = \underline{a}(\omega)$ holds for any $\omega \in U^*_{\underline{N}}$.

A partial realization problem of so-called linear systems is stated as follows:

< For any given partial time-invariant, affine input response $\underline{a} \in F(U^*_{\underline{N}}, Y)$, find a partial realization σ of $\underline{a}$ such that the dimensions of state space X of σ is minimum, where the σ is said to be a minimal partial realization of $\underline{a}$. Moreover, show when the minimal realizations are isomorphic.>

We have noted the representation for the time-invariant, affine input response maps. The representation says that any time-invariant, affine input response map can be characterized by the modified impulse response in Definition (4.1).

Note that the modified impulse response $I : \{0,1\} \to F(N, Y)$ can be represented by $(I(u)(t)) = a(u^{t+1}) - a(u^t)$ for $u \in \{0,1\}$, $t \in N$ and the time-invariant, affine input response map $a \in F(U^*, Y)$.

For any given partial time-invariant, affine input response $\underline{a} \in F(U^*_{\underline{N}}, Y)$, this correspondence can determine a partial modified impulse response $\underline{I}$: $\{0,1\} \to F(N_{\underline{N}}, Y); u \mapsto [t \mapsto (\underline{I}(u))(t) = \underline{a}(u^{t+1}) - \underline{a}(u^t)$, where $N_{\underline{N}} := \{1, 2, \cdots, \underline{N}$; for some $\underline{N} \in N\}$.

For a partial time-invariant, affine input response map $\underline{a} \in F(U^*_{\underline{N}}, Y)$, the following matrix $(I/O)_{\underline{a}\ (p,\underline{N}-p)}$ is said to be a finite-sized Input/output matrix of $\underline{a}$.

$$(I/O)_{\underline{a}\ (p,\underline{N}-p)} = \begin{pmatrix} & & (s,u) \\ & & \vdots \\ & & \vdots \\ & & \vdots \\ t\ \cdots & \cdots & a(u^{s+t+1}) - a(u^{s+t}) \\ & & \end{pmatrix},$$

where $0 \leq s \leq p$, $0 \leq t \leq \underline{N} - p$ and $u \in \{0,1\}$.

Since $I_{\underline{a}}(u)(i+j) = \underline{a}(u^{i+j+1}) - \underline{a}(u^{i+j})$ holds for $u \in \{0,1\}$, column vectors of $(I/O)_{\underline{a}}$ denote $S_l^i I_{\underline{a}}(u)$.

Let a matrix $(I/O)_{\underline{a}\ (p,\underline{N}-p)}(v,w)$ denote $(I/O)_{\underline{a}\ (p,\underline{N}-p)}(v,w)$ $= [I_{\underline{a}}(0), S_l I_{\underline{a}}(0), \cdots, S_l^{v-1} I_{\underline{a}}(0), I_{\underline{a}}(1), S_l I_{\underline{a}}(1), \cdots, S_l^{w-1} I_{\underline{a}}(1)]$.

When we actually treat approximate and noisy realization problems, we will use a notation $H_{\underline{a}\ (n_1+n_2, \underline{N}-n_1-n_2)}(n_1, n_2)$ expressed as follows:

$$H_{\underline{a}\ (n_1+n_2,\underline{N}-n_1-n_2)}(n_1,n_2) = [I_{\underline{a}}(0), \cdots, S_l^{n_1-1} I_{\underline{a}}(0), I_{\underline{a}}(1), \cdots, S_l^{n_2-1} I_{\underline{a}}(1)].$$

Theorem 4.14. *Let a time-invariant, affine input response map* $\underline{a} \in F(U^*_{\underline{N}}, Y)$ *satisfy rank* $(I/O)_{\underline{a}\ (p,\underline{N}-p)} = n$ *such that the rank value becomes the* *maximum value for some* $p \in N$. *Then the real time standard system* $\sigma_r = ((\boldsymbol{R}^n, F_r), g_r^0, g_r, h_r, a(1))$ *which realizes* $\underline{a}$ *is obtained by the following procedure:*

1) Select n_1 *independent vectors on the vectors* $\{S_l^s I_{\underline{a}}(0) : 0 \le s \le n\}$.
Select n_2 *independent vectors in* $\{S_l^s I_{\underline{a}}(1) : 0 \le s \le n\}$.
2) Let the state space be $\boldsymbol{R}^n$. *And let* g_r^0 *and* g_r *be as follows:* $g_r^0 = \mathbf{e}_1$,
$g_r = \mathbf{e}_{n_1+1}$. *Moreover,* $n = n_1 + n_2$ *and* $\mathbf{e}_i = [0, \cdots, 0, \overset{i}{1}, 0, \cdots, 0]^T$ *hold.*
3) $F_r \in \boldsymbol{R}^{n \times n}$ *is given in Definition (4.11),*
where $S_l^{n_1} I_{\underline{a}}(0) = \sum_{i=1}^{n_1} \alpha_{1i} S_l^{i-1} I_{\underline{a}}(0)$,
$S_l^{n_2} I_{\underline{a}}(1) = \sum_{i=1}^{n_1} \alpha_{2i} S_l^{i-1} I_{\underline{a}}(0) + \sum_{j=1}^{n_2} \alpha_{2n_1+j} S_l^{j-1} I_{\underline{a}}(1)$.
4) Let h_r *be* $h_r = [\underline{a}(0) - \underline{a}(1), \underline{a}(0^2) - \underline{a}(0), \cdots, \underline{a}(0^{n_1}) - \underline{a}(0^{n_1-1}),$
$\underline{a}(1) - \underline{a}(1), \underline{a}(0|1) - \underline{a}(0), \cdots, \underline{a}(0^{n_1-1}|1) - \underline{a}(0^{n_1-1})]$.
5) Let h^0 *be* $h^0 = a(1)D$

[proof] For the selected value $p \in N$, a linear combination obtained by $S_l^{n_1} I_{\underline{a}}(0)$ and $S_l^{n_2} I_{\underline{a}}(1)$ produces $a \in F(U^*, Y)$ obtained from the linear combination. Then the obtained time-invariant, affine input response map $a \in F(U^*, Y)$ satisfies rank $(I/O)_a = n$. Hence, the almost linear system $((\ll S_l^N(\chi(U)) \gg, S_l), \chi^0, \chi, 0, a(1))$ which realizes a is a canonical n-dimensional almost linear system by theorem.

Therefore, based on the $a \in F(U^*, Y)$, we obtain the real time standard system $\sigma_r = ((\boldsymbol{R}^n, F_r), g_r^0, g_r, h_r, a(1))$ which realizes a is obtained by the same procedure in theorem (4.13).

4.4 Real-Time Partial Realization of Almost Linear Systems

In general, it is well known that non-linear systems can only be determined by multi-experiments. The condition that a single experiment may pretend to produce the same effects is very hard for us to find. However, we can look for special single-experiments that simulate multi-experiments for any almost linear system.

In this section, based on the results of partial realization theory in the reference [Matsuo and Hasegawa, 2003], we will state a single-experiment for so-called linear systems.

Problem 4.15. Real time partial realization problem
Let a physical object, that is, $\underline{a} \in F(U_{\underline{N}}^*, Y)$ be a finite dimensional so-called linear system. Then, for any given finite data $\{a(\bar{\omega}); \bar{\omega}$ is a finite length input $\}$, find a so-called linear system $\sigma = ((K^n, F), x^0, g, h)$ and an input $\bar{\omega} \in U^*$ such that $a_\sigma(\omega) = a(\omega)$ for any $\omega \in U^*$.

Definition 4.16. For a finite dimensional almost linear system, if there exists a solution of the real time partial realization problem, then an input $\bar{\omega} \in U^*$ of the solution is said to be a (real time partial) realization signal.

Lemma 4.17. *Let a given time-invariant, affine input response map $a \in F(U^*, Y)$ have the behavior of an almost linear system whose state space is less than L-dimensional. Then there exists an input of finite length $\bar{\omega} \in U^*$ such that the following algorithm provides a finite Input/output matrix, where $p := max\{L_1, L_2\}$.*

1) Find an integer L_1 such that row vectors $\{\underline{S_l}^i \chi^0 \in K^L; 0 \le i \le L_1 - 1\}$ are linearly independent and $\{\underline{S_l}^i \chi^0 \in K^L; 0 \le i \le L_1\}$ are linearly dependent. Namely, feed an input $\omega_1 := 0^{L_1+L+1}$ into the plant, where $\underline{S_l}^i \chi^0 = [a(0^{n+1}) - a(0^n), a(0^n) - a(0^{n-1}), \cdots, a(0^{L+i+1}) - a(0^{L+i})]^T$.
2) Find an integer L_2 such that row vectors $\{\underline{S_l}^i \chi^0, \underline{S_l}^i(\chi^0 + \bar{\chi}) \in K^L; 0 \le i \le L_j - 1, 1 \le j \le 2\}$ are linearly independent and $\{\underline{S_l}^i \cdot \chi^0\ ,\ \underline{S_l}^i(\chi^0 + \bar{\chi} \cdot u \in K^L; 0 \le i \le L_j, 1 \le j \le 2\}$ are linearly dependent. Namely, feed a further input $\omega_2 := 0^{L_1+L-1}|1$ into the plant. Let $\bar{\omega} = \omega_2|\omega_1$.
Making the row vectors of a matrix from the row vectors $\{\underline{S_l}^i(\chi^0 + \bar{\chi} \cdot u)) \in K_L; 0 \le i \le L_j, 1 \le j \le 2, u \in \{0,1\}\}$ obtained by the above iterations, we will obtain a finite-sized Input/output matrix $(I/O)_{a\ (L-1,p)}$, where $\underline{S_l}^i \bar{\chi}$
$= [a(0^i|1) - a(0^{i+1}), a(0^{i+1}|1) - a(0^{i+2}), \cdots, a(0^{i+L}|1) - a(0^{i+L+1})]^T$.
And $a(0^i|1)$ is given by $a(0^j|1) = a(0^{i+1}|1|0^t) - a(0^{t+1}) + a(0^t)$ for any $i, t \in N$.

Theorem 4.18. *Let a given time-invariant, affine input response map $a \in F(U^*, Y)$ have the behavior of an almost linear system whose state space is less than L-dimensional. Then there exists a realization signal such that the real-time standard system $\sigma_s = ((K^n, F_s), g_s^0, g_s, h_s, h^0)$ which realizes a is obtained by the following algorithm:*
1) Find a finite Input/output matrix $(I/O)_{a\ (L-1,p)}$ based on the algorithm given in Lemma (4.17).
2) Apply the algorithm given in Theorem (4.14) to the above finite Input/output matrix $(I/O)_{a\ (L-1,p)}$.

4.5 Algebraically Approximate Realization of So-called Linear Systems

Here, we will discuss the algebraically approximate realization problem of so-called linear systems, which is stated as follows:

<For any given finite-length modified impulse response, find, using only algebraic calculations, a so-called linear system which approximates it.>

The algebraically approximate realization of non-linear systems is presented here for the first time.

In order to make our discussion simple, we assume that the set Y of output is the set $\boldsymbol{R}$ of real numbers, namely 1-output.

Theorem 4.19. *Algebraic algorithm for approximate realization*

Let a partial input response map $\underline{a}$ *be considered an object which is a so-called linear system. Then an approximate realization* $\sigma = ((\boldsymbol{R}^n, F_r), g_r^0, g_r, h_r)$ *of* $\underline{a}$ *is given by the following algorithm:*

1) *Based on the ratio of the square root of eigenvalues for a matrix* $H_{\underline{a}\ (p,\bar{p})}(p,0)H_{\underline{a}\ (p,\bar{p})}(p,0)^T$, *determine the value* n_1 *of rank for the Input/output matrix* $H_{\underline{a}\ (p,\bar{p})}(p,0)$, *where* $n_1 \leq p$.
   Namely, determine the value n_1 *of rank for the matrix* $H_{\underline{a}\ (p,\bar{p})}(p,0)$ *such that the ratio of the square root of eigenvalues for the covariance matrix becomes very small. The small ratio defines the nearness of approximation degree.*
2) *We use the algebraic CLS method as follows:*
   ① *Based on Proposition (2.14), determine coefficients* $\{\alpha_{1i} : 1 \leq i \leq n_1\}$. *The Q in Proposition (2.14) can be considered as the matrix composed from the eigenvectors of* $H_{\underline{a}\ (n_1+1,L)}(n_1+1,0)H^T_{\underline{a}\ (n_1+1,L)}(n_1+1,0)$.
   Let a matrix $A_1 \in \boldsymbol{R}^{1\times(n_1+1)}$ *be* $A_1 = [\alpha_{11}, \alpha_{12}, \cdots, \alpha_{1n_1}, -1]$.
   ② *Determine the error vectors* $\{\underline{S_l^i}\bar{I}_{\underline{a}} \in \boldsymbol{R}^{L\times 1} : 0 \leq i \leq n_1\}$ *by using the equation* $[\bar{I}_{\underline{a}}(0), \underline{S_l}\bar{I}_{\underline{a}}(0), \cdots, \underline{S_l^{n_1}}\bar{I}_{\underline{a}}(0)]^T :=$ $A_1^T[A_1A_1^T]^{-1}A_1H^T_{\underline{a}\ (n_1+1,L)}(n_1+1,0)$ *and* $H^T_{\underline{a}\ (n_1,L)}(n_1,0) := [I_{\underline{a}}(0), \cdots, S_l^{n_1-1}I_{\underline{a}}(0), S_l^{n_1}I_{\underline{a}}(0)]$.
   ③ *Let* $h_{1r} \in \boldsymbol{R}^{1\times n_1}$ *be* h_{1r} $= [(I_{\underline{a}}(0))(0) - (\bar{I}_{\underline{a}}(0))(0), (S_lI_{\underline{a}}(0))(0) - (S_l\bar{I}_{\underline{a}}(0))(0), \cdots,$ $(S_l^{n_1-1}I_{\underline{a}}(0))(0) - (S_l^{n_1-1}\bar{I}_{\underline{a}}(0))(0)]$.
3) *Based on the ratio of the square root of eigenvalues for a matrix* $H_{\underline{a}\ (n_1+p,\bar{p})}(n_1,p)H_{\underline{a}\ (n_1+p,\bar{p})}(n_1,p)^T$, *determine the value* n_2 *of rank for the matrix* $H_{\underline{a}\ (n_1+p,\bar{p})}(n_1,p)$, *where* $n_2 \leq p$.
   Namely, determine the value n_2 *of rank for the matrix* $H_{\underline{a}\ (p,\bar{p})}(p,0)$ *such that the ratio of the square root of eigenvalues for the covariance matrix becomes very small. The small ratio defines the nearness of approximation degree.*
4) *The algebraic CLS method is used as follows:*
   ① *Based on Proposition (2.14), determine coefficients* $\{\alpha_{2i} : 1 \leq i \leq n_1+n_2\}$.
   The Q in Proposition (2.14) can be considered as the matrix composed from the eigenvectors of $H_{\underline{a}\ (n_1,L)}(n_1,n_2+1)H^T_{\underline{a}\ (n_1,L)}(n_1,n_2+1)$.
   Let a matrix $A_2 \in \boldsymbol{R}^{1\times(n_1+n_2+1)}$ *be* $A_2 = [\alpha_{21}, \alpha_{22}, \cdots, \alpha_{2n_1+n_2}, -1]$.
   ② *Determine the error vectors* $\{\underline{S_l^i}\bar{I}_{\underline{a}} \in \boldsymbol{R}^{L\times 1} : 0 \leq i \leq n_1+n_2\}$ *by using the equation*$[\bar{I}_{\underline{a}}, \underline{S_l}\bar{I}_{\underline{a}}, \cdots, \underline{S_l^{n_1+n_2}}\bar{I}_{\underline{a}}]^T :=$ $A_2^T[A_2A_2^T]^{-1}A_2H^T_{\underline{a}\ (n_1+n_2,L)}(n_1,n_2+1)$ *and* $H^T_{\underline{a}\ (n_1,L)}(n_1,n_2+1) :=$ $[I_{\underline{a}}(0), \cdots, S_l^{n_1-1}I_{\underline{a}}(0), I_{\underline{a}}(1), \cdots, S_l^{n_2-1}I_{\underline{a}}(1), S_l^{n_2}I_{\underline{a}}(1)]$.
   ③ *Let* $F_r \in \boldsymbol{R}^{(n_1+n_2)\times(n_1+n_2)}$ *be given as the same as in Definition (4.11). Let* g_r^0 *be* $g_r^0 = \mathbf{e}_1$ *and* g_r *be* $g_r = \mathbf{e}_{n_1+1} - \mathbf{e}_1$,

where $\mathbf{e}_i = [0, \cdots, 0, \overset{i}{1}, 0, \cdots, 0]^T \in \boldsymbol{R}^{n_1+n_2}$.
④ *Let* $h_r \in \boldsymbol{R}^{1\times(n_1+n_2)}$ *be* h_r
$= [h_{1r}, (I_{\underline{a}}(1))(0) - (\bar{I}_{\underline{a}}(1))(0), (S_l I_{\underline{a}}(1))(0) - (S_l \bar{I}_{\underline{a}}(1))(0), \cdots,$
$(S_l^{n_2-1} I_{\underline{a}}(1))(0) - (S_l^{n_2-1} \bar{I}_{\underline{a}}(1))(0)]$.

[proof] By 1) and 3), the reduction part in the data can be excluded in the sense of the number of dimensions by using the ratio of the matrix norm, which produces a degree of information loss. The matrices A_1 in 2) and A_2 in 4) correspond to the matrix A in Lemma (2.17). Hence, if we determine the coefficients $\{\alpha_{ij} : i \leq i \leq 2,\ 1 \leq j \leq n_i\}$, we can obtain the approximate part of the finite-sizes Input/output matrices $H_{\underline{a}\ (n_1+1,\bar{p})}(n_1+1, 0)$ and $H_{\underline{a}\ (n_1+n_2+1,\bar{p})}(n_1, n_2+1)$ by using Lemma (2.17).

Therefore, we obtain the approximate Input/output matrices $\hat{H}_{\underline{a}\ (n_1+1,\bar{p})}(n_1+1, 0)$ and $\hat{H}_{\underline{a}\ (n_1+n_2+1,\bar{p})}(n_1, n_2+1)$. Finally, we apply Theorem (4.14) to the $\hat{H}_{\underline{a}\ (n_1+1,\bar{p})}(n_1+1, 0)$ and $\hat{H}_{\underline{a}\ (n_1+n_2+1,\bar{p})}(n_1, n_2+1)$.

Example 4.20. Let the signals be the modified impulse responses of the following 3-dimensional so-called linear system: $\sigma = ((\boldsymbol{R}^3, F), x^0, g, h)$, where

$$F = \begin{bmatrix} 0 & 0 & -0.7 \\ 1 & 0 & 0.4 \\ 0 & 1 & 0.7 \end{bmatrix},\ x^0 = [-0.83,\ -0.5,\ -1.7]^T,\ h = [12,\ 4,\ -3],\ g = [1, 0, 0]^T.$$

Then the algebraically approximate realization problem is solved as follows:

covariance matrix	eigenvalues			
	1	2	3	4
$H^T_{\underline{a}\ (2,50)}(2,0)H_{\underline{a}\ (2,50)}(2,0)$	2056	216		
$H^T_{\underline{a}\ (3,50)}(3,0)H_{\underline{a}\ (3,50)}(3,0)$	2338	742	7.4	
$H^T_{\underline{a}\ (4,50)}(4,0)H_{\underline{a}\ (4,50)}(4,0)$	2355	1534	7.6	0
covariance matrix	square root of eigenvalues			
$H^T_{\underline{a}\ (3,50)}(3,0)H_{\underline{a}\ (3,50)}(3,0)$	48.4	27.2	2.7	
$H^T_{\underline{a}\ (4,50)}(4,0)H_{\underline{a}\ (4,50)}(4,0)$	48.5	39.2	2.8	0
covariance matrix	eigenvalues			
	1	2	3	4
$H^T_{\underline{a}\ (3,50)}(2,1)H_{\underline{a}\ (3,50)}(2,1)$	5775	226	0.38	
$H^T_{\underline{a}\ (4,50)}(3,1)H_{\underline{a}\ (4,50)}(3,1)$	5959	853	8	0
covariance matrix	square root of eigenvalues			
$H^T_{\underline{a}\ (3,50)}(2,1)H_{\underline{a}\ (3,50)}(2,1)$	75	15	0.6	
$H^T_{\underline{a}\ (4,50)}(3,1)H_{\underline{a}\ (4,50)}(3,1)$	77.2	29.2	2.8	0

1) Since the ratio $\frac{2.7}{48.4} = 0.06$ obtained by the square root of $H^T_{\underline{a}\ (3,50)}(3,0) \times H_{\underline{a}\ (3,50)}(3,0)$ and $\frac{0.6}{75} = 0.01$ obtained by the square root of $H^T_{\underline{a}\ (3,50)}(2,1) \times H_{\underline{a}\ (3,50)}(2,1)$ are not so small, the approximate so-called linear system obtained by the algebraic CLS method may not be good.
2) After determining the numbers n_1 and n_2 of dimensions which are 2 and

0, we execute the algebraically approximate realization algorithm.

A so-called linear system $\sigma = ((\boldsymbol{R}^2,\ F_2),\ x^0,\ g_2,\ h_2)$ obtained by the algebraic CLS method is expressed as follows:

$$F_2 = \begin{bmatrix} 0 & -0.82 \\ 1 & 1.5 \end{bmatrix},\ h_2 = [19.5,\ 12],\ g_2 = [1,\ 0]^T, g_2^0 = [0.28,\ 0.55]^T.$$

In the case that $n_1 = 3$ and $n_2 = 0$, a 3-dimensional so-called linear system $\sigma_3 = ((\boldsymbol{R}^3,\ F_3),\ x_3^0,\ g_3,\ h_3)$ obtained by the algebraic CLS method is also expressed as follows:

$$F_3 = \begin{bmatrix} 0 & 0 & -0.7 \\ 1 & 0 & 0.4 \\ 0 & 1 & 0.7 \end{bmatrix},\ h_3 = [12,\ 4,\ -3], x_3^0 = [-0.83,\ -0.5,\ -1.7]^T, g_3 = [1,0,0]^T.$$

We can show that the algorithm for approximate realization given by the analytic CLS method in the reference [Hasegawa, 2008] produces the same systems as the above ones in the sense of the numerical calculation.

In this example, the original signals are considered as modified impulse responses of a 3-dimensional so-called linear system and the desirable modified impulse responses are obtained by the algebraic CLS method. The model

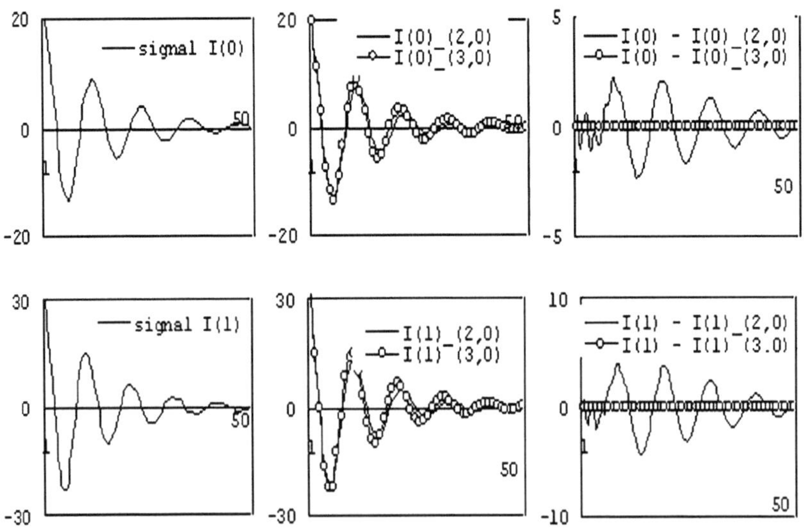

Fig. 4.1 The left are the original modified impulse responses $I(0)$ and $I(1)$. The middle are the obtained modified impulse responses $I(0)\_(2,0)$, $I(0)\_(3,0)$, $I(1)\_(2,0)$ and $I(1)\_(3,0)$ by the algebraic CLS method. The right are the difference between $I(0)$ and $I(0)\_(2,0)$ or $I(0)\_(3,0)$ and the difference between $I(1)$ and $I(1)\_(2,0)$ or $I(1)\_(3,0)$ in Example (4.20).

obtained by the algebraic CLS method is a 2-dimensional so-called linear system. Therefore, a good approximate realization could not be obtained. For reference, a 3-dimensional so-called linear system obtained by the algebraic CLS method is also given.

This table indicates that the 2-dimensional so-called linear system reconstructs the original signal with 21 and 21 % error to signal ratio and 0.06 and 0.01 ratio of matrices, and the 3-dimensional so-called linear system completely reconstructs the original system.

Just as we thought, the following table and Fig. 4.1 truly indicate that the two-dimensional so-called linear system is not a good approximation for the given system.

dimension	ratio of matrices	mean values of square root for sum of signal	signal by CLS	error	cosine ① and ②	error ratio
		①	②	③	$\cos\theta$	③/①
I(0)_(2,0)	0.06	0.73	0.69	0.15	0.979	0.21
I(1)_(2,0)	0.01	1.22	1.16	0.26	0.975	0.21
I(0)_(3,0)	0	0.73	0.73	0	1	0
I(1)_(3,0)	0	1.22	1.22	0	1	0

For the notations $I(0)\_(n_1, n_2)$ and $I(1)\_(n_1, n_2)$, see Definition (4.11).

Example 4.21. Let the signals be the modified impulse responses of the following 4-dimensional so-called linear system: $\sigma = ((\boldsymbol{R}^4, F), x^0, g, h)$, where

$$F = \begin{bmatrix} 0 & 0 & 0 & 0.6 \\ 1 & 0 & 0 & 0.4 \\ 0 & 1 & 0 & -0.3 \\ 0 & 0 & 1 & 0.1 \end{bmatrix}, x^0 = [-1,\ 0,\ 0,\ 0]^T,\ h = [8,\ 3,\ 0,\ -4],\ g = [1,\ 0,\ 0,\ 0]^T.$$

Then the algebraically approximate realization problem is solved by the following algorithm:

covariance matrix	eigenvalues				
	1	2	3	4	5
$H^T_{\underline{a}\ (3,50)}(3,0)H_{\underline{a}\ (2,50)}(3,0)$	1316	733	75		
$H^T_{\underline{a}\ (4,50)}(4,0)H_{\underline{a}\ (4,50)}(4,0)$	1468	1232	131	0.16	
$H^T_{\underline{a}\ (5,50)}(5,0)H_{\underline{a}\ (5,50)}(5,0)$	1909	1407	131	0.17	0
covariance matrix	square root of eigenvalues				
$H^T_{\underline{a}\ (4,50)}(4,0)H_{\underline{a}\ (4,50)}(4,0)$	38.3	35	11.4	0.4	
$H^T_{\underline{a}\ (5,50)}(5,0)H_{\underline{a}\ (5,50)}(5,0)$	43.7	37.5	11.4	0.4	0
covariance matrix	eigenvalues				
	1	2	3	4	5
$H^T_{\underline{a}\ (5,50)}(3,2)H_{\underline{a}\ (5,50)}(3,2)$	3852	2874	164	5.3	0
$H^T_{\underline{a}\ (5,50)}(4,1)H_{\underline{a}\ (5,50)}(4,1)$	3737	1686	205	4.6	0
covariance matrix	square root of eigenvalues				
$H^T_{\underline{a}\ (5,50)}(3,2)H_{\underline{a}\ (5,50)}(3,2)$	62.1	53.6	12.8	2.3	0
$H^T_{\underline{a}\ (5,50)}(4,1)H_{\underline{a}\ (5,50)}(4,1)$	61.1	41	14.3	2.1	0

1) Since the ratio $\frac{0.4}{38.3} = 0.01$ obtained by the square root of $H^T_{\underline{a}\,(4,50)}(4,0) \times H_{\underline{a}\,(4,50)}(4,0)$ is a little small and the ratio $\frac{2.3}{62.1} = 0.04$ obtained by the square root of $H^T_{\underline{a}\,(5,50)}(3,2)H_{\underline{a}\,(5,50)}(3,2)$ is somewhat small, an approximate almost linear system obtained by the algebraic CLS method is obtained as follows:
2) After determining the numbers n_1 and n_2 of dimensions which are 3 and 0, we execute the approximate realization algorithm by the algebraic CLS method.

$$\sigma_2 = ((\boldsymbol{R}^3,\ F_2),\ x_2^0,\ g_2,\ h_2, h^0),\ \text{where}\ F_2 = \begin{bmatrix} 0 & 0 & -0.62 \\ 1 & 0 & -1.1 \\ 0 & 1 & -0.83 \end{bmatrix},\ h_2 = [4.97,\ 2.95,\ 3.96],\ x_2^0 = [-0.82,\ -0.52,\ -0.28]^T,\ g_2 = [0.92\ 0.54\ 0.38]^T.$$

In the case that $n_1 = 4$ and $n_2 = 0$, a 4-dimensional so-called linear system $\sigma_4 = ((\boldsymbol{R}^4,\ F_4),\ x_4^0,\ g,\ h_4)$ obtained by the algebraic CLS method is also expressed as follows:

$$F_4 = \begin{bmatrix} 0 & 0 & 0 & 0.6 \\ 1 & 0 & 0 & 0.4 \\ 0 & 1 & 0 & -0.3 \\ 0 & 0 & 1 & 0.1 \end{bmatrix},\ h_4 = [5,\ 3,\ 4,\ -9.6],\ x_4^0 = [-4,\ -6,\ -4.5,\ -5]^T,$$
$$g_4 = [4,\ 6,\ 4.5,\ 5]^T.$$

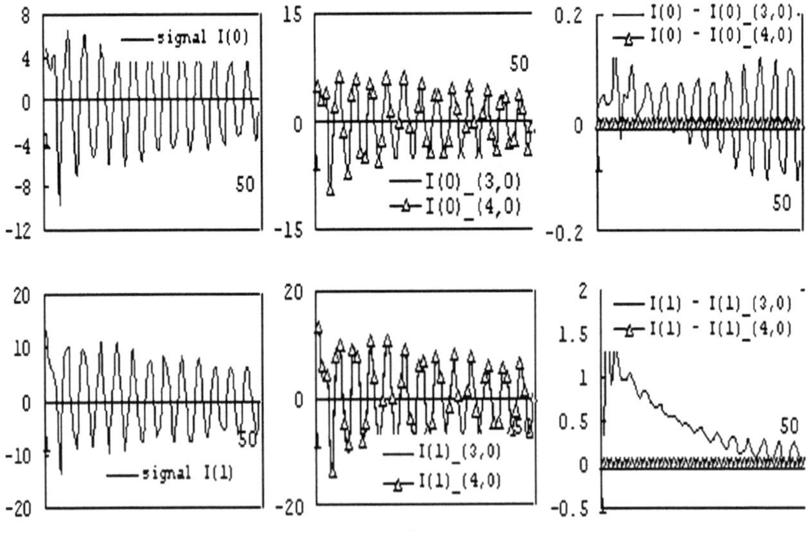

Fig. 4.2 The left are original modified impulse responses $I(0)$ and $I(1)$. The middle are the obtained modified impulse responses $I(0)\_(3,0)$, $I(0)\_(4,0)$, $I(1)\_(3,0)$, and $I(1)\_(4,0)$ by the algebraic CLS method. The right are the difference between $I(0)$ and $I(0)\_(3,0)$ or $I(0)\_(4,0)$ and the difference between $I(1)$ and $I(1)\_(3,0)$ or $I(1)\_(4,0)$ in Example (4.21).

We can show that the algorithm for approximate realization given by the analytic CLS method in the reference [Hasegawa, 2008] produces the same systems as the above ones in the sense of the numerical calculation.

In this example, the original signals are considered as the modified impulse responses of a 4-dimensional so-called linear system and the desirable modified impulse responses are obtained by the algebraic CLS method. The following table shows that the 3-dimensional so-called linear system reconstructs the original signal with 2 and 8 % error to signal ratio and with 0.01 and 0.04 ratio of matrices. The model obtained by the algebraic CLS method is a 3-dimensional so-called linear system. Therefore, a somewhat good approximate realization was obtained. For reference, a 4-dimensional so-called linear system is shown.

Fig. 4.2 also indicates that the 4-dimensional so-called linear system completely reconstructs the original system.

dimension	ratio of matrices	mean values of square root for sum of signal	signal by CLS	error	cosine ① and ②	error ratio
		①	②	③	$\cos\theta$	③/①
I(0)_(3,0)	0.01	0.587	0.582	0.01	0.999	0.017
I(1)_(3,0)	0.02	0.945	0.939	0.08	0.996	0.08
I(0)_(4,0)	0	0.587	0.42	0	1	0
I(1)_(4,0)	0	0.945	0.66	0	1	0

For the notations $I(0)_-(n_1, n_2)$ and $I(1)_-(n_1, n_2)$, see Definition (4.11).

Example 4.22. Let the signals be the modified impulse responses of the following 5-dimensional so-called linear system: $\sigma = ((\boldsymbol{R}^5, F), x^0, g, h)$, where

$$F = \begin{bmatrix} 0\,0\,0\,0 & 0.2 \\ 1\,0\,0\,0 & -0.04 \\ 0\,1\,0\,0 & -0.2 \\ 0\,0\,1\,0 & 0.5 \\ 0\,0\,0\,1 & -0.4 \end{bmatrix}, \; x^0 = [-10,\ 0,\ 1,\ 0,\ 0]^T,\ h = [10,\ 2,\ -5,\ -1,\ 2],$$

$g = [1,\ 0,\ 0,\ 0,\ 0]^T$.

Then the algebraically approximate realization problem is solved by the following algorithm:

covariance matrix	eigenvalues					
	1	2	3	4	5	6
$H^T_{\underline{a}\ (5,50)}(5,0)H_{\underline{a}\ (5,50)}(5,0)$	20769	8585	3192	790	31.7	
$H^T_{\underline{a}\ (6,50)}(6,0)H_{\underline{a}\ (6,50)}(6,0)$	20986	8636	4259	862	32	0
covariance matrix	square root of eigenvalues					
$H^T_{\underline{a}\ (5,50)}(5,0)H_{\underline{a}\ (5,50)}(5,0)$	145	93	56.5	28	5.6	
$H^T_{\underline{a}\ (6,50)}(6,0)H_{\underline{a}\ (6,50)}(6,0)$	145	93	65.3	28.4	5.7	0

covariance matrix	eigenvalues					
	1	2	3	4	5	6
$H^T_{\underline{a}\ (5,50)}(4,1)H_{\underline{a}\ (5,50)}(4,1)$	39362	8474	2889	350	0.6	
$H^T_{\underline{a}\ (6,50)}(5,1)H_{\underline{a}\ (6,50)}(5,1)$	39382	9034	3193	841	35	0
covariance matrix	square root of eigenvalues					
$H^T_{\underline{a}\ (5,50)}(4,1)H_{\underline{a}\ (5,50)}(4,1)$	198	92.1	53.7	18.7	0.8	
$H^T_{\underline{a}\ (6,50)}(5,1)H_{\underline{a}\ (6,50)}(5,1)$	198	95	56.5	29	5.9	0

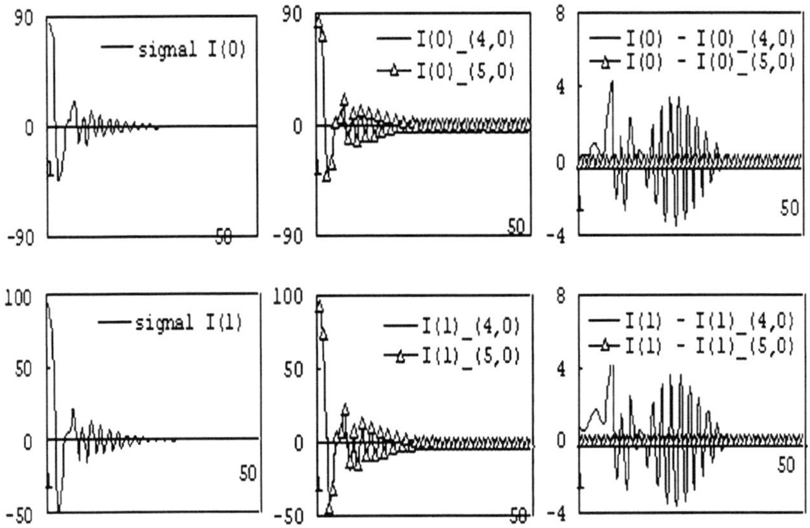

Fig. 4.3 The left are original modified impulse responses $I(0)$ and $I(1)$. The middle are obtained modified impulse responses $I(0)\_(4,0)$, $I(0)\_(5,0)$, $I(1)\_(4,0)$ and $I(1)\_(5,0)$ by the algebraic CLS method, The right are the difference between $I(0)$ and $I(0)\_(4,0)$ or $I(0)\_(5,0)$ and the difference between $I(1)$ and $I(1)\_(4,0)$ or $I(1)\_(5,0)$ in Example (4.22).

1) Since the ratio $\frac{5.6}{145} = 0.04$ obtained by the square root of $H^T_{\underline{a}\ (5,50)}(5,0) \times H_{\underline{a}\ (5,50)}(5,0)$ is somewhat small and the ratio $\frac{0.8}{198} \approx 0$ obtained by the square root of $H^T_{\underline{a}\ (5,50)}(4,1)H_{\underline{a}\ (5,50)}(4,1)$ is very small, the approximation of the original so-called linear system may be somewhat good.
2) After determining the numbers n_1 and n_2 of dimensions which are 4 and 0, we execute the algebraically approximate realization algorithm.

The approximate so-called linear system $\sigma_3 = ((\boldsymbol{R}^4,\ F_3),\ x_3^0,\ g_3,\ h_3)$ obtained by the algebraic CLS method is obtained as follows:

$$F_3 = \begin{bmatrix} 0 & 0 & 0 & -0.33 \\ 1 & 0 & 0 & -0.41 \\ 0 & 1 & 0 & -0.46 \\ 0 & 0 & 1 & -1.28 \end{bmatrix},$$

$h_3 = [83.8,\ 72.7,\ -40.7,\ -31.3]$, $g_3 = [0.09,\ 0.073,\ 0.06,\ 0.03]^T, x_3^0 = [-0.9,\ -0.79,\ -0.66,\ -0.29]^T$.

In the case that $n_1 = 5$, $n_2 = 0$, a 5-dimensional so-called linear system $\sigma = ((\boldsymbol{R}^5,\ F_4),\ x_4^0,\ g_4,\ h_4)$ obtained by the algebraic CLS method is expressed as follows:

$$F_4 = \begin{bmatrix} 0\,0\,0\,0 & 0.2 \\ 1\,0\,0\,0 & -0.04 \\ 0\,1\,0\,0 & -0.2 \\ 0\,0\,1\,0 & 0.5 \\ 0\,0\,0\,1 & -0.4 \end{bmatrix},\ h_4 = [84,\ 73,\ -40.4,\ -30.5,\ 1.6],$$

$x_4^0 = [1.21,\ -1.17,\ -0.96,\ -1.5,\ -1.1]^T, g_4 = [0.12,\ 0.12,\ 0.1,\ 0.16,\ 0.12]^T$.

We can show that the algorithm for approximate realization given by the analytic CLS method in the reference [Hasegawa, 2008] produces the same systems as the above ones in the sense of the numerical calculation.

In this example, original signals are considered as modified impulse responses of a 5-dimensional so-called linear system and the desirable modified impulse responses are obtained by the algebraic CLS method.

The following table shows that the 4-dimensional so-called linear system reconstructs the original signal with 9 and 9 % error to signal ratio and with 0.04 and 0 ratio of matrices. Therefore, a somewhat good approximate realization was obtained. For reference, a 5-dimensional so-called linear system is also obtained by the algebraic CLS method. The system completely reconstructs the original system.

Just as we expected, the following table and Fig. 4.3 truly indicate that the 4-dimensional so-called linear system is a somewhat good approximation for a given 5-dimensional so-called linear system.

dimen-sion	ratio of matrices	mean values of square root for sum of signal	signal by CLS	error	cosine ① and ②	error ratio
		①	②	③	$\cos\theta$	③/①
I(0)-(4,0)	0.04	2.56	2.57	0.24	0.996	0.09
I(1)-(4,0)	0	2.77	2.78	0.26	0.996	0.09
I(0)-(5,0)	0	2.56	2.56	0	1	0
I(1)-(5,0)	0	2.77	2.77	0	1	0

For the notations $I(0)\_(n_1, n_2)$ and $I(1)\_(n_1, n_2)$, see Definition (4.11).

Example 4.23. Let the signals be the modified impulse responses of the following 6-dimensional so-called linear system: $\sigma = ((\boldsymbol{R}^6, F), x^0, g, h)$, where $F =$

$$\begin{bmatrix} 0\,0\,0\,0\,0 & 0.3 \\ 1\,0\,0\,0\,0 & -0.3 \\ 0\,1\,0\,0\,0 & -0.1 \\ 0\,0\,1\,0\,0 & 0.1 \\ 0\,0\,0\,1\,0 & 0.4 \\ 0\,0\,0\,0\,1 & -0.1 \end{bmatrix},\ x^0 = [-10,\ 0,\ 1,\ 0,\ 0, 0]^T, h = [5,\ 2,\ -5,\ -1,\ 2,\ -1],$$

$g = [1,\ 0,\ 0,\ 0,\ 0,\ 0]^T$.

Then the algebraically approximate realization problem is solved as follows:

covariance matrix	eigenvalues						
	1	2	3	4	5	6	7
$H^T_{\underline{a}\ (6,50)}(6,0)H_{\underline{a}\ (6,50)}(6,0)$	20261	18140	2359	854	832	18	
$H^T_{\underline{a}\ (7,50)}(7,0)H_{\underline{a}\ (7,50)}(7,0)$	21290	18709	2367	1026	841	18	0
covariance matrix	square root of eigenvalues						
$H^T_{\underline{a}\ (6,50)}(6,0)H_{\underline{a}\ (6,50)}(6,0)$	142	135	48.6	29.2	28.8	4.2	
$H^T_{\underline{a}\ (7,50)}(7,0)H_{\underline{a}\ (7,50)}(7,0)$	146	137	49	32	29	4.2	0

covariance matrix	eigenvalues						
	1	2	3	4	5	6	7
$H^T_{\underline{a}\ (6,50)}(5,1)H_{\underline{a}\ (6,50)}(5,1)$	32716	17194	2842	877	508	0.43	
$H^T_{\underline{a}\ (7,50)}(6,1)H_{\underline{a}\ (7,50)}(6,1)$	33090	19427	2843	900	857	21	0
covariance matrix	square root of eigenvalues						
$H^T_{\underline{a}\ (6,50)}(5,1)H_{\underline{a}\ (6,50)}(5,1)$	181	131	53.3	29.6	22.5	0.7	
$H^T_{\underline{a}\ (7,50)}(6,1)H_{\underline{a}\ (7,50)}(6,1)$	182	139	53.3	30	29.3	4.6	0

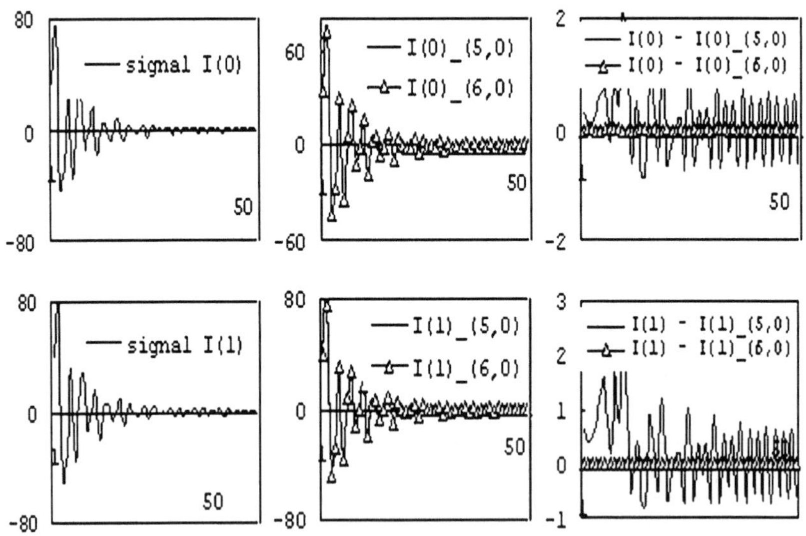

Fig. 4.4 The left are original modified impulse responses $I(0)$ and $I(1)$. The middle are obtained modified impulse responses $I(0)\_(5,0)$, $I(0)\_(6,0)$, $I(1)\_(5,0)$ and $I(1)\_(6,0)$ by the algebraic CLS method, The right are the difference between $I(0)$ and $I(0)\_(5,0)$ or $I(0)\_(6,0)$ and the difference between $I(1)$ and $I(1)\_(5,0)$ or $I(1)\_(6,0)$ in Example (4.23).

1) Since the ratio $\frac{4.2}{142} = 0.03$ obtained by the square root of $H^T_{\underline{a}\ (6,50)}(6,0) \times H_{\underline{a}\ (6,50)}(6,0)$ is a little small and the ratio $\frac{0.7}{181} = 0.003$ obtained by the square root of $H^T_{\underline{a}\ (6,50)}(5,1)H_{\underline{a}\ (6,50)}(5,1)$ is very small, the approximate almost

system obtained by the CLS method may be good.
2) After determining the numbers n_1 and n_2 of dimensions which are 5 and 0, we execute the algebraically approximate realization algorithm.

The approximate so-called linear system $\sigma_3 = ((\boldsymbol{R}^5,\ F_3),\ x_3^0,\ g_3,\ h_3)$ obtained by the algebraic CLS method is expressed as follows;

$$F_3 = \begin{bmatrix} 0\,0\,0\,0 & -0.4 \\ 1\,0\,0\,0 & -0.2 \\ 0\,1\,0\,0 & -0.19 \\ 0\,0\,1\,0 & -0.36 \\ 0\,0\,0\,1 & -0.88 \end{bmatrix},\ h_3 = [33.7,\ 72.9,\ -43.1,\ -27.1,\ 29],$$

$g_3 = [0.09, 0.07, 0.07, 0.06, 0.04]^T, x_3^0 = [-0.87, -0.8, -0.74, -0.62, -0.33]^T$.

In the case that $n_1 = 6$ and $n_2 = 0$, a 6-dimensional so-called linear system $\sigma_4 = ((\boldsymbol{R}^6,\ F_4),\ x_4^0,\ g_4,\ h_4)$ obtained by the algebraic CLS method is expressed as follows:

$$F_4 = \begin{bmatrix} 0\,0\,0\,0\,0 & 0.3 \\ 1\,0\,0\,0\,0 & -0.3 \\ 0\,1\,0\,0\,0 & -0.1 \\ 0\,0\,1\,0\,0 & 0.1 \\ 0\,0\,0\,1\,0 & 0.4 \\ 0\,0\,0\,0\,1 & -0.1 \end{bmatrix},\ h_4 = [34,\ 73,\ -43,\ -26.8,\ 29.6,\ -34.6],$$

$x_4^0 = [-1.4, -1, -0.86, -1, -1.6, -1.4]^T, g_4 = [0.15, 0.1, 0.09, 0.11, 0.17, 0.16]^T$.

We can show that the algorithm for approximate realization given by the analytic CLS method in the reference [Hasegawa, 2008] produces the same systems as the above ones in the sense of the numerical calculation.

In this example, original signals are considered as modified impulse responses of a 6-dimensional so-called linear system and the desirable modified impulse responses are obtained by the algebraic CLS method.

The following table shows that the 5-dimensional so-called linear system reconstructs the original signal with 4 and 4 % error to signal ratio and with 0.03 and 0.003 ratio of matrices. Therefore, a somewhat good approximate realization was obtained. For reference, a 6-dimensional so-called linear system obtained by the algebraic CLS method is also shown.

Just as we expected, the following table and Fig. 4.4 truly indicate that the 5-dimensional so-called linear system is a somewhat good approximation for the given 6-dimensional so-called linear system.

dimension	ratio of matrices	mean values of square root for sum of signal			cosine ① and ②	error ratio
		signal	signal by CLS	error		
		①	②	③	$\cos\theta$	③/①
I(0)_(5,0)	0.03	2.27	2.26	0.1	0.999	0.04
I(1)_(5,0)	0.003	2.41	2.4	0.11	0.999	0.04
I(0)_(6,0)	0	2.27	2.27	0	1	0
I(1)_(6,0)	0	2.41	2.41	0	1	0

For the notations $I(0)\_(n_1, n_2)$ and $I(1)\_(n_1, n_2)$, see Definition (4.11).

4.6 Algebraically Noisy Realization of So-called Linear Systems

In this section, we discuss an algebraically noisy realization problem of so-called linear systems which are non-linear systems. The noisy realization of nonlinear is presented here for the first time.

In order to make our discussion simple, we assume that the set Y of output is the set $\boldsymbol{R}$ of real numbers, namely 1-output.

For noise $\{\bar{\gamma}(t) : t \in N\}$ added to an unknown so-called linear system σ, we will obtain the observed data $\{\hat{\gamma}(|\omega|) + \bar{\gamma}(|\omega|) : \omega \in U^*\}$.

For any given data $\{\hat{\gamma}(|\omega|) + \bar{\gamma}(|\omega|) : \omega \in U^*\}$, σ which satisfies $\{a_\sigma(\omega) = \hat{\gamma}(|\omega|) : \omega \in U^*\}$ is called a noisy realization of an input response map a.

We can propose the following algebraically noisy realization problem:

For any given $\{\hat{\gamma}(|\omega|) + \bar{\gamma}(|\omega|) : \omega \in U^*\}$, find, using only algebraic calculations, a so-called linear system σ which satisfies $a_\sigma(\omega) \approx \hat{\gamma}(|\omega|)$ for any $\omega \in U^*$.

A situation for algebraically noisy realization problem 4.24
Let the observed object be a so-called linear system and noise be added to the output. Then we will obtain the data $\{\gamma(t) = \hat{\gamma}(t) + \bar{\gamma}(t) : 0 \leq t \leq \underline{N}\}$ for some integer $\underline{N} \in N$, where $\hat{\gamma}(t)$ is the exact signal which comes from the observed so-called linear system and $\bar{\gamma}(t)$ is the noise added at the time of observation.

Problem 4.25. Problem statement of noisy realization for so-called linear systems

Let $H_{\underline{a}\ (p,\bar{p})}$ be the measured finite-sized Input/output matrix. Then find, using only algebraic calculations, the cleaned-up Input/output matrix $\hat{H}_{\underline{a}\ (p,\bar{p})}$ such that $H_{\underline{a}\ (p,\bar{p})} = \hat{H}_{\underline{a}\ (p,\bar{p})} + \bar{H}_{\underline{a}\ (p,\bar{p})}$ holds.

Namely, find, using only algebraic calculations, a minimal dimensional so-called linear system $\sigma = ((\boldsymbol{R}^n, F), g, h))$ which realizes $\hat{H}_{\underline{a}\ (p,\bar{p})}$.

Theorem 4.26. *Algebraically algorithm for noisy realization*

Let a partial input response map $\underline{a}$ be a considered object which is a so-called linear system. Then an algebraically noisy realization $\sigma_r = ((\boldsymbol{R}^n, F_r), g_r^0, g_r, h_r, h^0)$ of $\underline{a}$ is given by the following algorithm:

1) *Based on the square root of eigenvalues for a matrix $H_{\underline{a}\ (p,\bar{p})}(p,0)H_{\underline{a}\ (p,\bar{p})}(p,0)^T$, determine the value n_1 of rank for the Input/output matrix $H_{\underline{a}\ (p,\bar{p})}(p,0)$, where $n_1 \leq p$.*
   Namely, determine the value n_1 of rank for the matrix $H_{\underline{a}\ (p,\bar{p})}(p,0)$ such that a set of the square root of eigenvalues for the covariance matrix composed of relatively small and equally-sized numbers may be found, in which the signal part may be divided from the noisy part.

2) We use the algebraic CLS method as follows:

① *Based on Proposition (2.14), determine coefficients* $\{\alpha_{1i} : 1 \le i \le n_1\}$. *The Q in Proposition (2.14) can be considered as the matrix composed from the eigenvectors of* $H_{\underline{a}\ (n_1+1,L)}(n_1+1,0)H^T_{\underline{a}\ (n_1+1,L)}(n_1+1,0)$.
Let a matrix $A_1 \in \boldsymbol{R}^{1\times(n_1+1)}$ *be* $A_1 = [\alpha_{11}, \alpha_{12}, \cdots, \alpha_{1n_1}, -1]$.
② *Determine the error vectors* $\{\underline{S}_l^i \bar{I}_{\underline{a}} \in \boldsymbol{R}^{L\times 1} : 0 \le i \le n_1\}$ *by using the equation* $[\bar{I}_{\underline{a}}(0), \underline{S}_l\bar{I}_{\underline{a}}(0), \cdots, \underline{S}_l^{n_1}\bar{I}_{\underline{a}}(0)]^T :=$
$A_1^T[A_1A_1^T]^{-1}A_1H^T_{\underline{a}\ (n_1+1,L)}(n_1+1,0)$ *and*
$H^T_{\underline{a}\ (n_1,L)}(n_1,0) := [I_{\underline{a}}(0), \cdots, S_l^{n_1-1}I_{\underline{a}}(0), S_l^{n_1}I_{\underline{a}}(0)]$.
③ *Let* $h_{1r} \in \boldsymbol{R}^{1\times n_1}$ *be* $h_{1r} = [I_{\underline{a}}(1) - \bar{I}_{\underline{a}}(1), I_{\underline{a}}(2) - \bar{I}_{\underline{a}}(2), \cdots, I_{\underline{a}}(n_1) - \bar{I}_{\underline{a}}(n_1)]$.

3) Based on the square root of eigenvalues for a matrix

$H_{\underline{a}\ (n_1+p,\bar{p})}(n_1,p)H_{\underline{a}\ (n_1+p,\bar{p})}(n_1,p)^T$, *determine the value* n_2 *of rank for the matrix* $H_{\underline{a}\ (n_1+p,\bar{p})}(n_1,p)$, *where* $n_2 \le p$.
Namely, determine the value n_2 *of rank for the matrix* $H_{\underline{a}\ (n_1+p,\bar{p})}(n_1,p)$ *such that a set of the square root of eigenvalues for the covariance matrix composed of relatively small and equally-sized numbers is excluded, where the signal part effected by the set may be the noisy part.*

4) The algebraic CLS method is used as follows:
① *Based on Proposition (2.14), determine coefficients* $\{\alpha_{2i} : 1 \le i \le n_1+n_2\}$.
The Q in Proposition (2.14) can be considered as the matrix composed from the eigenvectors of $H_{\underline{a}\ (n_1,L)}(n_1,n_2+1)H^T_{\underline{a}\ (n_1,L)}(n_1,n_2+1)$.
Let a matrix $A_2 \in \boldsymbol{R}^{1\times(n_1+n_2+1)}$ *be* $A_2 = [\alpha_{21}, \alpha_{22}, \cdots, \alpha_{2n_1+n_2}, -1]$.
② *Determine the error vectors* $\{\underline{S}_l^i \bar{I}_{\underline{a}} \in \boldsymbol{R}^{L\times 1} : 0 \le i \le n_1+n_2\}$ *by using the equation*$[\bar{I}_{\underline{a}}, \underline{S}_l\bar{I}_{\underline{a}}, \cdots, \underline{S}_l^{n_1+n_2}\bar{I}_{\underline{a}}]^T :=$
$A_2^T[A_2A_2^T]^{-1}A_2H^T_{\underline{a}\ (n_1+n_2,L)}(n_1,n_2+1)$ *and* $H^T_{\underline{a}\ (n_1,L)}(n_1,n_2+1) :=$
$[I_{\underline{a}}(0), \cdots, S_l^{n_1-1}I_{\underline{a}}(0), I_{\underline{a}}(1), \cdots, S_l^{n_2-1}I_{\underline{a}}(1), S_l^{n_2}I_{\underline{a}}(1)]$.
③ *Let* $F_r \in \boldsymbol{R}^{(n_1+n_2)\times(n_1+n_2)}$ *be given as the same as in Definition (4.11). Let* g_r^0 *be* $g_r^0 = \mathbf{e}_1$ *and* g_r *be* $g_r = \mathbf{e}_{n_1+1} - \mathbf{e}_1$, *where*
$\mathbf{e}_i = [0, \cdots, 0, \overset{i}{1}, 0, \cdots, 0]^T \in \boldsymbol{R}^{n_1+n_2}$.
④ *Let* h_r *be* $h_s =$
$[h_{1r}, I_{\underline{a}}(1) - \bar{I}_{\underline{a}}(1), I_{\underline{a}}(1^2) - \bar{I}_{\underline{a}}(1^2), \cdots, I_{\underline{a}}(1^{n_2-1}) - \bar{I}_{\underline{a}}(1^{n_2-1})]$.

[proof] By 1) and 3), the noisy part in the data can be excluded in the sense of the number of dimensions by checking what part is the noisy part. The matrices A_1 in 2) and A_2 in 4) correspond to the matrix A in Lemma (2.17). Hence, if we determine the coefficients $\{\alpha_{ij} : 0 \le i \le 1,\ 1 \le j \le n_{i+1}\}$, we can obtain the noisy part of the finite Input/output matrices $H_{\underline{a}\ (n_1+1,\bar{p})}(n_1+1,0)$ and $H_{\underline{a}\ (n_1+n_2+1,\bar{p})}(n_1,n_2+1)$ by using Lemma (2.17).

Therefore, we obtain the cleaned-up Input/output matrices $\hat{H}_{\underline{a}\ (n_1+1,\bar{p})}(n_1+1,0)$ and $\hat{H}_{\underline{a}\ (n_1+n_2+1,\bar{p})}(n_1,n_2+1)$. Finally, we apply Theorem (4.14) to the $\hat{H}_{\underline{a}\ (n_1+1,\bar{p})}(n_1+1,0)$ and $\hat{H}_{\underline{a}\ (n_1+n_2+1,\bar{p})}(n_1,n_2+1)$.

Remark 1: Let S and N be the norm of a signal and a noise. Then the selected ratio of matrices in the algorithm may be considered as $\frac{N}{S+N}$.
Remark 2: This algebraically noisy realization method is very new.

Example 4.27. Let signals be the modified impulse responses of the following 3-dimensional so-called linear system $\sigma = ((R^3, F), x^0, g, h)$,

where $F = \begin{bmatrix} 0 & 0 & -0.5 \\ 1 & 0 & -0.4 \\ 0 & 1 & 0.7 \end{bmatrix}$, $h = [12,\ 6,\ -1]$, $x^0 = [-1.1,\ 1.9,\ -0.7]^T$,
$g = [1,\ 0,\ 0]^T$.

The almost linear system which corresponds to the so-called linear system is given by $\sigma = ((R^3, F), g^0, g, h, h^0)$, where $g^0 = [1.4,\ -2.7,\ 2.1]^T$, $h^0 = -0.73$.

Let added noises be given in Fig. 4.5.

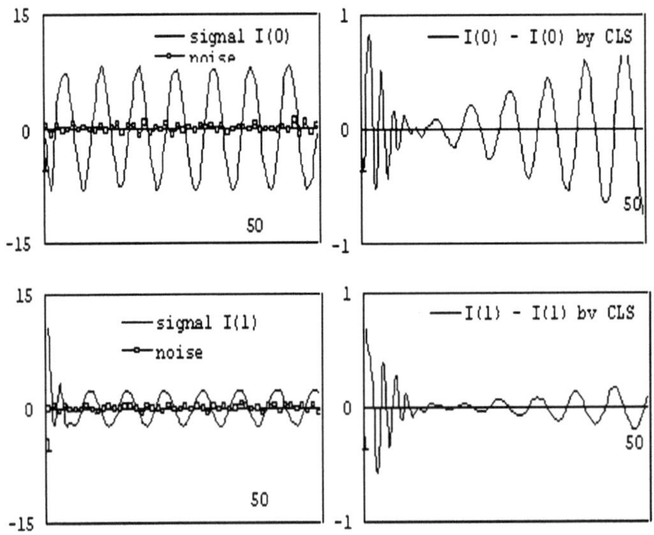

Fig. 4.5 The left are the exact modified impulse responses $I(0)$ and $I(1)$ and noises added to $I(0)$ and $I(1)$. The right are the difference between the exact ones and the obtained modified impulse responses by the algebraic CLS in Example (4.27).

Then the algebraically noisy realization problem is solved as follows:

covariance matrix	eigenvalues					
	1	2	3	4	5	6
$H^T_{\underline{a}\ (4,50)}(4,0)H_{\underline{a}\ (4,50)}(4,0)$	3969	2813	101	12		
$H^T_{\underline{a}\ (5,50)}(5,0)H_{\underline{a}\ (5,50)}(5,0)$	5232	3227	103	16	3.9	
$H^T_{\underline{a}\ (6,50)}(6,0)H_{\underline{a}\ (6,50)}(6,0)$	5680	4463	105	18	8	1.4
covariance matrix	square root of eigenvalues					
	1	2	3	4	5	6
$H^T_{\underline{a}\ (6,50)}(6,0)H_{\underline{a}\ (6,50)}(6,0)$	77	67	10	4.2	2.8	1.2

covariance matrix	eigenvalues					
	1	2	3	4	5	6
$H^T_{\underline{a}\ (4,50)}(3,1)H_{\underline{a}\ (4,50)}(3,1)$	2876	2351	212	9.4		
$H^T_{\underline{a}\ (5,50)}(3,2)H_{\underline{a}\ (5,50)}(3,2)$	2929	2435	234	11	4.5	
$H^T_{\underline{a}\ (6,50)}(3,3)H_{\underline{a}\ (6,50)}(3,3)$	3077	2436	243	11	7.8	2.6
covariance matrix	square root of eigenvalues					
	1	2	3	4	5	6
$H^T_{\underline{a}\ (6,50)}(3,3)H_{\underline{a}\ (6,50)}(3,3)$	55.4	49.4	15.6	3.3	2.8	1.6

1) Since a set $\{4.2,\ 2.8,\ 1.2\}$ is composed of relatively small and equally-sized numbers in the square root of $H^T_{\underline{a}\ (6,50)}(6,0)H_{\underline{a}\ (6,50)}(6,0)$, the noisy realization of a so-called linear system obtained by the algebraic CLS method may be good for a 3-dimensional space.
2) After determining the number n_1 of dimensions which is 3, we will continue the algebraically noisy realization algorithm.

Therefore, the modified impulse response $I(0)$ of a so-called linear system obtained by the algebraic CLS method is constructed by a 3-dimensional so-called linear system.
3) Since a set $\{3.3,\ 2.8,\ 1.6\}$ is composed of relatively small and equally-sized numbers in the square root of $H^T_{\underline{a}\ (6,50)}(3,3)H_{\underline{a}\ (6,50)}(3,3)$, an almost linear system obtained by the algebraic CLS method may be somewhat good by adding nothing.
4) After determining the number n_2 of dimensions which is 0, we will continue the algebraically noisy realization algorithm.

Therefore, the modified impulse response $I(1)$ of a so-called linear system obtained by the algebraic CLS method is constructed by adding nothing.

Therefore, the modified impulse responses $I(0)$ and $I(1)$ of an almost linear system obtained by the algebraic CLS method is constructed by a 3-dimensional so-called linear system.

The system is given by $\sigma_o = ((\boldsymbol{R}^3, F_o), x_o^0, g_o, h_o)$, where $F_o = \begin{bmatrix} 0 & 0 & -0.58 \\ 1 & 0 & -0.31 \\ 0 & 1 & 0.62 \end{bmatrix}$,
$x_o^0 = [-0.54, -0.29, -0.79]^T$, $g_o = [-0.5, -0.72, 0.83]^T$, $h_o = [-1.2, -8.8, 4.9]$. The obtained modified impulse responses $I(0)$ and $I(1)$ are illustrated in Fig. 4.5.

We can show that the algorithm for noisy realization given by the analytic CLS method in the reference [Hasegawa, 2008] produces the same systems as the above one in the sense of the numerical calculation.

In this example, original signals are considered as the modified impulse responses of a 3-dimensional so-called linear system and the desirable modified impulse responses are obtained by the algebraic CLS method. A model obtained by the algebraic CLS method is a 3-dimensional so-called linear system which has the same number of dimensions as the number of the original system. The following table indicates that the 3-dimensional so-called linear system reconstructs the original signal with a 6 and 6 % error to signal ratio and with 0.05 and 0.06 noise to signal ratio, please refer to Remark 1 in Theorem 4.26 for the noise to signal ratio. Just as we expected, the following table and Fig. 4.5 indicate that the model obtained by the algebraic CLS method is a somewhat good 3-dimensional system for the original 3-dimensional system.

dimension	ratio of matrices	mean values of square root for sum of signal	signal by CLS	error	cosine ① and ②	error ratio
		①	②	③	$\cos\theta$	③/①
I(0)_(3,0)	0.05	0.819	0.83	0.05	0.998	0.06
I(1)_(3,0)	0.06	0.325	0.323	0.02	0.997	0.06

For the notations $I(0)\_(n_1, n_2)$ and $I(1)\_(n_1, n_2)$, see Definition (4.11).

Example 4.28. Let signals be the modified impulse responses of the following 4-dimensional so-called linear system $\sigma = ((R^4, F), x^0, g, h)$,

where $F = \begin{bmatrix} 0\,0\,0 & -0.6 \\ 1\,0\,0 & -0.5 \\ 0\,1\,0 & 0.9 \\ 0\,0\,1 & 0.3 \end{bmatrix}$, $h = [10,\ 3,\ 0,\ -2]$, $x^0 = [-1,\ 0,\ 0.1,\ 0]^T$,
$g = [1,\ 0,\ 0,\ 0]^T$.

The almost linear system which corresponds to the so-called linear system is given by $\sigma = ((R^4, F), g^0, g, h, h^0)$, where $g^0 = [1,\ -1,\ -0.1,\ 0.1]^T$, $h^0 = -10$

Let added noises be given in Fig. 4.6.

Then the algebraically noisy realization problem is solved as follows:

covariance matrix	eigenvalues						
	1	2	3	4	5	6	7
$H^T_{\underline{a}\ (5,80)}(5,0)H_{\underline{a}\ (5,80)}(5,0)$	12368	9380	151	58	3.5		
$H^T_{\underline{a}\ (6,80)}(6,0)H_{\underline{a}\ (6,80)}(6,0)$	14660	12127	158	91	4.4	2.7	
$H^T_{\underline{a}\ (7,80)}(7,0)H_{\underline{a}\ (7,80)}(7,0)$	18785	13070	164	93	5.9	3.4	1.8
covariance matrix	square root of eigenvalues						
	1	2	3	4	5	6	7
$H^T_{\underline{a}\ (7,80)}(7,0)H_{\underline{a}\ (7,80)}(7,0)$	137	114	12.8	9.6	2.4	1.8	1.3

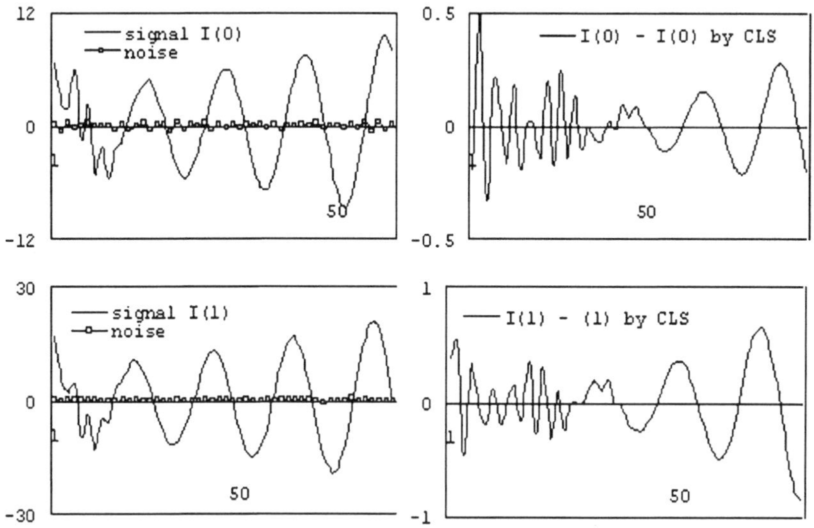

Fig. 4.6 The left are exact modified impulse responses $I(0)$ and $I(1)$ and noises added to $I(0)$ and $I(1)$. The right are the difference between the exact ones and the obtained modified impulse responses by the algebraic CLS in Example (4.28).

covariance matrix	eigenvalues						
	1	2	3	4	5	6	7
$H^T_{\underline{a}\ (5,80)}(4,1)H_{\underline{a}\ (5,80)}(4,1)$	31839	5439	162	65	2.9		
$H^T_{\underline{a}\ (6,80)}(4,2)H_{\underline{a}\ (6,80)}(4,2)$	49573	9508	213	67	3.6	1.7	
$H^T_{\underline{a}\ (7,80)}(4,3)H_{\underline{a}\ (7,80)}(4,3)$	62981	19530	394	75	4.3	2	1.2
covariance matrix	square root of eigenvalues						
	1	2	3	4	5	6	7
$H^T_{\underline{a}\ (7,80)}(4,3)H_{\underline{a}\ (7,80)}(4,3)$	251	140	19.8	8.7	2	1.4	1.1

1) Since a set $\{2.4,\ 1.8,\ 1.3\}$ is composed of relatively small and equally-sized numbers in the square root of $H^T_{\underline{a}\ (7,80)}(7,0)H_{\underline{a}\ (7,80)}(7,0)$, the algebraically noisy realization of a so-called linear system may be good for a 4-dimensional space.
2) After determining the number n_1 of dimensions which is 4, we will continue the algebraically noisy realization algorithm.

Therefore, the modified impulse response $I(0)$ of a so-called linear system obtained by the algebraic CLS method is constructed for a 4-dimensional space.
3) Since a set $\{2,\ 1.4,\ 1.1\}$ is composed of relatively small and equally-sized numbers in the square root of $H^T_{\underline{a}\ (7,80)}(4,3)H_{\underline{a}\ (7,80)}(4,3)$, an almost linear system obtained by the algebraic CLS method may be somewhat good by adding nothing.
4) After determining the number n_2 of dimensions which is 0, we will continue the noisy realization algorithm by the algebraic CLS method.

Therefore, the modified impulse response $I(1)$ of a so-called linear system obtained by the algebraic CLS method is constructed by adding nothing.

Therefore, the modified impulse responses $I(0)$ and $I(1)$ of the so-called linear system obtained by the algebraic CLS method is realized by a 4-dimensional so-called linear system.

The system is given by $\sigma_o = ((\boldsymbol{R}^4, F_o), x_o^0, g_o, h_o)$, where $F_o = \begin{bmatrix} 0 & 0 & 0 & -0.65 \\ 1 & 0 & 0 & -0.46 \\ 0 & 1 & 0 & 0.92 \\ 0 & 0 & 1 & 0.26 \end{bmatrix}$,

$x_o^0 = [-0.3,\ 0.2,\ -0.8,\ -1.1]^T$, $g_o = [0.17,\ -0.29,\ 0.91,\ 1.13]^T$,
$h_o = [7,\ 1.9,\ 2.54,\ 5.69]$.

The obtained modified impulse responses $I(0)$ and $I(1)$ are illustrated in Fig. 4.6.

We can show that the algorithm for noisy realization given by the analytic CLS method in the reference [Hasegawa, 2008] produces the same system as the above one in the sense of the numerical calculation.

In this example, original signals are considered as the modified impulse responses of a 4-dimensional so-called linear system and the desirable modified impulse responses are obtained by the algebraic CLS method. The model obtained by the algebraic CLS method is a 4-dimensional so-called linear system which has the same number of dimensions as the number of the original system.

The following table indicates that the 4-dimensional so-called linear system reconstructs the original signal with a 3 and 3 % error to signal ratio and with 0.02 and 0.01 noise to signal ratio, please refer to Remark 1 in Theorem 4.26 for the noise to signal ratio.

Just as we expected, the following table and Fig. 4.6 indicate that the model obtained by the algebraic CLS method is a somewhat good 4-dimensional system for the original 4-dimensional system.

dimen-sion	ratio of matrices	mean values of square root for sum of			cosine	error
		signal	signal by CLS	error	① and ②	ratio
		①	②	③	$\cos\theta$	③/①
I(0)_(4,0)	0.02	0.687	0.686	0.02	0.999	0.03
I(1)_(4,0)	0.01	1.501	1.502	0.05	0.999	0.03

For the notations $I(0)\_(n_1, n_2)$ and $I(1)\_(n_1, n_2)$, see Definition (4.11).

Example 4.29. Let signals be the modified impulse responses of the following 5-dimensional so-called linear system $\sigma = ((R^5, F), x^0, g, h)$, where

$$F = \begin{bmatrix} 0 & 0 & 0 & 0 & 0.2 \\ 1 & 0 & 0 & 0 & -0.4 \\ 0 & 1 & 0 & 0 & -0.1 \\ 0 & 0 & 1 & 0 & 0.5 \\ 0 & 0 & 0 & 1 & -0.4 \end{bmatrix},\ h = [12,\ 2,\ -5,\ -1,\ 2],\ x^0 = [-1,\ 0,\ 0,\ 1,\ 0]^T,$$

$g = [1,\ 0,\ 0,\ 0,\ 0]^T$.

The almost linear system which corresponds to the so-called linear system is given by $\sigma = ((R^4, F), g^0, g, h, h^0)$, where $g^0 = [10, -10, -1, 1, 0]^T$, $h^0 = -13$.

Let added noises be given in Fig. 4.7.

Then the algebraically noisy realization problem is solved as follows:

covariance matrix	eigenvalues						
	1	2	3	4	5	6	7
$H^T_{\underline{a}\ (5,50)}(5,0)H_{\underline{a}\ (5,50)}(5,0)$	28820	26149	11525	2061	138		
$H^T_{\underline{a}\ (6,50)}(6,0)H_{\underline{a}\ (6,50)}(6,0)$	28826	28446	17097	2064	138	20.2	
$H^T_{\underline{a}\ (7,50)}(7,0)H_{\underline{a}\ (7,50)}(7,0)$	30226	28653	22654	2321	144	28	12
covariance matrix	square root of eigenvalues						
	1	2	3	4	5	6	7
$H^T_{\underline{a}\ (7,50)}(7,0)H_{\underline{a}\ (7,50)}(7,0)$	174	169	151	45.4	12	5.3	3.5

covariance matrix	eigenvalues						
	1	2	3	4	5	6	7
$H^T_{\underline{a}\ (6,50)}(5,1)H_{\underline{a}\ (6,50)}(5,1)$	58427	27305	11526	2073	146	7.8	
$H^T_{\underline{a}\ (7,50)}(5,2)H_{\underline{a}\ (7,50)}(5,2)$	58743	41003	14679	2335	148	8.5	7.3

covariance matrix	square root of eigenvalues						
	1	2	3	4	5	6	7
$H^T_{\underline{a}\ (7,50)}(5,2)H_{\underline{a}\ (7,50)}(5,2)$	242	202	121	15.3	12.1	2.9	2.7

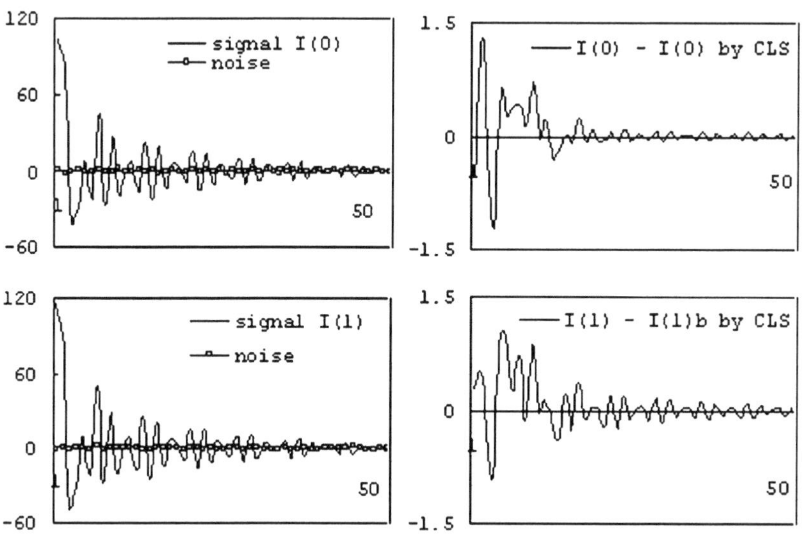

Fig. 4.7 The left are exact modified impulse responses $I(0)$ and $I(1)$ and noises added to $I(0)$ and $I(1)$. The right are the difference between the exact ones and the obtained modified impulse responses by the algebraic CLS in Example (4.29).

1) Since a set $\{5.3,\ 3.5\}$ is composed of relatively small and equally-sized numbers in the square root of $H^T_{\underline{a}\ (7,50)}(7,0)H_{\underline{a}\ (7,50)}(7,0)$, a noisy realization of a so-called linear system obtained by the algebraic CLS method may be good for a 5-dimensional space.
2) After determining the number n_1 of dimensions which is 5, we will continue the algebraically noisy realization algorithm.

Therefore, the modified impulse response $I(0)$ of a so-called linear system obtained by the algebraic CLS method is constructed for a 5-dimensional space.
3) Since a set $\{2.9,\ 2.7\}$ is composed of relatively small and equally-sized numbers in the square root of $H^T_{\underline{a}\ (7,50)}(5,2)H_{\underline{a}\ (7,50)}(5,2)$, an almost linear system obtained by the algebraic CLS method may be somewhat good by adding nothing.
4) After determining the number n_2 of dimensions which is 0, we will continue the algebraically noisy realization algorithm.

Therefore, the modified impulse response $I(1)$ of a so-called linear system obtained by the algebraic CLS method is constructed by adding nothing.

Therefore, the modified impulse responses $I(0)$ and $I(1)$ of an almost linear system obtained by the algebraic CLS method is realized by a 5-dimensional so-called linear system.
The system is given by $\sigma_o = ((\boldsymbol{R}^5,\ F_o),\ g_o^0, g_o, h_o,\ h^0)$, where

$$F_o = \begin{bmatrix} 0 & 0 & 0 & 0 & 0.17 \\ 1 & 0 & 0 & 0 & -0.42 \\ 0 & 1 & 0 & 0 & -0.12 \\ 0 & 0 & 1 & 0 & 0.42 \\ 0 & 0 & 0 & 1 & -0.47 \end{bmatrix},\ x_o^0 = [-1.12,\ -0.83,\ -0.74,\ -1.04,\ -0.7]^T,$$

$g_o = [0.11,\ 0.06,\ 0.07,\ 0.09,\ 0.062]^T$, $h_o = [104.5,\ 71.7,\ -41,\ -26.4,\ 6.56]$ and $h^0 = -15$.

The obtained modified impulse responses $I(0)$ and $I(1)$ are illustrated in Fig. 4.7.

We can show that the algorithm for noisy realization given by the analytic CLS method in the reference [Hasegawa, 2008] produces the same system as the above one in the sense of the numerical calculation.

In this example, original signals are considered as the modified impulse responses of a 5-dimensional so-called linear system and the desirable modified impulse responses are obtained by the algebraic CLS method. The model obtained by the algebraic CLS method is a 5-dimensional so-called linear system which has the same number of dimensions as the number of the original system.

The following table indicates that the 5-dimensional so-called linear system reconstructs the original signal with a 1 and 1 % error to signal ratio and with 0.03 and 0.01 noise to signal ratio, please refer to Remark 1 in Theorem 4.26 for the noise to signal ratio.

Just as we expected, the following table and Fig. 4.7 indicate that the model obtained by the algebraic CLS method is a good 5-dimensional system for the original 5-dimensional system.

dimen- sion	ratio of matrices	mean values of square root for sum of signal ①	signal by CLS ②	error ③	cosine ① and ② $\cos\theta$	error ratio ③/①
I(0)_(5,0)	0.03	3.255	3.248	0.04	0.9999	0.01
I(1)_(5,0)	0.01	3.512	3.51	0.05	0.9999	0.01

For the notations $I(0)\_(n_1, n_2)$ and $I(1)\_(n_1, n_2)$, see Definition (4.11).

Example 4.30. Let signals be the modified impulse responses of the following 6-dimensional so-called linear system $\sigma = ((R^6, F), x^0, g, h)$, where

$$F = \begin{bmatrix} 0 & 0 & 0 & 0 & 0 & -0.5 \\ 1 & 0 & 0 & 0 & 0 & -0.4 \\ 0 & 1 & 0 & 0 & 0 & -0.2 \\ 0 & 0 & 1 & 0 & 0 & 0.2 \\ 0 & 0 & 0 & 1 & 0 & 0.5 \\ 0 & 0 & 0 & 0 & 1 & -0.7 \end{bmatrix}, \ h = [6, 4, -3, -1, 3, -2], x^0 = [-10, 0, 1, 0, 0, 0]^T,$$
$$g = [1,\ 0,\ 0,\ 0,\ 0,\ 0]^T.$$

The almost linear system which corresponds to the so-called linear system is given by $\sigma = ((R^6, F), g^0, g, h, h^0)$, where $g^0 = [10, -10, -1, 1, 0, 0]^T$, $h^0 = -63$.

Let added noises be given in Fig. 4.8.

Then the algebraically noisy realization problem is solved as follows:

covariance matrix	eigenvalues								
	1	2	3	4	5	6	7	8	9
$H^T_{\underline{a}\ (8,50)}(8,0)H_{\underline{a}\ (8,50)}(8,0)$	19040	17079	13029	4083	292	139	10	7.4	
$H^T_{\underline{a}\ (9,50)}(9,0)H_{\underline{a}\ (9,50)}(9,0)$	20072	17079	13338	5201	317	151	10.8	8.4	4.7
covariance matrix	square root of eigenvalues								
	1	2	3	4	5	6	7	8	9
$H^T_{\underline{a}\ (9,50)}(9,0)H_{\underline{a}\ (9,50)}(9,0)$	142	131	115	72	17.8	12.2	3.3	2.9	2.2

covariance matrix	eigenvalues								
	1	2	3	4	5	6	7	8	9
$H^T_{\underline{a}\ (7,50)}(6,1)H_{\underline{a}\ (7,50)}(6,1)$	32072	17258	12098	1724	235	131	4.8		
$H^T_{\underline{a}\ (8,50)}(6,2)H_{\underline{a}\ (8,50)}(6,2)$	36523	27018	12160	1741	280	132	5	4.4	
$H^T_{\underline{a}\ (9,50)}(6,3)H_{\underline{a}\ (9,50)}(6,3)$	36639	32789	14252	1890	374	134	5.3	4.9	3.8
covariance matrix	square root of eigenvalues								
$H^T_{\underline{a}\ (9,50)}(6,3)H_{\underline{a}\ (9,50)}(6,3)$	191	181	119	43.4	19.3	11.6	2.3	2.2	1.9

1) Since a set $\{3.3,\ 2.9,\ 2.2\}$ is composed of relatively small and equally-sized numbers in the square root of $H^T_{\underline{a}\ (9,50)}(9,0)H_{\underline{a}\ (9,50)}(9,0)$, a noisy realization of a so-called linear system obtained by the algebraic CLS method may be good for a 6-dimensional space.
2) After determining the number n_1 of dimensions which is 6, we will continue the algebraically noisy realization algorithm.

Therefore, the modified impulse response $I(0)$ of a so-called linear system obtained by the algebraic CLS method is constructed for a 6-dimensional space.
3) Since a set $\{2.3,\ 2.2,\ 1.9\}$ is composed of relatively small and equally-sized numbers in the square root of $H^T_{\underline{a}\ (9,50)}(6,3)H_{\underline{a}\ (9,50)}(6,3)$, a so-called linear system obtained by the algebraic CLS method may be somewhat good by adding nothing.
4) After determining the number n_2 of dimensions which is 0, we will continue the algebraically noisy realization algorithm.

Therefore, the modified impulse response $I(1)$ of a so-called linear system obtained by the algebraic CLS method is constructed by adding nothing.

Therefore, the modified impulse responses $I(0)$ and $I(1)$ of a so-called linear system obtained by the algebraic CLS method is realized by a 6-dimensional so-called linear system.

In particular, $I(0)$ has a 6-dimensional space and $I(1)$ has a 0-dimensional space.

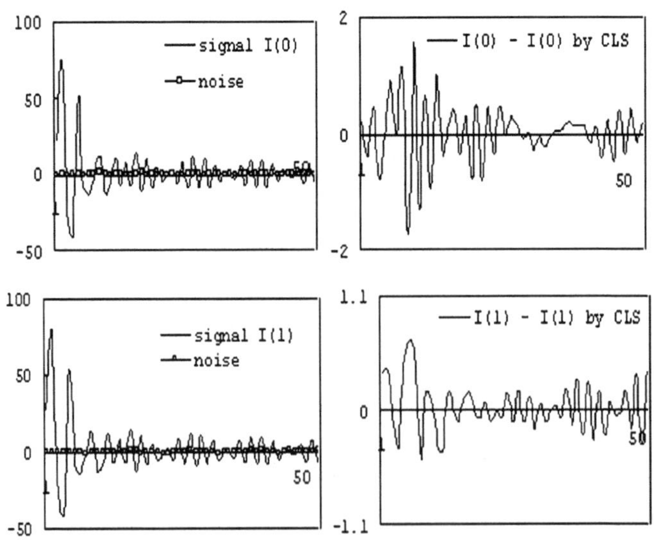

Fig. 4.8 The left are exact modified impulse responses $I(0)$ and $I(1)$ and noises added to $I(0)$ and $I(1)$. The right are the difference between the exact ones and the obtained modified impulse responses by the algebraic CLS in Example (4.30).

The system is given by $\sigma_o = ((\boldsymbol{R}^6,\ F_o),\ x_o^0, g_o, h_o)$, where

$$F_o = \begin{bmatrix} 0\,0\,0\,0\,0 & -0.5 \\ 1\,0\,0\,0\,0 & -0.38 \\ 0\,1\,0\,0\,0 & -0.16 \\ 0\,0\,1\,0\,0 & 0.23 \\ 0\,0\,0\,1\,0 & 0.49 \\ 0\,0\,0\,0\,1 & -0.7 \end{bmatrix},\ x_o^0 = [-0.75,\ -0.57,\ -0.49,\ -0.6,\ -0.85,\ -0.5]^T,$$

$$g_o = [0.09,\ 0.07,\ 0.08,\ 0.05,\ 0.043,\ 0.03]^T,$$
$$h_o = [21.8, 74.4, -25.4, -38.5, 51.3, -7.5].$$

The obtained modified impulse responses $I(0)$ and $I(1)$ are illustrated in Fig. 4.8.

We can show that the algorithm for noisy realization given by the analytic CLS method in the reference [Hasegawa, 2008] produces the same system as the above one in the sense of the numerical calculation.

In this example, original signals are considered as the modified impulse responses of a 6-dimensional so-called linear system and desirable modified impulse responses are obtained by the algebraic CLS method. The model obtained by the algebraic CLS method is a 6-dimensional so-called linear system which has the same number of dimensions as the number of the original system.

The following table indicates that the 6-dimensional so-called linear system reconstructs the original signal with a 3 and 3 % error to signal ratio and with 0.02 and 0.01 noise to signal ratio, please to Remark 1 in Theorem 4.26 for the noise to signal ratio.

Just as we expected with observed data of a short length, the following table and Fig. 4.8 indicate that the model obtained by the algebraic CLS method is a good 6-dimensional system for the original 6-dimensional system even though the signal has been a little damped.

dimen-sion	ratio of matrices	mean values of square root for sum of signal	signal by CLS	error	cosine ① and ②	error ratio
		①	②	③	$\cos\theta$	③/①
I(0)_(6,0)	0.02	2.303	2.31	0.08	0.999	0.03
I(1)_(6,0)	0.01	2.453	2.456	0.08	0.999	0.03

For the notations $I(0)\_(n_1, n_2)$ and $I(1)\_(n_1, n_2)$, see Definition (4.11).

4.7 Historical Notes and Concluding Remarks

Algebraically approximate realization and noisy realization problems of so-called linear systems were studied from the viewpoint of Input/output matrix norm and the algebraic CLS method. The matrix norm is used for determining the dimension number of state space and the algebraic CLS method is used for determining the parameters of so-called linear systems, which are sort of non-linear systems.

For our treatment of the approximate and noisy realization problems, as we said, there may be a time for using singular value decomposition and the algebraic Constrained Least Square (CLS) in Kalman [1997]. In the reference, Kalman also pointed out that the identification problem from noisy data should be treated without any prejudice, hence, should be viewed in a statistical sense, not a probabilistic sense. Here, we only insist that the signal and the noise are not correlated. Then, we could discuss algebraically noisy realization problems of non-linear system that could not be treated up to this point.

In order to ascertain that our method for algebraically approximate and noisy realization is effective for non-linear cases, we provided several examples. Based on the result of the examples, we have shown that the ratio of the square root of singular values implies the degree of approximation. For our algebraically noisy realization problem, we have shown that we can determine the dimension number of so-called linear systems when a set of relatively small and equally-sized numbers of the square root of singular values can be found.

As stated in the Historical notes and concluding remarks of chapter 3, our methods can be roughly summarized as follows:

Intuitively, our several examples for algebraically approximate realization problems demonstrate that the smaller the ratio of matrices is, the smaller the error to signal ratio is. The ratio 0.01 Input/output matrix ratio implies a range of 1 to 4 % error to signal ratio.

The several examples suggest that three unique features can be expressed as follows:

(1) : The ratio of the matrix norm determines the degree of the crossed angle between directions of the approximated signal and the original signal.
(2) : We could propose a new law which says that so-called linear systems obtained by the algebraic CLS method are the same as ones obtained by the analytic CLS method proposed in the reference [Hasegawa, 2008]. The law is said to be a law of a constrained least square.
(3) : The algebraic CLS method determines the coefficients of linearly dependent vectors such that the error between the approximate signal and the original signal has a minimum value in the sense of a square norm while conserving the crossed angle.

Intuitively, our several examples for algebraically noisy realization problem show that the smaller the ratio of matrices is, the smaller the error to signal ratio is. The ratio within a 0.05 input/output matrix ratio implies an error to signal ratio within 6 %.

The several examples suggest that three unique features can be expressed as follows:

(1) : The ratio of matrices determines the degree of the crossed angle between directions of the obtained signal and the original signal.
(2) : We could propose a new law which says that so-called linear systems obtained by the algebraic CLS method are the same as ones obtained by the analytic CLS method proposed in the reference [Hasegawa, 2008].
(3) : The algebraic CLS method determines the coefficients of linearly dependent vectors such that the error between the obtained signal and original signal has a minimum value in the sense of a square norm while conserving the crossed angle.

The analytic CLS method for determing n variables is reduced to the minimization of the following rational polynomial $p(x_1, x_2, \cdots, x_n)$ in n variables:

$$p(x_1, x_2, \cdots, x_n) = \frac{\sum_{c_1, c_2, \cdots, c_n} \alpha(c_1, c_2, \cdots, c_n) \times x_1^{c_1} x_2^{c_2} \cdots x_n^{c_n}}{(1 + x_1^2 + x_2^2 + \cdots + x_n^2)^2},$$

where $\sum_{c_1, c_2, \cdots, c_n}$ means all summations of any combination with the conditions $0 \leq c_i \leq 4$ and $c_1 + c_2 + \cdots + c_n \leq 4$ for any $1 \leq i \leq n$ and $\alpha(c_1, c_2, \cdots, c_n) \in \boldsymbol{R}$.

Therefore, our new Law demonstrates that approximate and noisy problems can be solved using only algebraic calculations, namely, without treating partial differential equations.

Chapter 5
Algebraically Approximate and Noisy Realization of Almost Linear Systems

Let the set of output's values Y be a linear space over the real number field $\boldsymbol{R}$. In the reference [Matsuo and Hasegawa, 2003], we introduced almost linear systems that are in a subclass of pseudo linear systems, which are very close to linear systems.

At first, their realization theory was stated. Namely, it was shown that any almost linear systems can be characterized by time-invariant, affine input response maps and any time-invariant, affine input response maps, that is, any input/output maps with causality, time-invariance and affinity can be completely characterized by two modified impulse responses, where the modified impulse response may be a slightly revised version of an impulse response in linear systems. An existence theorem and a uniqueness theorem were also proved.

Secondly, details of finite dimensional almost linear systems were investigated. A criterion for the canonical finite dimensional almost linear systems and representation theorems of isomorphic classes for canonical almost linear systems were given. Moreover, a criterion for the behavior of finite dimensional almost linear systems and a procedure to obtain the canonical almost linear systems were given. The criterion is the finite rank condition of an Input/output matrix, which is a natural extension of a finite rank of a Hankel matrix in linear systems.

Thirdly, their partial realization was discussed according to the above results. An algorithm to obtain an almost linear system from the given partial input response map was given.

We can easily understand that the above results of our systems are the same as ones obtained in linear system theory.

In chapter 4, we stated fundamental facts about so-called linear systems for preparation of their approximate and noisy realization problems. Since so-called linear systems are in a subclass of almost linear systems, the problems were discussed in the sense of almost linear systems in order to make a unified treatment as much as we could. Therefore, the fundamental facts about almost linear systems were stated. Hence, please refer to the facts needed for

Y. Hasegawa: Algebra. Approx. & Noisy Reali. of Discrete-Time Sys., LNEE 50, pp. 87–109.
springerlink.com

our discussion about algebraically approximate and noisy realization problems in chapter 4.

5.1 Basic Facts of Almost Linear Systems

Definition 5.1. Almost Linear System

1) A system given by the following system equations is written as a collection $\sigma = ((X, F), g^0, g, h, h^0)$ and is said to be an almost linear system.

$$\begin{cases} x(t+1) &= Fx(t) + g^0 + g\omega(t+1) \\ x(0) &= 0 \\ \hat{\gamma}(t) &= h^0 + hx(t) \end{cases}$$

for any $t \in N$, $x(t) \in X$, $\gamma(t) \in Y$, and X is a linear space over the field $\boldsymbol{R}$, F is a linear operator on X, $g^0,\ g \in X$ and $h : X \to Y$ is a linear operator.

2) The input response map $a_\sigma : U^* \to Y; \omega \mapsto h^0 + h(\sum_{j=1}^{|\omega|} F^{|\omega|-j}(g^0 + g\omega(j))$ is said to be the behavior of σ.

3) For the almost linear system σ and any $i \geq 1$, $I_\sigma(1)(i) := a_\sigma(0^{i-1}|1^1) - a_\sigma(0^{i-1}) = hF^{i-1}(g^0 + g)$ and $I_\sigma(0)(i) := a_\sigma(0^i) - a_\sigma(0^{i-1}) = hF^{i-1}g^0$ are said to be modified impulse responses of σ, where $0^0 := 1$.

   Note that there is a one-to-one correspondence between the behavior of σ and the modified impulse responses $I_\sigma(0)$ and $I_\sigma(1) \in F(N, Y)$ of σ by the relations $a_\sigma(\omega) = (\sum_{j=1}^{|\omega|}(I_\sigma(0)(|\omega| - j + 1) + I_\sigma(1)(|\omega| - j + 1) \times \omega(j))$.

4) An almost linear system σ is said to be quasi-reachable if the linear hull of the reachable set $\{\sum_{j=1}^{|\omega|} F^{|\omega|-j}(g^0 + g\omega(j)); \omega \in U^*\}$ is equal to X and an almost linear system σ is called to be observable if $hF^i x_1 = hF^i x_2$ for any $i \in N$ implies $x_1 = x_2$.

   Especially, an almost linear σ is called reachable if the reachable set $\{\sum_{j=1}^{|\omega|} F^{|\omega|-j}(g^0 + g\omega(j)); \omega \in U^*\}$ is equal to X.

5) An almost linear system σ is called canonical if σ is reachable and observable.

6) A system σ is said to be intrinsically canonical if σ is reachable and observable.

Example 5.2. $A(N \times \{0, 1\}, K) := \{\lambda = \sum_{n,u} \lambda(n, u)\mathbf{e}_{(\mathbf{n},\mathbf{u})}$ (finite sum) $; n \in N, u \in \{0, 1\}\}$, where $\mathbf{e}_{(\mathbf{n},\mathbf{u})}$ is given by the following equations for $n, n' \in N$ and $u, u' \in \{0, 1\}$. If $n = n'$ and $u = u'$ imply $\mathbf{e}_{(\mathbf{n},\mathbf{u})}(n', u') = 1$. If $n \neq n'$ or $u \neq u'$ imply $\mathbf{e}_{(\mathbf{n},\mathbf{u})}(n', u') = 0$. Then $A(N \times \{0, 1\}, K)$ is clearly a linear space. Let S_r be $S_r\mathbf{e}_{(\mathbf{n},\mathbf{u})} = \mathbf{e}_{(\mathbf{n+1},\mathbf{u})}$. Then $S_r \in L(A(N \times \{0, 1\}, K))$ and S_r is irrelevant to the input value's set $\{0, 1\}$. S_r is a right shift operator. Let $\bar{\eta} := e_{(0,1)} - e_{(0,0)}$ and let a linear map $\bar{a} : A(N \times \{0, 1\}, K) \to Y$ be $\bar{a}(\mathbf{e}_{(\mathbf{n},\mathbf{u})}) = a(u^{n+1}) - a(u^n)$ for any time-invariant, affine input response map

$a \in F(U^*, Y)$. Then a collection $((A(N \times \{0,1\}, K), S_r), \mathbf{e}_{(\mathbf{0},\mathbf{0})}, \bar{\eta}, \bar{a}, a(1))$ is a quasi-reachable almost linear system that realizes a.

Let $F(N, Y) := \{$ any function $f : N \to Y\}$ and let $S_l\gamma(t) = \gamma(t+1)$ for any $\gamma \in F(N, Y)$ and $t \in N$. Then $S_l \in L(F(N, Y))$. Let a map $\chi^0 \in F(N, Y)$ be $(\chi^0)(t) := a(\omega|0) - a(\omega)$ and $\bar{\chi} \in F(N, Y)$ be $(\bar{\chi})(t) := a(\omega|1) - a(\omega|0)$ for any $t \in N$, a time-invariant, affine input response map $a \in F(U^*, Y)$ and $\omega \in U^*$ such that $|\omega| = t$. Moreover, let a linear map 0 be $F(N, Y) \to Y; \gamma \mapsto \gamma(0)$. Then a collection $((F(N, Y), S_l), \chi^0, \bar{\chi}, 0, a(1))$ is an observable almost linear system that realizes a.

Theorem 5.3. *The following two almost linear systems are canonical realizations of any time-invariant, affine input response map $a \in F(U^*, Y)$.*

1) $(A(N \times \{0,1\}, K)/_{=a}, \hat{S}_r), [e_{(0,0)}], \hat{\bar{\eta}}, \hat{\bar{a}}, a(1))$,
where $A(N \times \{0,1\}, K)/_{=a}$ is a quotient space obtained by equivalence relation $\sum_{(n,u)} \lambda_1(n,u)\mathbf{e}_{(\mathbf{n},\mathbf{u})} = \sum_{(n',u')} \lambda_2(n',u')\mathbf{e}_{(\mathbf{n}',\mathbf{u}')} \iff \sum_{(n,u)} \lambda(n,u)(a(u^{n+1}) - a(u^n)) = \sum_{(n,u)} \lambda(n,u)(a(u^{n+1}) - a(u^n))$.
Moreover, $\hat{S}_r \in L(A(N \times \{0,1\}, K)/_{=a})$ *is given by* $\hat{S}_r[e(n,u)] = [e_{(n+1,u)}]$ *for* $[e_{(n,u)}] \in A(N \times \{0,1\}, K)/_{=a}$, *and* $\hat{\bar{\eta}} = [e_{(0,1)}] - [e_{(0,0)}]$, $\hat{\bar{a}}$ *is given by* $\hat{\bar{a}} : A(N \times \{0,1\}, K)/_{=a} \to Y; [e(n,u)] \mapsto a(u^{n+1}) - a(u^n)$.

2) $((\ll S_l^N(\chi(U)) \gg, S_l), \chi^0, \bar{\chi}, 0, a(1))$,
where $\ll S_l^N(\chi(U)) \gg$ *is the smallest linear space that contains* $S_l^N(\chi(U)) := \{S_l^i(\chi^0 + \bar{\chi}u); u \in K, i \in N, S_l^i(\chi^0 + \bar{\chi}u)(t) = (\chi(u)(i+t+1) = a(\omega|u) - a(\omega)$ *for* $\omega \in U^*, \ |\omega| = i + t\}$.

Theorem 5.4. *Realization Theorem of almost linear systems*

Existence: For any time-invariant, affine input response map $a \in F(U^*, Y)$, there exist at least two canonical almost linear systems that realize a.

Uniqueness: Let σ_1 and σ_2 be any two canonical almost linear systems which realize a time-invariant, affine input response map $a \in F(U^*, Y)$. Then there exists an isomorphism $T : \sigma_1 \to \sigma_2$.

For the isomorphism of almost linear systems, see Definition (4.7).

5.2 Finite Dimensional Almost Linear Systems

Based on the realization theory (5.4), we want to review the fundamental facts about almost linear systems in this section. The facts are as follows:

1) when an almost linear system is finite dimensional.
2) when a finite dimensional almost linear system is canonical.
3) how we find a standard almost linear system.
4) a criterion for an Input/output relation to be the behavior of finite dimensional almost linear systems.

5) a procedure to obtain the standard system which realizes a given input response map.
6) how to find a partial realization σ from a given partial input/output data.
7) how to find a partial realization σ from a given partial input/output data in real time.

In chapter 4, since the above facts were stated when discussing algebraically approximate and noisy realization problems in the sense of almost linear systems, they are omitted here.

There is a fact about finite dimensional linear spaces that an n-dimensional linear space over the field $\boldsymbol{R}$ is isomorphic to $\boldsymbol{R}^n$, and $L(\boldsymbol{R}^n, \boldsymbol{R}^m)$ is isomorphic to $\boldsymbol{R}^{m\times n}$ (See Halmos [1958]). Therefore, without loss of generality, we can consider a n-dimensional almost linear system as $\sigma = ((\boldsymbol{R}^n, F), g^0, g, h, h^0)$, where $F \in \boldsymbol{R}^{n\times n}$, $g,\ g^0 \in \boldsymbol{R}^n$ and $h \in \boldsymbol{R}^{p\times n}$.

Proposition 5.5. *An almost linear system $\sigma = ((\boldsymbol{R}^n, F), g^0, g, h, h^0)$ is intrinsically canonical if and only if the following two conditions hold.*
rank $[g, Fg, F^2g, \cdots, F^{n-1}g] = n$
rank $[h^T, (hF)^T, \cdots, (hF^{n-1})^T] = n$.

Definition 5.6. For any time-invariant, affine input response map $a \in F(U^*, Y)$, the corresponding linear input/output map $A : (A(N \times \{0,1\}, \boldsymbol{R}), S_r) \to (F(N, Y), S_l)$ satisfies $A(\mathbf{e}_{(\mathbf{s},\mathbf{u})})(t) = a(u^{s+t+1}) - a(u^{s+t})$ for any $u \in \{0,1\}$.

Therefore, the map A can be represented by the infinite matrix $(I/O)_a$. This $(I/O)_a$ is said to be an Input/output matrix of a. For the Input/output matrix $(I/O)_a$, see Definition (4.9).

For a partial time-invariant, affine input response map $\underline{a} \in F(U^*_{\underline{N}}, Y)$, the matrix $(I/O)_{\underline{a}\ (p, \underline{N}-p)}$ is said to be a finite-sized Input/output matrix of $\underline{a}$, where $0 \leq s \leq p$, $0 \leq t \leq \underline{N} - p$ and $u \in \{0,1\}$. For $(I/O)_{\underline{a}\ (p,\underline{N}-p)}$, see section (4.4).

Since $I_{\underline{a}}(u)(i+j) = \underline{a}(u^{i+j+1}) - \underline{a}(u^{i+j})$ holds for $u \in \{0,1\}$, column vectors of $(I/O)_{\underline{a}}$ are denoted by $S_l^i I_{\underline{a}}(u)$.

Let a matrix $(I/O)_{\underline{a}\ (p,\underline{N}-p)}(v,w)$ denote $(I/O)_{\underline{a}\ (p,\underline{N}-p)}(v,w) := [I_{\underline{a}}(0), S_l I_{\underline{a}}(0), \cdots, S_l^{v-1} I_{\underline{a}}(0), I_{\underline{a}}(1), S_l I_{\underline{a}}(1), \cdots, S_l^{w-1} I_{\underline{a}}(1)]$.

When we actually treat approximate and noisy realization problems, we will use a notation $H_{\underline{a}\ (n_1+n_2, \underline{N}-n_1-n_2)}(n_1, n_2)$ expressed as follows:
$H_{\underline{a}\ (n_1+n_2, \underline{N}-n_1-n_2)}(n_1, n_2) = [I_{\underline{a}}(0), \cdots, S_l^{n_1-1} I_{\underline{a}}(0), I_{\underline{a}}(1), \cdots, S_l^{n_2-1} I_{\underline{a}}(1)]$.

5.3 Algebraically Approximate Realization of Almost Linear Systems

In this section, we discuss algebraically approximate realization problems of almost linear systems.

Here, we will discuss the algebraically approximate realization problem of almost linear systems, which is stated as follows:

<For any given finite-length modified impulse response of an almost linear system, find, using only algebraic calculations, an almost linear system which approximates it.>

The algebraically approximate realization of an almost linear system is presented here for the first time.

In order to make our discussion simple, we assume that the set Y of output is the set $\boldsymbol{R}$ of real numbers, namely 1-output.

Theorem 5.7. *Algorithm for algebraically approximate realization*

Let a partial input response map $\underline{a}$ be a considered object which is an almost linear system. Then an approximate realization $\sigma_r = ((\boldsymbol{R}^n, F_r), g_r^0, g_r, h_r, h^0)$ of $\underline{a}$ is given by the following algorithm:

1) Based on the ratio of the square root of eigenvalues for a matrix $H_{\underline{a}\ (p,\bar{p})}(p,0)H_{\underline{a}\ (p,\bar{p})}(p,0)^T$, determine the value n_1 of rank for the matrix $H_{\underline{a}\ (p,\bar{p})}(p,0)$, where $n_1 \leq p$.
Namely, determine the value n_1 of rank for the matrix $H_{\underline{a}\ (p,\bar{p})}(p,0)$ such that the ratio of the square root of eigenvalues for the covariance matrix becomes very small. The small ratio means the nearness of approximation degree.

2) We use the algebraic CLS method as follows:
① Based on Proposition (2.14), determine coefficients $\{\alpha_{1i} : 1 \leq i \leq n_1\}$. The Q in Proposition (2.14) can be considered as the matrix composed from the eigenvectors of $H_{\underline{a}\ (n_1+1,L)}(n_1+1,0)H^T_{\underline{a}\ (n_1+1,L)}(n_1+1,0)$.
Let a matrix $A_1 \in \boldsymbol{R}^{1\times(n_1+1)}$ be $A_1 = [\alpha_{11}, \alpha_{12}, \cdots, \alpha_{1n_1}, -1]$.
② Determine the error vectors $\{\underline{S}_l^i \bar{I}_{\underline{a}} \in \boldsymbol{R}^{L\times 1} :\ 0 \leq i \leq n_1\}$ by using the equation $[\bar{I}_{\underline{a}}(0), \underline{S}_l\bar{I}_{\underline{a}}(0), \cdots, \underline{S}_l^{n_1}\bar{I}_{\underline{a}}(0)]^T :=$
$A_1^T[A_1A_1^T]^{-1}A_1H^T_{\underline{a}\ (n_1+1,L)}(n_1+1,0)$ and
$H^T_{\underline{a}\ (n_1,L)}(n_1,0) := [I_{\underline{a}}(0), \cdots, S_l^{n_1-1}I_{\underline{a}}(0), S_l^{n_1}I_{\underline{a}}(0)]$.
③ Let $h_{1r} \in \boldsymbol{R}^{1\times n_1}$ be h_{1r}
$= [(I_{\underline{a}}(0))(0) - (\bar{I}_{\underline{a}}(0))(0), (S_lI_{\underline{a}}(0))(0) - (S_l\bar{I}_{\underline{a}}(0))(0), \cdots,$
$(S_l^{n_1-1}I_{\underline{a}}(0))(0) - (S_l^{n_1-1}\bar{I}_{\underline{a}}(0))(0)]$.

3) Based on the ratio of the square root of eigenvalues for a matrix $H_{\underline{a}\ (n_1+p,\bar{p})}(n_1,p)H_{\underline{a}\ (n_1+p,\bar{p})}(n_1,p)^T$, determine the value n_2 of rank for the matrix $H_{\underline{a}\ (n_1+p,\bar{p})}(n_1,p)$, where $n_2 \leq p$.
Namely, determine the value n_2 of rank for the matrix $H_{\underline{a}\ (p,\bar{p})}(p,0)$ such that the ratio of the square root of eigenvalues for the covariance matrix becomes very small. The small ratio means the nearness of approximation degree.

4) The algebraic CLS method is used as follows:
① Based on Proposition (2.14), determine coefficients $\{\alpha_{2i} : 1 \leq i \leq n_1 + n_2\}$.

The Q in Proposition (2.14) can be considered as the matrix composed from the eigenvectors of $H_{\underline{a}\ (n_1+n_2,L)}(n_1,n_2+1)H^T_{\underline{a}\ (n_1+n_2,L)}(n_1,n_2+1)$.
Let a matrix $A_2 \in \boldsymbol{R}^{1\times(n_1+n_2+1)}$ *be* $A_2 = [\alpha_{21}, \alpha_{22}, \cdots, \alpha_{2n_1+n_2}, -1]$.
② *Determine the error vectors* $\{\underline{S}_l^i \bar{I}_{\underline{a}} \in \boldsymbol{R}^{L\times 1} : 0 \le i \le n_1+n_2\}$ *by using the equation*$[\bar{I}_{\underline{a}}, \underline{S}_l \bar{I}_{\underline{a}}, \cdots, \underline{S}_l^{n_1+n_2}\bar{I}_{\underline{a}}]^T :=$
$A_2^T[A_2A_2^T]^{-1}A_2H^T_{\underline{a}\ (n_1+n_2,L)}(n_1,n_2+1)$ *and* $H^T_{\underline{a}\ (n_1,L)}(n_1,n_2+1) :=$
$[I_{\underline{a}}(0), \cdots, S_l^{n_1-1}I_{\underline{a}}(0), I_{\underline{a}}(1), \cdots, S_l^{n_2-1}I_{\underline{a}}(1), S_l^{n_2}I_{\underline{a}}(1)]$.
③ *Let* $F_r \in \boldsymbol{R}^{(n_1+n_2)\times(n_1+n_2)}$ *be given as the same as in Definition (4.11). Let* g_r^0 *be* $g_r^0 = \mathbf{e}_1$ *and* g_r *be* $g_r = \mathbf{e}_{n_1+1} - \mathbf{e}_1$,
where $\mathbf{e}_i = [0, \cdots, 0, \overset{i}{1}, 0, \cdots, 0]^T \in \boldsymbol{R}^{n_1+n_2}$.
④ *Let* h_r *be* $h_r = [h_{1r}, (I_{\underline{a}}(1))(0) - (\bar{I}_{\underline{a}}(1))(0), (S_lI_{\underline{a}}(1))(0) - (S_l\bar{I}_{\underline{a}}(1))(0), \cdots, (S_l^{n_2-1}I_{\underline{a}}(1))(0) - (S_l^{n_2-1}\bar{I}_{\underline{a}}(1))(0)]$.

For the real time standard system $\sigma_r = ((\boldsymbol{R}^n, F_r), g_r^0, g_r, h_r, h^0)$, *its modified impulse responses* $I(0)(i) := h_rF_r^ig_r^0$ *and* $I(1)(i) := h_rF_r^i(g_r^0+g_r)$ *are written by* $I(0)\_(n_1,n_2)$ *and* $I(1)\_(n_1,n_2)$ *respectively, where* $n := n_1+n_2$.

[proof] This is the same as the Algorithm for algebraically approximate realization (4.19).

Example 5.8. Let the signals be the modified impulse responses of the following 4-dimensional almost linear system: $\sigma = ((\boldsymbol{R}^4, F), g, g^0, h, h^0)$, where $F = \begin{bmatrix} 0 & 0 & 0.7 & 0 \\ 1 & 0 & 0.4 & 0 \\ 0 & 1 & -0.2 & 0 \\ 0 & 0 & 0.1 & 0.1 \end{bmatrix}$, $g^0 = [1,0,0,0]^T, h = [16, 8, -5, -1], g = [0,0,0,1]^T, h^0 = 1$.

Then the algebraically approximate realization problem is solved as follows:

covariance matrix	eigenvalues				
	1	2	3	4	5
$H^T_{\underline{a}\ (5,50)}(5,0)H_{\underline{a}\ (5,50)}(5,0)$	2164	741	599	0.003	0
covariance matrix	square root of eigenvalues				
$H^T_{\underline{a}\ (5,50)}(5,0)H_{\underline{a}\ (5,50)}(5,0)$	46.5	27.2	24.5	0.05	0
covariance matrix	eigenvalues				
	1	2	3	4	5
$H^T_{\underline{a}\ (5,50)}(3,1)H_{\underline{a}\ (5,50)}(3,1)$	2270	635	485	0.3	
$H^T_{\underline{a}\ (5,50)}(3,2)H_{\underline{a}\ (5,50)}(3,2)$	2492	1156	486	0.3	0
covariance matrix	square root of eigenvalues				
$H^T_{\underline{a}\ (5,50)}(3,2)H_{\underline{a}\ (5,50)}(3,2)$	50	34	22	0.5	0

1) Since the ratio $\frac{0.05}{46.5} = 0.001$ obtained by the square root of $H^T_{\underline{a}\ (5,50)}(5,0)H_{\underline{a}\ (5,50)}(5,0)$ and the ratio $\frac{0.5}{50} = 0.01$ obtained by the square root of $H^T_{\underline{a}\ (5,50)}(3,2)H_{\underline{a}\ (5,50)}(3,2)$ are small, the approximate almost linear system obtained by the algebraic CLS method may be good.

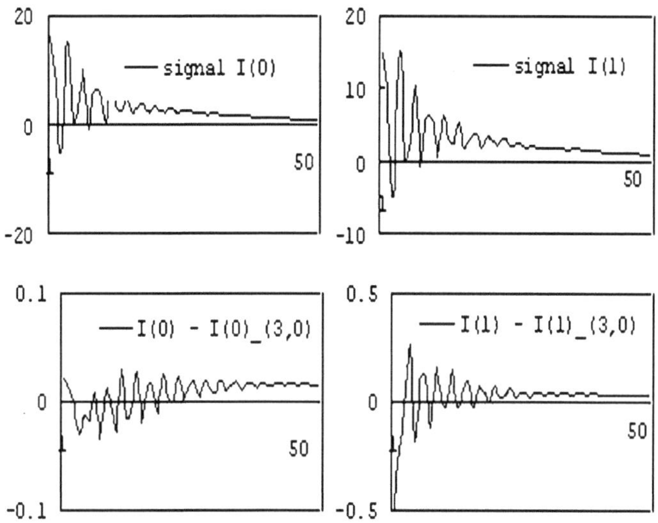

Fig. 5.1 The left are the original modified impulse response $I(0)$ of a 4-dimensional almost linear system and the difference between the original signal and the obtained one by a 4-dimensional almost linear system. The right are the original modified impulse response $I(1)$ of a 4-dimensional almost linear system and the difference between the original signal and the obtained one by a 3-dimensional almost linear system in Example (5.8).

2) After determining the numbers n_1 and n_2 of dimensions which are 3 and 0, we execute the algebraically approximate realization algorithm.

The almost linear system $\sigma_1 = ((\boldsymbol{R}^3,\ F_1),\ g_1^0,\ g_1,\ h_1, h^0)$ obtained by the algebraic CLS method is expressed as follows:

$$F_1 = \begin{bmatrix} 0 & 0 & 0.7 \\ 1 & 0 & 0.4 \\ 0 & 1 & -0.2 \end{bmatrix},\ h_1 = [16,\ 8,\ -5],\ g_1 = [-0.02,\ -0.006,\ 0.02]^T, g_1^0 = [1,\ 0,\ 0]^T,\ h^0 = 1.$$

For reference, a 4-dimensional almost linear system $\sigma_2 = ((\boldsymbol{R}^4,\ F_2),\ g_2^0,\ g_2,\ h_2, h^0)$ obtained by the algebraic CLS method is expressed as follows:

$$F_2 = \begin{bmatrix} 0 & 0 & 0 & -0.07 \\ 1 & 0 & 0 & 0.66 \\ 0 & 1 & 0 & 0.42 \\ 0 & 0 & 1 & -0.1 \end{bmatrix},\ h_2 = [16,\ 8,\ -5,\ 15.3],\ g_2 = [-7,\ -4,\ 2,\ 10]^T,$$
$$g_2^0 = [1,\ 0,\ 0,\ 0]^T, h^0 = 1.$$

We can show that the algorithm for approximate realization given by the analytic CLS method in the reference [Hasegawa, 2008] produces the same systems as the above ones in the sense of the numerical calculation.

In this example, the original signals are considered as the modified impulse responses of a 4-dimensional almost linear system and the desirable modified impulse responses are obtained by the algebraic CLS method.

The following table indicates that the 3-dimensional almost linear system reconstructs the original signal with 0.3 and 2 % error to signal ratio and with 0.001 and 0.01 ratio of matrices, and the 4-dimensional almost linear system completely reconstructs the original system.

Therefore, an approximate realization could be obtained.

Just as we thought, the following table and Fig. 5.1 truly indicate that the 3-dimensional almost linear system given by the algebraic CLS method is a good approximation for the original 4-dimensional system.

dimen-sion	ratio of matrices	mean values of square root for sum of signal	signal by CLS	error	cosine ① and ②	error ratio
		①	②	③	$\cos\theta$	③/①
I(0)_(3,0)	0.001	0.632	0.631	0.002	0.9999	0.003
I(1)_(3,0)	0.01	0.6217	0.621	0.014	0.999	0.02
I(0)_(4,0)	0	0.632	0.632	0	1	0
I(1)_(4,0)	0	0.6217	0.6217	0	1	0

For the notations $I(0)\_(n_1, n_2)$ and $I(1)\_(n_1, n_2)$, see Algorithm (5.7).

Example 5.9. Let the signals be the modified impulse responses of the following 5-dimensional almost linear system: $\sigma = ((\boldsymbol{R}^5, F), g, g^0, h, h^0)$, where

$$F = \begin{bmatrix} 0 & 0 & -0.2 & 0 & 0 \\ 1 & 0 & -0.3 & 0 & 0 \\ 0 & 1 & 0 & 0 & 0 \\ 0 & 0 & 1 & 0 & 0.5 \\ 0 & 0 & 0 & 0.5 & -0.4 \end{bmatrix}, \; g^0 = [1,\ 0,\ 0,\ 0,\ 0]^T,\; h = [10,\ 2,\ -5,\ -1,\ 3],$$

$g = [0,\ 0,\ 0,\ 1,\ 0]^T$, $h^0 = 1$.

Then the algebraically approximate realization problem is solved by the following algorithm:

covariance matrix	eigenvalues					
	1	2	3	4	5	6
$H^T_{\underline{a}\ (5,50)}(5,0) H_{\underline{a}\ (5,50)}(5,0)$	212	64	10.5	2.3	0.05	
$H^T_{\underline{a}\ (6,50)}(6,0) H_{\underline{a}\ (6,50)}(6,0)$	215	64	11	2.3	0.05	0
covariance matrix	square root of eigenvalues					
$H^T_{\underline{a}\ (5,50)}(5,0) H_{\underline{a}\ (5,50)}(5,0)$	14.6	8	3.2	1.5	0.2	
$H^T_{\underline{a}\ (6,50)}(6,0) H_{\underline{a}\ (6,50)}(6,0)$	14.7	8	3.3	1.5	0.2	0

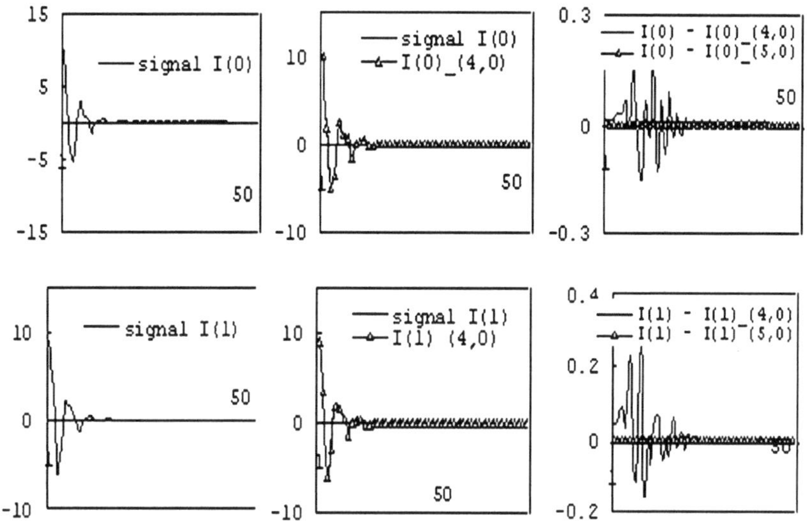

Fig. 5.2 The left are the original modified impulse response $I(0)$ and $I(1)$ of an original 5-dimensional almost linear system. The middle are the original modified impulse responses and the obtained ones by a 4-dimensional almost linear system. The right are the difference between the original signals and the obtained ones by a 4 or 5-dimensional almost linear system in Example (5.9).

covariance matrix	eigenvalues					
	1	2	3	4	5	6
$H^T_{\underline{a}\ (5,50)}(4,1)H_{\underline{a}\ (5,50)}(4,1)$	348	62.7	10	2	0	
$H^T_{\underline{a}\ (6,50)}(4,2)H_{\underline{a}\ (6,50)}(4,2)$	358	117	10	2.3	0.05	0
covariance matrix	square root of eigenvalues					
$H^T_{\underline{a}\ (5,50)}(4,1)H_{\underline{a}\ (5,50)}(4,1)$	18.7	7.9	3.2	1.4	0	
$H^T_{\underline{a}\ (6,50)}(4,2)H_{\underline{a}\ (6,50)}(4,2)$	18.9	10.8	3.2	1.5	0.2	0

1) Since the ratio $\frac{0.2}{14.6} = 0.01$ obtained by the square root of $H^T_{\underline{a}\ (5,50)}(5,0)H_{\underline{a}\ (5,50)}(5,0)$ is small and the ratio $\frac{0.2}{18.9} = 0.01$ obtained by the square root of $H^T_{\underline{a}\ (6,50)}(4,2)H_{\underline{a}\ (6,50)}(4,2)$ is small, the approximate almost linear system obtained by the algebraic CLS method may be good.
2) After determining the number n_1 and n_2 of dimensions which are 4 and 0, we execute the algebraically approximate realization algorithm.

The almost linear system $\sigma_1 = ((\boldsymbol{R}^4,\ F_1),\ g_1^0,\ g_1,\ h_1)$ obtained by the algebraic CLS method is expressed as follows:

$$F_1 = \begin{bmatrix} 0 & 0 & 0 & -0.27 \\ 1 & 0 & 0 & -0.5 \\ 0 & 1 & 0 & -0.51 \\ 0 & 0 & 1 & -1 \end{bmatrix},\ h_1 = [10,\ 2,\ -5,\ -3.6],\ g_1 = [0.2,\ 0.3,\ 0,\ 1]^T,$$

$g_1^0 = [1,\ 0,\ 0,\ 0]^T,\ h^0 = 1.$

For reference, a 5-dimensional almost linear system $\sigma_2 = ((\boldsymbol{R}^5,\ F_2),\ g_2^0,\ g_2,\ h_2, h^0)$ obtained by the algebraic CLS method is expressed as follows:

$$F_2 = \begin{bmatrix} 0\,0\,0\,0 & 0.05 \\ 1\,0\,0\,0 & -0.005 \\ 0\,1\,0\,0 & -0.32 \\ 0\,0\,1\,0 & -0.05 \\ 0\,0\,0\,1 & -0.4 \end{bmatrix},\ h_2 = [10,\ 2,\ -5,\ -3.6,\ 2.6],\ g_2 = [0.2,\ 0.3,\ 0,\ 1,\ 0]^T,$$

$g_2^0 = [1,\ 0,\ 0,\ 0,\ 0]^T,\ h^0 = 1.$

We can show that the algorithm for approximate realization given by the analytic CLS method in the reference [Hasegawa, 2008] produces the same systems as the above ones in the sense of the numerical calculation.

In this example, the original signals are considered as the modified impulse responses of a 5-dimensional almost linear system and the desirable modified impulse responses are obtained by the algebraic CLS method.

The following table indicates that the 4-dimensional almost linear system reconstructs the original signal with 3 and 4 % error to signal ratio and with 0.01 and 0.01 ratio of matrices, and the 5-dimensional almost linear system obtained by the algebraic CLS method completely reconstructs the original system.

Therefore, a somewhat good approximate realization could be obtained.

The following table and Fig. 5.2 truly indicate that the 4-dimensional almost linear system obtained by the algebraic CLS method is a somewhat good approximation within our expectations. Hence, there exists a somewhat good approximation for the given system except for the peak values in the modified impulse response $I(1)$.

dimen-sion	ratio of matrices	mean values of square root for sum of signal ①	signal by CLS ②	error ③	cosine ① and ② $\cos\theta$	error ratio ③/①
I(0)-(4,0)	0.01	0.2473	0.2476	0.007	0.999	0.03
I(1)-(4,0)	0.01	0.2405	0.2405	0.009	0.999	0.042
I(0)-(5,0)	0	0.2473	0.2473	0	1	0
I(1)-(5,0)	0	0.2405	0.2405	0	1	0

For the notations $I(0)\text{-}(n_1, n_2)$ and $I(1)\text{-}(n_1, n_2)$, see Algorithm (5.7).

Example 5.10. Let the signals be the modified impulse responses of the following 6-dimensional almost linear system: $\sigma = ((\boldsymbol{R}^6, F), g, g^0, h, h^0)$, where

$$F = \begin{bmatrix} 0 & 0 & -0.1 & 0 & 0 & 0.1 \\ 1 & 0 & -0.4 & 0 & 0 & -0.3 \\ 0 & 1 & 0 & 0 & 0 & 0.2 \\ 0 & 0 & 1 & 0 & 0 & 0.5 \\ 0 & 0 & 0 & 1 & 0.5 & -0.4 \\ 0 & 0 & 0 & 0 & 1 & 0.5 \end{bmatrix},\ g^0 = [1,0,0,0,0,0]^T, h = [10,2,-5,-1,3,2],$$

$g = [0,0,0,1,0,0]^T,\ h^0 = 1.$

Then the algebraically approximate realization problem is solved as follows:

covariance matrix	eigenvalues						
	1	2	3	4	5	6	7
$H^T_{\underline{a}\ (6,50)}(6,0)H_{\underline{a}\ (6,50)}(6,0)$	685	514	61	21	13	0.005	
$H^T_{\underline{a}\ (7,50)}(76,0)H_{\underline{a}\ (7,50)}(7,0)$	796	522	74	26	17	0.007	0
covariance matrix	square root of eigenvalues						
$H^T_{\underline{a}\ (6,50)}(6,0)H_{\underline{a}\ (6,50)}(6,0)$	26.2	22.7	7.8	4.6	1	0.07	
$H^T_{\underline{a}\ (7,50)}(7,0)H_{\underline{a}\ (7,50)}(7,0)$	28.2	22.8	8.6	5.1	4.1	0.08	0

covariance matrix	eigenvalues						
	1	2	3	4	5	6	7
$H^T_{\underline{a}\ (6,50)}(5,1)H_{\underline{a}\ (6,50)}(5,1)$	1140	461	92	22	2.1	0	
$H^T_{\underline{a}\ (7,50)}(6,1)H_{\underline{a}\ (7,50)}(6,1)$	1140	626	99	24	13	0.006	0
covariance matrix	square root of eigenvalues						
$H^T_{\underline{a}\ (6,50)}(5,1)H_{\underline{a}\ (6,50)}(5,1)$	33.7	21.5	9.6	4.7	1.4	0	
$H^T_{\underline{a}\ (7,50)}(6,1)H_{\underline{a}\ (7,50)}(6,1)$	33.7	25	9.9	4.9	1	0.08	0

1) Since the ratio $\frac{0.07}{26.2} = 0.003$ obtained by the square root of $H^T_{\underline{a}\ (6,50)}(6,0)H_{\underline{a}\ (6,50)}(6,0)$ is small and the ratio $\frac{0.08}{33.7} = 0.002$ obtained by the square root of $H^T_{\underline{a}\ (7,50)}(65,1)H_{\underline{a}\ (7,50)}(6,1)$ is also small, the approximate almost linear system obtained by the algebraic CLS method may be good.
2) After determining the number n_1 and n_2 of dimensions which are 5 and 0, we execute the algebraically approximate realization algorithm.

The almost linear system $\sigma_1 = ((\boldsymbol{R}^5,\ F_1),\ g_1^0,\ g_1,\ h_1)$ obtained by the algebraic CLS method is expressed as follows:

$$F_1 = \begin{bmatrix} 0\,0\,0\,0 & 0.33 \\ 1\,0\,0\,0 & -0.92 \\ 0\,1\,0\,0 & 1.81 \\ 0\,0\,1\,0 & -1.9 \\ 0\,0\,0\,1 & 1.53 \end{bmatrix},\ h_1 = [10,\ 2,\ -5,\ -2.8,\ 4.8],\ g_1 = [0.1,\ 0.4,\ 0,\ 1,\ 0]^T,$$

$g_1^0 = [1,\ 0,\ 0,\ 0,\ 0]^T,\ h^0 = 1.$

For reference, a 6-dimensional almost linear system $\sigma_2 = ((\boldsymbol{R}^6, F_2), g_2^0, g_2, h_2, h^0)$ obtained by the algebraic CLS method is expressed as follows:

$$F_2 = \begin{bmatrix} 0\,0\,0\,0\,0 & 0.15 \\ 1\,0\,0\,0\,0 & -0.17 \\ 0\,1\,0\,0\,0 & 0.04 \\ 0\,0\,1\,0\,0 & 0.8 \\ 0\,0\,0\,1\,0 & -1.05 \\ 0\,0\,0\,0\,1 & 1 \end{bmatrix},\ h_2 = [10,\ 2,\ -5,\ -2.8,\ 4.8,\ 5.12],$$

$g_2 = [0.1,\ 0.4,\ 0,\ 1,\ 0,\ 0]^T, g_2^0 = [1,\ 0,\ 0,\ 0,\ 0,\ 0]^T,\ h^0 = 1.$

We can show that the algorithm for approximate realization given by the analytic CLS method in the reference [Hasegawa, 2008] produces the same systems as the above ones in the sense of the numerical calculation.

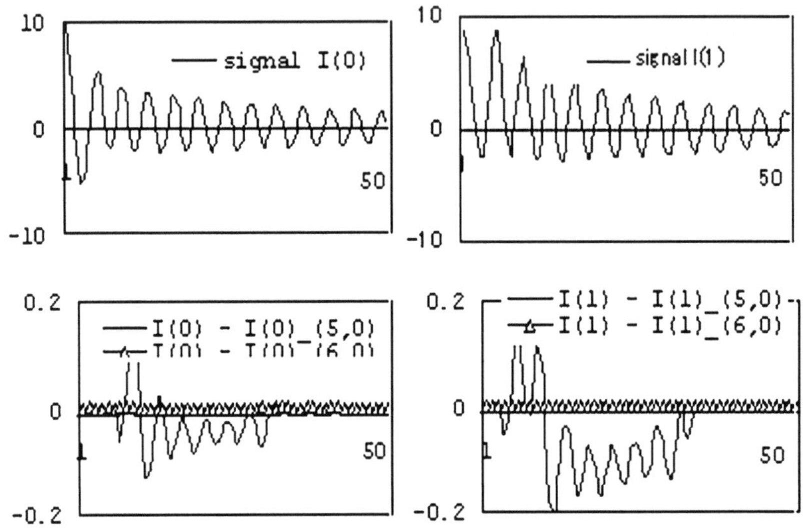

Fig. 5.3 The left are the original modified impulse response $I(0)$ of a 6-dimensional almost linear system and the difference between the original signal and the obtained one by a 5 or 6-dimensional almost linear system. The right are the original modified impulse response $I(1)$ of a 6-dimensional almost linear system and the difference between the original signal and the obtained one by a 5 or 6-dimensional almost linear system in Example (5.10).

In this example, the original signals are considered as the modified impulse responses of a 6-dimensional almost linear system and the desirable modified impulse responses are obtained by the algebraic CLS method.

The following table indicates that the 5-dimensional almost linear system reconstructs the original signal with 2 and 3 % error to signal ratio and with 0.003 and 0.002 ratio of matrices, and the 6-dimensional almost linear system completely reconstructs the original system.

The following table and Fig. 5.3 truly indicate that the 5-dimensional almost linear system obtained by the algebraic CLS method is a good approximation within our expectations. For reference, the modified impulse responses of the same dimensional almost linear system as the original system are shown. Hence, there exists a good approximation for the given system except for the peak values in the modified impulse response $I(1)$.

dimen- sion	ratio of matrices	mean values of square root for sum of			cosine	error
		signal	signal by CLS	error	① and ②	ratio
		①	②	③	$\cos\theta$	③/①
I(0)_(5,0)	0.003	0.3685	0.3682	0.008	0.999	0.02
I(1)_(5,0)	0.002	0.4370	0.43703	0.01	0.9998	0.03
I(0)_(6,0)	0	0.3684	0.3684	0	1	0
I(1)_(6,0)	0	0.4370	0.4370	0	1	0

For the notations $I(0)\_(n_1, n_2)$ and $I(1)\_(n_1, n_2)$, see Algorithm (5.7).

5.4 Algebraically Noisy Realization of Almost Linear Systems

In this section, we discuss an algebraically noisy realization of almost linear systems.

We will obtain the observed data $\{\hat{\gamma}(|\omega|)+\bar{\gamma}(|\omega|) : \omega \in U^*\}$ for noise $\{\bar{\gamma}(t) : t \in N\}$ added to the unknown almost linear system σ with the behavior a.

For any given $\{\hat{\gamma}(|\omega|) + \bar{\gamma}(|\omega|) : \omega \in U^*\}$, a system σ which satisfies $a_\sigma(\omega) \approx \hat{\gamma}(|\omega|)$ for any $\omega \in U^*$ is called a noisy realization of a.

In order to make our discussion simple, we assume that the set Y of output is the set $\boldsymbol{R}$ of real numbers, namely 1-output.

We can propose the following algebraically noisy realization problem:

For any given $\{\hat{\gamma}(|\omega|) + \bar{\gamma}(|\omega|) : \omega \in U^*\}$, find, using only algebraic calculations, an almost linear system σ which satisfies $a_\sigma(\omega) = \hat{\gamma}(|\omega|)$ for any $\omega \in U^*$.

A situation for the algebraically noisy realization problem 5.11
Let the observed object be an almost linear system and noise be added to its output. Then we will obtain the data $\{\gamma(t) = \hat{\gamma}(t) + \bar{\gamma}(t) : 0 \le t \le \underline{N}\}$ for some integer $\underline{N} \in N$, where $\hat{\gamma}(t)$ is the exact signal which comes from the observed almost linear system and $\bar{\gamma}(t)$ is the noise added at observation.

Problem 5.12. Problem statement of the algebraically noisy realization for almost linear systems.

Let $H_{\underline{a}\ (p,\bar{p})}$ be the measured finite-sized Input/output matrix. Then find the cleaned-up Input/output matrix $\hat{H}_{\underline{a}\ (p,\bar{p})}$ such that $H_{\underline{a}\ (p,\bar{p})} = \hat{H}_{\underline{a}\ (p,\bar{p})} + \bar{H}_{\underline{a}\ (p,\bar{p})}$ holds.

Namely, find, using only algebraic calculations, a minimal dimensional almost linear system $\sigma = ((\boldsymbol{R}^n, F), g^0, g, h, h^0))$ which realizes $\hat{H}_{\underline{a}\ (p,\bar{p})}$.

Definition 5.13. A canonical almost linear system $\sigma_r = ((\boldsymbol{R}^n, F_r), g_r^0, g_r, h_r, h^0)$ is said to be a real time standard system if $g_r^0 = \mathbf{e}_1$, $\mathbf{e}_i = F_r^{i-1}\mathbf{e}_1$ for $i \le n_1$ and $F_r^{n_1}\mathbf{e}_1 = \sum_{i=1}^{n_1} \alpha_{0i} F_r^{i-1}\mathbf{e}_1$ hold. $g_r = \mathbf{e}_{n_1+1} - \mathbf{e}_1$, $\mathbf{e}_{n_1+i} = F_r^{i-1}\mathbf{e}_{n_1+1}$ for $i \le n_2$ and $F_r^{n_2}\mathbf{e}_{n_1+1} = \sum_{i=1}^{n_2} \alpha_{1i} F_r^{i-1}\mathbf{e}_{n_1+1}$ hold. F_r is given by the following.

$$F_r = \begin{bmatrix} 0 \cdots & 0 & \alpha_{11} & 0 \cdots & \cdots 0 & \alpha_{21} \\ 1 \ddots & & \alpha_{12} & 0 \cdots & 0 & \alpha_{22} \\ \vdots \ddots & 0 & \vdots & \vdots & \vdots & \vdots \\ 0 & 1 & \alpha_{1n_1} & \vdots & \vdots & \alpha_{2n_1} \\ 0\ 0 & \cdots & 0 & 0 \cdots & \cdots 0 & \alpha_{2n_1+1} \\ 0\ 0 & \cdots & \vdots & 1 \ddots & \vdots & \alpha_{2n_1+2} \\ 0\ 0 & \cdots & \vdots & 0 \ddots & \ddots \vdots & \vdots \\ 0 \ddots & \cdots & \vdots & \vdots \ddots & 1\ 0 & \alpha_{2n-1} \\ 0\ 0 & \cdots & 0 & 0 \cdots & 0\ 1 & \alpha_{2n} \end{bmatrix}.$$

Theorem 5.14. *Algebraic algorithm for noisy realization*

Let $\underline{a}$ be a considered object which is an almost linear system. Then a noisy realization $\sigma_r = ((\boldsymbol{R}^n, F_r), g_r^0, g_r, h_r, h^0)$ of $\underline{a}$ is given by the following algorithm:

1) *Based on the square root of eigenvalues for a matrix $H_{\underline{a}(p,\bar{p})}(p,0)H_{\underline{a}(p,\bar{p})}(p,0)^T$, determine the value n_1 of rank for the matrix $H_{\underline{a}\ (p,\bar{p})}(p,0)$, where $n_1 \le p$. Namely, determine the value n_1 of rank for the matrix $H_{\underline{a}\ (p,\bar{p})}(p,0)$ such that a set of the square root of eigenvalues for the covariance matrix composed of relatively small and equally-sized numbers is excluded, where the signal part effected by the set may be a noisy part.*
2) *The algebraic CLS method is used as follows:*
   ① *Based on Proposition (2.14), determine coefficients $\{\alpha_{1i} : 1 \le i \le n_1\}$. The Q in Proposition (2.14) can be considered as the matrix composed from the eigenvectors of $H_{\underline{a}\ (n_1+1,L)}(n_1+1,0)H^T_{\underline{a}\ (n_1+1,L)}(n_1+1,0)$.*
   Let a matrix $A_1 \in \boldsymbol{R}^{1\times(n_1+1)}$ be $A_1 = [\alpha_{11}, \alpha_{12}, \cdots, \alpha_{1n_1}, -1]$.
   ② *Determine the error vectors $\{\underline{S_l^i}\bar{I}_{\underline{a}} \in \boldsymbol{R}^{L\times 1} :\ 0 \le i \le n_1\}$ by using the equation $[\bar{I}_{\underline{a}}(0), \underline{S_l}\bar{I}_{\underline{a}}(0), \cdots, \underline{S_l^{n_1}}\bar{I}_{\underline{a}}(0)]^T :=$*
   $A_1^T[A_1A_1^T]^{-1}A_1H^T_{\underline{a}\ (n_1+1,L)}(n_1+1,0)$ and
   $H^T_{\underline{a}\ (n_1,L)}(n_1,0) := [I_{\underline{a}}(0), \cdots, S_l^{n_1-1}I_{\underline{a}}(0), S_l^{n_1}I_{\underline{a}}(0)]$.
   ③ *Let $h_{1r} \in \boldsymbol{R}^{1\times n_1}$ be $h_{1r} = [I_{\underline{a}}(1) - \bar{I}_{\underline{a}}(1), I_{\underline{a}}(2) - \bar{I}_{\underline{a}}(2), \cdots, I_{\underline{a}}(n_1) - \bar{I}_{\underline{a}}(n_1)]$.*
3) *Based on the square root of eigenvalues for a matrix $H_{\underline{a}\ (n_1+p,\bar{p})}(n_1,p)H_{\underline{a}\ (n_1+p,\bar{p})}(n_1,p)^T$, determine the value n_2 of rank for the matrix $H_{\underline{a}\ (n_1+p,\bar{p})}(n_1,p)$, where $n_2 \le p$.*
   Namely, determine the value n_2 of rank for the matrix $H_{\underline{a}\ (n_1+p,\bar{p})}(n_1,p)$ such that a set of the square root of eigenvalues for the covariance matrix composed of relatively small and equally-sized numbers is excluded, where the signal part effected by the set may be a noisy part.

4) The algebraic CLS method is used as follows:
① *Based on Proposition (2.14), determine coefficients* $\{\alpha_{2i} : 1 \leq i \leq n_1 + n_2\}$.
The Q in Proposition (2.14) can be considered as the matrix composed from the eigenvectors of $H_{\underline{a}\ (n_1+n_2,L)}(n_1, n_2+1) H^T_{\underline{a}\ (n_1+n_2,L)}(n_1, n_2+1)$.
Let a matrix $A_2 \in \boldsymbol{R}^{1\times(n_1+n_2+1)}$ *be* $A_2 = [\alpha_{21}, \alpha_{22}, \cdots, \alpha_{2n_1+n_2}, -1]$.
② *Determine the error vectors* $\{\underline{S_l^i} \bar{I}_{\underline{a}} \in \boldsymbol{R}^{L\times 1} : 0 \leq i \leq n_1+n_2\}$ *by using the equation* $[\bar{I}_{\underline{a}}, \underline{S_l}\bar{I}_{\underline{a}}, \cdots, \underline{S_l^{n_1+n_2}}\bar{I}_{\underline{a}}]^T :=$
$A_2^T[A_2A_2^T]^{-1}A_2H^T_{\underline{a}\ (n_1+n_2,L)}(n_1, n_2+1)$ *and* $H^T_{\underline{a}\ (n_1,L)}(n_1, n_2+1) :=$
$[I_{\underline{a}}(0), \cdots, S_l^{n_1-1}I_{\underline{a}}(0), I_{\underline{a}}(1), \cdots, S_l^{n_2-1}I_{\underline{a}}(1), S_l^{n_2}I_{\underline{a}}(1)]$.
③ *Let* $F_r \in \boldsymbol{R}^{(n_1+n_2)\times(n_1+n_2)}$ *be given as the same as in Definition (4.11). Let* g_r^0 *be* $g_r^0 = \mathbf{e}_1$ *and* g_r *be* $g_r = \mathbf{e}_{n_1+1} - \mathbf{e}_1$, *where*
$\mathbf{e}_i = [0, \cdots, 0, \overset{i}{1}, 0, \cdots, 0]^T \in \boldsymbol{R}^{n_1+n_2}$.
④ *Let* h_r *be*
$h_r = [h_{1r}, I_{\underline{a}}(1) - \bar{I}_{\underline{a}}(1), I_{\underline{a}}(1^2) - \bar{I}_{\underline{a}}(1^2), \cdots, I_{\underline{a}}(1^{n_2-1}) - \bar{I}_{\underline{a}}(1^{n_2-1})]$.

For the real time standard system $\sigma_r = ((\boldsymbol{R}^n, F_r), g_r^0, g_r, h_r, h_r^0)$, *its modified impulse responses* $I(0)(i) := h_rF_r^ig_r^0$ *and* $I(1)(i) := h_rF_r^i(g_r^0 + g_r)$ *may be written by* $I(0)\_(n_1, n_2)$ *and* $I(1)\_(n_1, n_2)$ *repectively.*

[proof] This algorithm is the same as the algorithm for algebraically noisy realization (4.26) except 1).

Remark 1: A determination method of the degree n in the almost linear system $\sigma = ((\boldsymbol{R}^n, F_s), g, h_s)$ can be found in the Principal Component Method. This method is popular.
Remark 2: Let S and N be the norm of a signal and a noise. Then the selected ratio of matrices in the algorithm may be considered as $\frac{N}{S+N}$.
Remark 3: This algebraically noisy realization method is very new.
Remark 4: For a noisy case, the AIC is famous for determining only linear systems including dimensions of the state space.

Example 5.15. Let signals be the modified impulse responses of the following 4-dimensional almost linear system $\sigma = ((R^4, F), g^0, g, h, h^0)$,

$$\text{where } F = \begin{bmatrix} 0 & 0 & -0.5 & 0 \\ 1 & 0 & 0.2 & 0 \\ 0 & 1 & 0.5 & 0 \\ 0 & 0 & 0 & -0.9 \end{bmatrix}, \ h = [15,\ -2,\ 0,\ -10],\ g^0 = [1,\ 0,\ 0,\ 0]^T,$$

$g = [-1,\ 0,\ 0,\ 1]^T,\ h^0 = 1.$

Let added noises be given in Fig. 5.4.

Then the algebraically noisy realization problem is solved as follows:

covariance matrix	eigenvalues					
	1	2	3	4	5	6
$H_{\underline{a}\ (4,50)}(4,0)H^T_{\underline{a}\ (4,50)}(4,0)$	418	167	74.1	5.9		
$H_{\underline{a}\ (5,50)}(5,0)H^T_{\underline{a}\ (5,50)}(5,0)$	432	167	94.7	7.2	3.4	
$H_{\underline{a}\ (6,50)}(6,0)H^T_{\underline{a}\ (6,50)}(6,0)$	447	177	95	8.8	4.7	1.8
covariance matrix	square root of eigenvalues					
$H_{\underline{a}\ (6,50)}(6,0)H^T_{\underline{a}\ (6,50)}(6,0)$	21.1	13.3	9.7	3	2.2	1.3

covariance matrix	eigenvalues						
	1	2	3	4	5	6	7
$H_{\underline{a}\ (5,50)}(3,2)H^T_{\underline{a}\ (5,50)}(3,2)$	1114	222	130	30.2	4.8		
$H_{\underline{a}\ (6,50)}(3,3)H^T_{\underline{a}\ (6,50)}(3,3)$	1450	226	131	31	5	3.8	
$H_{\underline{a}\ (7,50)}(3,4)H^T_{\underline{a}\ (7,50)}(3,4)$	1723	229	130	31	5.4	4	3.4
covariance matrix	square root of eigenvalues						
$H_{\underline{a}\ (7,50)}(3,4)H^T_{\underline{a}\ (7,50)}(3,4)$	41.5	15.1	11.4	5.6	2.3	2	1.8

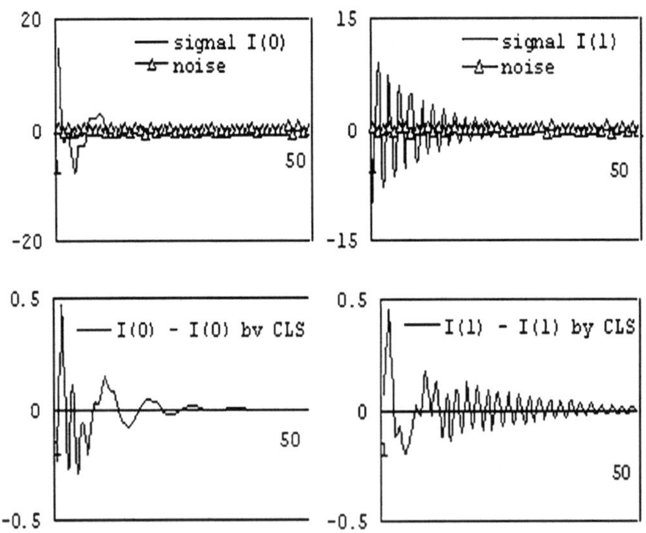

Fig. 5.4 The left are the exact modified impulse response $I(0)$ of a 4-dimensional almost linear system with noise and the difference between the original signal and the obtained one $I(0)_{(2,2)}$ by the algebraic CLS method. The right are the original modified impulse response $I(1)$ of a 4-dimensional almost linear system with noise and the difference between the original signal and the obtained one $I(1)_{(2,2)}$ by the algebraic CLS method in Example (5.15).

1) Since a set $\{3,\ 2.2,\ 1.3\}$ is composed of relatively small and equally-sized numbers in the square root of $H_{\underline{a}\ (6,50)}(6,0)H^T_{\underline{a}\ (6,50)}(6,0)$, the almost linear system obtained by the algebraic CLS method may be realized for a 3-dimensional space.
2) After determining the number n_1 of dimensions which is 3, we will continue the algebraically noisy realization algorithm.

Therefore, the modified impulse response $I(0)$ of an almost linear system obtained by the algebraic CLS method is characterized by a 3-dimensional almost linear system.
3) Since a set $\{2.3,\ 2,\ 1.8\}$ is composed of relatively small and equally-sized numbers in the square root of $H_{\underline{a}\ (7,50)}(3,4)H^T_{\underline{a}\ (7,50)}(3,4)$, the almost linear system obtained by the algebraic CLS method may be realized by adding another one-dimensional space.
4) After determining the number n_2 of dimensions which is 1, we will continue the algebraically noisy realization algorithm.

Therefore, the modified impulse response $I(1)$ of an almost linear system obtained by the algebraic CLS method is realized by adding another 1-dimensional space.

Therefore, the modified impulse responses $I(0)$ and $I(1)$ of an almost linear system obtained by the algebraic CLS method can be realized by a (3,1)-dimensional almost linear system.

The system is given as follows: $\sigma_o = ((\boldsymbol{R}^4,\ F_o),\ g_o^0, g_o, h_o,\ h^0)$,

where $F_o = \begin{bmatrix} 0 & 0 & -0.5 & -0.04 \\ 1 & 0 & 0.2 & -0.03 \\ 0 & 1 & 0.5 & 0.04 \\ 0 & 0 & 0 & -0.9 \end{bmatrix}$, $g_o^0 = \mathbf{e}_1$, $g_o = [-1,\ 0,\ 0,\ 1]^T$,

$h_o = [15.2,\ -2.5,\ 0.26,\ -10.1]$ and $h^0 = 1$.

The obtained modified impulse responses $I(0)$ and $I(1)$ are illustrated in Fig. 5.4.

We can show that the algorithm for noisy realization given by the analytic CLS method in the reference [Hasegawa, 2008] produces the same system as the above one in the sense of the numerical calculation.

In this example, the original signal $I(0)$ is considered as the modified impulse response of a 3-dimensional linear space and the original signal $I(1)$ is considered as the modified impulse response of an added 1-dimensional linear space. The desirable modified impulse responses have been attempted to be obtained by the algebraic CLS method. The model obtained by the algebraic CLS method is a (3,1)-dimensional almost linear system which has the same number of dimensions as the number of the original system.

Just as we expected, the following table and Fig. 5.4 indicate that the model obtained by the algebraic CLS method is a good (3,1)-dimensional almost linear system for the original (3,1)-dimensional system.

dimen-sion	ratio of matrices	mean values of square root for sum of signal ①	signal by CLS ②	error ③	cosine ① and ② $\cos\theta$	error ratio ③/①
I(0)_(3,1)	0.14	0.3622	0.3674	0.01	0.999	0.03
I(1)_(3,1)	0.06	0.45883	0.4592	0.01	0.999	0.02

For the notations $I(0)\_(n_1, n_2)$ and $I(1)\_(n_1, n_2)$, see Algorithm (5.14).

Example 5.16. Let signals be the modified impulse responses of the following 5-dimensional almost linear system $\sigma = ((R^5, F), x^0, g, h)$, where

$$F = \begin{bmatrix} 0 & 0 & 0.4 & 0 & -0.5 \\ 1 & 0 & -0.3 & 0 & 0.6 \\ 0 & 1 & 0.4 & 0 & -0.9 \\ 0 & 0 & 0 & 0 & -0.6 \\ 0 & 0 & 0 & 1 & 0.8 \end{bmatrix}, \; h = [15, \; -4, \; 2, \; -3, \; 5], \; g^0 = [1, \; 0, \; 0, \; 0, \; 0]^T,$$

$$g = [-1, \; 0, \; 0, \; 1, \; 0]^T, \; h^0 = 1.$$

Let added noises be given in Fig. 5.5.

Then the algebraically noisy realization problem is solved as follows:

covariance matrix	eigenvalues						
	1	2	3	4	5	6	7
$H_{\underline{a}\,(6,50)}(6,0)H^T_{\underline{a}\,(6,50)}(6,0)$	405	108	88	3.5	1.2	0.4	
$H_{\underline{a}\,(7,50)}(7,0)H^T_{\underline{a}\,(7,50)}(7,0)$	408	110	93	3.7	1.8	0.6	0.3
covariance matrix	square root of eigenvalues						
$H_{\underline{a}\,(7,50)}(7,0)H^T_{\underline{a}\,(7,50)}(7,0)$	20.2	10.5	9.6	1.9	1.3	0.8	0.5

covariance matrix	eigenvalues						
	1	2	3	4	5	6	7
$H_{\underline{a}\,(5,50)}(3,2)H^T_{\underline{a}\,(5,50)}(3,2)$	465	335	104	30	9.8		
$H_{\underline{a}\,(6,50)}(3,3)H^T_{\underline{a}\,(6,50)}(3,3)$	469	363	259	31	11.1	2.6	
$H_{\underline{a}\,(7,50)}(3,4)H^T_{\underline{a}\,(7,50)}(3,4)$	531	364	351	32	11.3	2.6	0.8
covariance matrix	square root of eigenvalues						
$H_{\underline{a}\,(7,50)}(3,4)H^T_{\underline{a}\,(7,50)}(3,4)$	23	19	18.7	5.7	3.4	1.3	0.9

1) Since a set $\{1.9, \; 1.3, \; 0.8, \; 0.5\}$ is composed of relatively small and equally-sized numbers in the square root of $H_{\underline{a}\,(7,50)}(7,0)H^T_{\underline{a}\,(7,50)}(7,0)$, the almost linear system obtained by the algebraic CLS method may be good for a 3-dimensional space.
2) After determining the number n_1 of dimensions which is 3, we will continue the algebraically noisy realization algorithm.
Therefore, the modified impulse response $I(0)$ of an almost linear system obtained by the algebraic CLS method is characterized by 3-dimensional almost linear system.
3) Since a set $\{1.3, \; 0.9\}$ is composed of relatively small and equally-sized numbers in the square root of $H_{\underline{a}\,(7,50)}(3,4)H^T_{\underline{a}\,(7,50)}(3,4)$, the almost linear system obtained by the algebraic CLS method may be somewhat good by adding another 2-dimensional space.

4) After determining the number n_2 of dimensions which is 2, we will continue the algebraically noisy realization algorithm.

Therefore, the modified impulse response $I(1)$ of an almost linear system obtained by the algebraic CLS method is realized by adding another 2-dimensional space.

The system is given by $\sigma_o = ((\boldsymbol{R}^5,\ F_o),\ g_o^0, g_o, h_o,\ h^0)$,

where $F_o = \begin{bmatrix} 0 & 0 & 0.39 & 0 & -0.46 \\ 1 & 0 & -0.28 & 0 & 0.53 \\ 0 & 1 & 0.4 & 0 & -0.94 \\ 0 & 0 & 0 & 0 & -0.6 \\ 0 & 0 & 0 & 1 & 0.75 \end{bmatrix}$, $g_o^0 = \mathbf{e}_1$, $g_o = [-1,\ 0,\ 0,\ 1,\ 0]^T$,

$h_o = [15.1,\ -4.3,\ 2.1,\ -2.97,\ 4.8]$ and $h^0 = 1$.

The obtained modified impulse responses $I(0)$ and $I(1)$ are illustrated in Fig. 5.5.

We can show that the algorithm for noisy realization given by the analytic CLS method in the reference [Hasegawa, 2008] produces the same system as the above one in the sense of the numerical calculation.

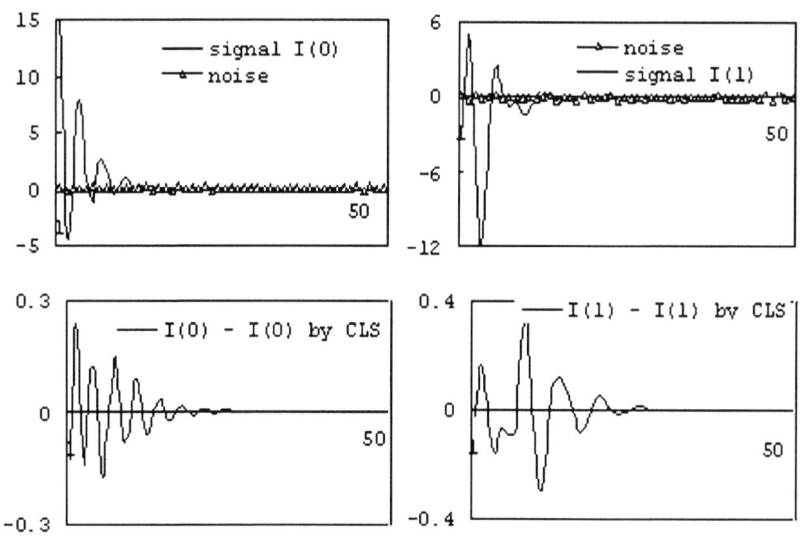

Fig. 5.5 The left are the exact modified impulse response $I(0)$ of a 5-dimensional almost linear system with noise and the difference between the original signal and the obtained one $I(0)_{(3,2)}$ by the algebraic CLS method. The right are the original modified impulse response $I(1)$ of a 5-dimensional almost linear system with noise and the difference between the original signal and the obtained one $I(1)_{(3,2)}$ by the algebraic CLS method in Example (5.16).

In this example, the original signal $I(0)$ is characterized by the modified impulse response of a 3-dimensional linear space and the original signal $I(1)$ is characterized as the modified impulse responses of an added 2-dimensional linear space. The desirable modified impulse responses have been attempted to be obtained by the algebraic CLS method. The model obtained by the algebraic CLS method is a (3,2)-dimensional almost linear system which has the same number of dimensions as the number of the original system.

Just as we expected, the following table and Fig. 5.5 indicate that the model obtained by the algebraic CLS method is a somewhat good (3,2)-dimensional system for the original (3,2)-dimensional system.

dimen-sion	ratio of matrices	mean values of square root for sum of signal	signal by CLS	error	cosine ① and ②	error ratio
		①	②	③	$\cos\theta$	③/①
I(0)_(3,2)	0.09	0.3587	0.3604	0.009	0.9997	0.03
I(1)_(3,2)	0.06	0.2999	0.2960	0.01	0.9992	0.03

For the notations $I(0)\_(n_1, n_2)$ and $I(1)\_(n_1, n_2)$, see Algorithm (5.14).

Example 5.17. Let signals be the modified impulse responses of the following 6-dimensional almost linear system $\sigma = ((\boldsymbol{R}^6, F), g^0, g, h, h^0)$,

$$\text{where } F = \begin{bmatrix} 0 & 0 & 0 & -0.7 & 0 & 0 \\ 1 & 0 & 0 & 0.5 & 0 & 0 \\ 0 & 1 & 0 & -0.7 & 0 & 0 \\ 0 & 0 & 1 & 0.5 & 0 & 0 \\ 0 & 0 & 0 & 0 & 0 & -1 \\ 0 & 0 & 0 & 0 & 1 & 0.8 \end{bmatrix}, \ h = [12,\ 2,\ -1,\ -1,\ 4,\ 2],$$

$g^0 = [1,\ 0,\ 0,\ 0,\ 0,\ 0]^T$, $g = [-1,\ 0,\ 0,\ 0,\ 1,\ 0]^T$, $h^0 = 1$.

Let added noises be given in Fig. 5.6.

Then the algebraically noisy realization problem is solved as follows:

covariance matrix	eigenvalues						
	1	2	3	4	5	6	7
$H_{\underline{a}\ (5,50)}(5,0)H^T_{\underline{a}\ (5,50)}(5,0)$	419	196	153	108	12.4		
$H_{\underline{a}\ (6,50)}(6,0)H^T_{\underline{a}\ (6,50)}(6,0)$	463	208	159	134	15	2.3	
$H_{\underline{a}\ (7,50)}(7,0)H^T_{\underline{a}\ (7,50)}(7,0)$	470	253	172	137	15	6.5	0.7
covariance matrix	square root of eigenvalues						
$H_{\underline{a}\ (7,50)}(7,0)H^T_{\underline{a}\ (7,50)}(7,0)$	21.7	15.9	12.6	11.7	3.9	2.5	0.8

covariance matrix	eigenvalues							
	1	2	3	4	5	6	7	8
$H_{\underline{a}\ (7,50)}(4,3)H^T_{\underline{a}\ (7,50)}(4,3)$	1214	944	324	189	150	65	3.2	
$H_{\underline{a}\ (8,50)}(4,4)H^T_{\underline{a}\ (8,50)}(4,4)$	1838	1039	324	189	149	66	4.2	2.9
covariance matrix	square root of eigenvalues							
$H_{\underline{a}\ (8,50)}(4,4)H^T_{\underline{a}\ (8,50)}(4,4)$	42.9	32.2	18	13.7	12.2	8.1	2	1.7

1) Since a set $\{3.9,\ 2.5,\ 0.8\}$ is composed of relatively small and equally-sized numbers in the square root of $H_{\underline{a}\ (7,50)}(7,0)H^T_{\underline{a}\ (7,50)}(7,0)$, the almost linear system obtained by the algebraic CLS method may be good for a 4-dimensional space.
2) After determining the number n_1 of dimensions which is 4, we will continue the algebraically noisy realization algorithm.

Therefore, the modified impulse response $I(0)$ of an almost linear system obtained by the algebraic CLS method is characterized by a 4-dimensional almost linear system.
3) Since a set $\{2,\ 1.7\}$ is composed of relatively small and equally-sized numbers in the square root of $H_{\underline{a}\ (8,50)}(4,4)H^T_{\underline{a}\ (8,50)}(4,4)$, the almost linear system obtained by the algebraic CLS method may be somewhat good by adding another 2-dimensional space.
4) After determining the number n_2 of dimensions which is 2, we will continue the algebraically noisy realization algorithm.

Therefore, the modified impulse response $I(1)$ of an almost linear system obtained by the algebraic CLS method is constructed by adding another 2-dimensional space.

The system is given by $\sigma_o = ((\boldsymbol{R}^6,\ F_o),\ g_o^0, g_o, h_o,\ h^0)$,

$$\text{where } F_o = \begin{bmatrix} 0 & 0 & 0 & -0.67 & 0 & 0.06 \\ 1 & 0 & 0 & 0.48 & 0 & -0.1 \\ 0 & 1 & 0 & -0.69 & 0 & 0.06 \\ 0 & 0 & 1 & 0.47 & 0 & -0.01 \\ 0 & 0 & 0 & 0 & 0 & -1 \\ 0 & 0 & 0 & 0 & 1 & 0.8 \end{bmatrix},\ g_o^0 = \mathbf{e}_1,\ g_o = [-1,\ 0,\ 0,\ 0,\ 1,\ 0]^T,$$

$h_o = [12.2,\ 1.54,\ -0.7,\ -1.22,\ 4.1,\ 1.6]$ and $h^0 = 1$.

The obtained modified impulse responses $I(0)$ and $I(1)$ are illustrated in Fig. 5.6.

We can show that the algorithm for noisy realization given by the analytic CLS method in the reference [Hasegawa, 2008] produces the same system as the above one in the sense of the numerical calculation.

In this example, the original signal $I(0)$ is characterized as the modified impulse response of a 4-dimensional linear space and the original signal $I(1)$ is characterized as the modified impulse response of an added 2-dimensional linear space. The desirable modified impulse responses have been attempted to be obtained by the algebraic CLS method. The model obtained by the algebraic CLS method is a (4,2)-dimensional almost linear system which has the same number of dimensions as the number of the original system.

Just as we expected, the following table and Fig. 5.6 indicate that the model obtained by the algebraic CLS method is a somewhat good (4,2)-dimensional system for the original (4,2)-dimensional system.

dimen-sion	ratio of matrices	mean values of square root for sum of signal	signal by CLS	error	cosine ① and ②	error ratio
		①	②	③	$\cos\theta$	③/①
I(0)_(4,2)	0.18	0.3412	0.3388	0.03	0.994	0.09
I(1)_(4,2)	0.04	0.4067	0.40577	0.02	0.998	0.05

For the notations $I(0)\_(n_1, n_2)$ and $I(1)\_(n_1, n_2)$, see Algorithm (5.14).

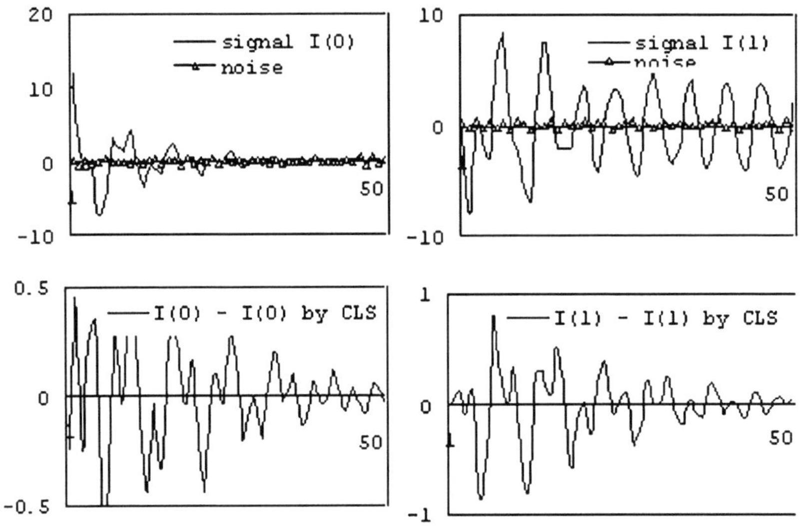

Fig. 5.6 The left are the exact modified impulse response $I(0)$ of a 6-dimensional almost linear system with noise and the difference between the original signal and the obtained one $I(0)_{(4,2)}$ by the algebraic CLS method. The right are the original modified impulse response $I(1)$ of a 6-dimensional almost linear system with noise and the difference between the original signal and the obtained one $I(1)_{(4,2)}$ by the algebraic CLS method in Example (5.17).

5.5 Historical Notes and Concluding Remarks

Algebraically approximate and noisy realization problems of almost linear systems have been studied with the notion of Input/output matrix norm and the algebraic CLS method. The matrix norm is used for determining the dimensions of the state space and the algebraic CLS method is used for determining the parameters of almost linear systems, which are a sort of non-linear system.

For the approximate and noisy realization problems, as has been mentioned, there may be an indication for using singular value decomposition and the Constrained Least Square (CLS) method as shown in the reference [Kalman, 1997]. In the reference, Kalman also pointed out that the identification problem from noisy data should be treated without any prejudice, hence, should be described in a statistical sense, not a probabilistic sense. Here, we only insist that the signal and the noise are not correlated. Then we discussed algebraically approximate and noisy realization problems of non-linear systems that could not be treated using existing methods.

In order to insist that our method for approximate and noisy realization is also effective for non-linear cases, we gave several examples. As shown, the numerical results of the examples have demonstrated that the ratio of the square root of singular values implies a degree of approximation in the sense of the square norm. For our algebraically noisy realization problem, we demonstrated that we can determine the dimensions of almost linear systems when a set of equally-sized numbers of the square root of singular values can be found.

In a similar manner in Chapters 3 and 4, our several examples of both algebraically approximate and noisy realization problems in almost linear systems suggest that our three features can also be expressed as follows:

(1) : The ratio of the matrix norm determines the degree of the crossed angle between directions of the approximated signal and the original signal.
(2) : We could propose a new law which says that almost linear systems obtained by the algebraic CLS method are the same as ones obtained by the analytic CLS method proposed in the reference [Hasegawa, 2008]. The law is said to be a law of a constrained least square.
(3) : The algebraic CLS method determines the coefficients of linearly dependent vectors such that the error between the approximate signal and the original signal has a minimum value in the sense of a square norm while conserving the crossed angle.

In particular, our several examples of algebraically approximate realization have shown that the changing relations among the ratio of matrices and the error to signal ratio are proportional relations and the ratio 0.01 of Input/output matrix ratio ranges from 0.06 to 0.02 for the error to signal ratio.

In addition, our several examples of algebraically noisy realization have shown that the changing relations among the ratio of matrices and the error to signal ratio are proportinal relations and the ratio 0.01 of Input/output matrix ratio ranges from 0.001 to 0.006 for the error to signal ratio.

The analtic CLS method for determing n variables is reduced to the minimization of the following rational polynomial $p(x_1, x_2, \cdots, x_n)$ in n variables:

$$p(x_1, x_2, \cdots, x_n) = \frac{\sum_{c_1, c_2, \cdots, c_n} \alpha(c_1, c_2, \cdots, c_n) \times x_1^{c_1} x_2^{c_2} \cdots x_n^{c_n}}{(1 + x_1^2 + x_2^2 + \cdots + x_n^2)^2},$$

where $\sum_{c_1, c_2, \cdots, c_n}$ means all summation of any combination with the conditions $0 \leq c_i \leq 4$ and $c_1 + c_2 + \cdots + c_n \leq 4$ for any $1 \leq i \leq n$ and $\alpha(c_1, c_2, \cdots, c_n) \in \boldsymbol{R}$.

Therefore, our new Law shows that approximate and noisy problems can be solved using only algebraic calculations, namely, without treating partial differential equations.

Chapter 6
Algebraically Approximate and Noisy Realization of Pseudo Linear Systems

Let the set Y of output's values be a linear space over the real number field $\boldsymbol{R}$. In the reference [Matsuo and Hasegawa, 2003], pseudo linear systems were presented with a main theorem, which says that for any time-invariant input response map, there exist at least two canonical (quasi-reachable and distinguishable) pseudo linear systems which realize, that is, faithfully describe it, and any two canonical pseudo linear systems with the same behavior are isomorphic.

As previously described, the fundamental facts about pseudo linear systems are stated for preparation of their algebraically approximate and noisy realization problems.

Firstly, their realization theory is stated.

Secondly, the main facts about finite dimensional pseudo linear systems are stated. A criterion for the canonical finite dimensional pseudo linear systems, representation theorems of isomorphic classes for canonical pseudo linear systems and a procedure to obtain a canonical one are stated.

Thirdly, their partial realization is discussed according to the above results. The main are the following:

An algorithm to obtain a natural partial realization from a given partial time-invariant input response map is given.

We can easily understand that the above results of our systems are the same as ones obtained in linear system theory.

Moreover, for the time-invariant input response map, we can discuss a real time partial realization problem. Namely, by a single experiment, we find a mathematical model from on-line data. An algorithm to obtain a partial realization from the data is given if a physical object is finite dimensional.

6.1 Basic Facts about Pseudo Linear Systems

Definition 6.1. Pseudo Linear System

1) A system given by the following equations is written as a collection $\sigma = ((X, F), g, h, h^0)$ and it is said to be a pseudo linear system.

Y. Hasegawa: Algebra. Approx. & Noisy Reali. of Discrete-Time Sys., LNEE 50, pp. 111–145.
springerlink.com

$$\begin{cases} x(t+1) &= Fx(t) + g(\omega(t+1)) \\ x(0) &= 0 \\ \gamma(t) &= h^0 + hx(t) \end{cases},$$

where X is a linear space over the field $\boldsymbol{R}$, F is a linear operator on X and $\omega(t) \in U$ for any $t \in N$. And g is a function : $U \mapsto X$, and h is a linear operator : $X \to Y$ and $h^0 \in Y$.

2) The input response map $a_\sigma : U^* \to Y; \omega \mapsto h^0 + h(\sum_{j=1}^{|\omega|}\{((F^{|\omega|-j})g(\omega(j)))$ is said to be a behavior of σ.

For a time-invariant input response map $a \in F(U^*, Y)$, σ that satisfies $a_\sigma = a$ is called a realization of a.

3) For the pseudo linear system σ and any $u \in U$, $i \in N$, $I_\sigma(u)(i) := hF^{i-1}g(u)$ is said to be a modified impulse response of σ, where $u^0 := 1$. The relation $I_\sigma(u)(i) = a_\sigma(u^i) - a_\sigma(u^{i-1})$ holds.

Note that there is a one-to-one correspondence between the behavior of σ and the modified impulse responses $I_\sigma(u) \in F(N, Y)$ of σ by the relations $a_\sigma(\omega) = a_\sigma(1) + \sum_{j=1}^{|\omega|} I_\sigma(\omega(i))(|\omega| - j + 1)$.

4) A pseudo linear system σ is said to be quasi-reachable if the linear hull of the reachable set $\{\sum_{j=1}^{|\omega|}\{((F^{|\omega|-j})g(\omega(j))); \omega \in U^*\}$ is equal to X.

A pseudo linear system σ is called observable if $hF^m x_1 = hF^m x_2$ for any $m \in N$ implies $x_1 = x_2$.

5) A pseudo linear system σ is said to be canonical if σ is quasi-reachable and observable.

Example 6.2. $A(N \times U, \boldsymbol{R}) := \{\lambda = \sum_{n,u} \lambda(n,u)e_{(n,u)}(\text{finite sum}); n \in N, u \in U\}$, where $e_{(n,u)}$ is given by the following equations for $n, n' \in N$ and $u, u' \in U$. If $n = n'$ and $u = u'$, it implies $e_{(n,u)}(n', u') = 1$. If $n \neq n'$ or $u \neq u'$, it implies $e_{(n,u)}(n', u') = 0$. Then $A(N \times U, \boldsymbol{R})$ is clearly a linear space. Let S_r be $S_r(e_{(n,u)}) = S_r(e_{(n+1,u)})$, then $S_r \in L(A(N \times U, \boldsymbol{R}))$ and S_r is irrelevant to the input value's set U. S_r is a right shift operator. Let a map $\eta : U \to A(N \times U, \boldsymbol{R}); u \mapsto e_{(0,u)}$ and let a linear map $\bar{a} : A(N \times U, \boldsymbol{R}) \to Y$ be $\bar{a}(e_{(n,u)}) = a(u^{n+1}) - a(u^n)$ for any time-invariant input response map $a \in F(U^*, Y)$. Then a collection $((A(N \times U, \boldsymbol{R}), S_r), \eta, \bar{a}, a(1))$ is a quasi-reachable pseudo linear system that realizes a.

Let $F(N, Y) := \{$ any function $f : N \to Y\}$. Let $S_l\, \gamma(t) = \gamma(t+1)$ for any $\gamma \in F(N, Y)$ and $t \in N$, then $S_l \in L(F(N, Y))$. Let a map $\chi : U \to F(N, Y)$ be $(\chi(u))(t) := a(\omega|u) - a(\omega)$ for any $u \in U, t \in N$, a time-invariant input response map $a \in F(U^*, Y)$ and ω such that $|\omega| = t$. Moreover, let a linear map 0 be $F(N, Y) \to Y; \gamma \mapsto \gamma(0)$. Then a collection $((F(N, Y), S_l), \chi, 0, a(1))$ is a distinguishable pseudo linear system that realizes a.

Theorem 6.3. *The following two pseudo linear systems are canonical realizations of any time-invariant input response map $a \in F(U^*, Y)$.*

1) $((A(N \times U, \boldsymbol{R})/_{=a}, \tilde{S}_r), \tilde{\eta}, \tilde{\tilde{a}}, a(1))$,
where $A(N \times U, \boldsymbol{R})/_{=a}$ *is a quotient space obtained by equivalence relation* $\sum_{n,u} \lambda_1(n,u)e_{(n,u)} = \sum_{\bar{n},\bar{u}} \lambda_2(\bar{n},\bar{u})e_{(\bar{n},\bar{u})} \iff \sum_{n,u}(a(u^{n+1} - a(u^n)) = \sum_{\bar{n},\bar{u}}(a(\bar{u}^{\bar{n}+1} - a(\bar{u}^{\bar{n}}))$.
And $\tilde{S}_r \in L(A(N \times U, \boldsymbol{R})/_{=a})$ *is given by* $\tilde{S}_r[e_{(n,u)}] = [e_{(n+1,u)}]$ *for* $[e_{(n,u)}] \in A(N \times U, \boldsymbol{R})/_{=a}$, *and* $\tilde{\eta}$ *is a map* : $U \to A(N \times U, \boldsymbol{R})/_{=a}; u \mapsto [e_{(0,u)}]$, *and* $\tilde{\tilde{a}}$ *is given by* : $\tilde{\tilde{a}} \to Y; [e_{(n,u)}] \mapsto a(u^{n+1}) - a(u^n)$.
2) $((\ll S_l^N(\chi(U)) \gg, S_l), \chi, 0, a(1))$,
where $\ll S_l^N(\chi(U)) \gg$ *is the smallest linear space which contains* $S_l^N(\chi(U)) := \{S_l^i(\chi(u)); u \in U, i \in N,\ S_l^i(\chi(u))(t) = (\chi(u))(t+i) = a(\omega|u) - a(\omega), \omega \in U^*, |\omega| = t+i\}$.

Definition 6.4. Let $\sigma_1 = ((X_1, F_1, g_1, h_1, h^0)$ and $\sigma_2 = ((X_2, F_2, g_2, h_2, h^0)$ be pseudo linear systems, then a linear operator $T : X_1 \to X_2$ is said to be a pseudo linear system morphism $T : \sigma_1 \to \sigma_2$ if T satisfies $TF_1 = F_2T$, $Tg_1 = g_2$ and $h_1 = h_2T$.

If $T : X_1 \to X_2$ is bijective, then $T : \sigma_1 \to \sigma_2$ is said to be an isomorphism.

Theorem 6.5. *Realization Theorem of Pseudo Linear Systems*

Existence : For any time-invariant input response map $a \in F(U^*, Y)$,
there exist at least two canonical pseudo linear systems which realize a.
Uniqueness : Let σ_1 *and* σ_2 *be any two canonical pseudo linear systems that realize a time-invariant input response map* $a \in F(U^*, Y)$.
Then there exists an isomorphism $T : \sigma_1 \to \sigma_2$.

6.2 Finite Dimensional Pseudo Linear Systems

Based on the realization theory (6.5), we will state facts about finite dimensional pseudo linear systems as previously described.

To state clear facts, we assume that the set U of input values is finite , i.e., $U := \{u_i; 1 \leq i \leq m\}$ for some $m \in N\}$. This assumption will imply that the g of a pseudo linear system $\sigma = ((X, F), g, h, h^0)$ is completely determined by the finite vectors $\{g(u_i); 1 \leq i \leq m, m \in N\}$, and it was presented that the assumption is not so special in the reference [Matsuo and Hasegawa, 2003].

We only state the following four facts needed for this chapter.

① : The condition for the finite dimensional pseudo linear system is to be canonical.
② : A representation theorem for finite dimensional canonical pseudo linear systems, i.e., we show the real time standard system as a representative.
③ : The criterion for the behavior of finite dimensional pseudo linear systems is to be given by the rank condition of an Input/output matrix.
④ : The procedure to obtain the quasi-reachable standard system that realizes a given time-invariant input response map.

Corollary 6.6. *Let T be a pseudo linear system morphism $T : \sigma_1 \to \sigma_2$, then $a_{\sigma_1} = a_{\sigma_2}$ holds.*

The following is a fact about finite dimensional linear spaces:

FACT : < An n-dimensional linear space over the field $\boldsymbol{R}$ is isomorphic to $\boldsymbol{R}^n$ and $L(\boldsymbol{R}^n, \boldsymbol{R}^m)$ is isomorphic to $\boldsymbol{R}^{m\times n}$. (See Halmos [1958]).>

Therefore, without loss of generality, we can consider a n-dimensional pseudo linear system as $\sigma = ((\boldsymbol{R}^n, F), g, h, h^0)$, where $F \in \boldsymbol{R}^{n\times n}$, $g(u) \in \boldsymbol{R}^n$ and $h \in \boldsymbol{R}^{p\times n}$.

Theorem 6.7. *A pseudo linear system $\sigma = ((\boldsymbol{R}^n, F), g, h, h^0)$ is canonical if and only if the following conditions 1) and 2) hold:*

1) rank $[g(u_1), Fg(u_1), \cdots, F^{n-1}g(u_1), g(u_2), Fg(u_2), \cdots, F^{n-1}g(u_2), \cdots, g(u_m), Fg(u_m), \cdots, F^{n-1}g(u_m)] = n$
2) rank $[h^T, (hF)^T, \cdots, \cdots, (hF^{n-1})^T] = n.$

Definition 6.8. A canonical pseudo linear system $\sigma_s = ((\boldsymbol{R}^n, F_s), g_s, h_s, h^0)$ is said to be a real time standard system if a set $\{(i, u_j) \in N \times U, 1 \le j \le m\}$ given by $\mathbf{e}_{m_1+\cdots+m_{j-1}+i} = F_s^{i-1}g_s(u_j)$ satisfies the following conditions:

1) $g_s(u_j) = e_{m_1+\cdots+m_{j-1}+1}$ and $\mathbf{e}_{m_1+\cdots+m_{j-1}+i} = F_s^{i-1}g_s(u_j)$ hold for any $i\ (1 \le i \le m_j,\ j\ (1 \le j \le m)$.
2) $F_s^{m_p}g_s(u_p) = \sum_{i=1}^{m_1+\cdots+m_p} \alpha_{p,i}\mathbf{e}_i$ holds for any $1 \le p \le m$, where $\alpha_{p,i} \in \boldsymbol{R}$ and $\mathbf{e}_i = [0, 0, \cdots, 0, \overset{i}{1}, 0, \cdots, 0]^T$.
3) $n = \sum_{i=1}^{m} m_i$ holds.
4) F_s is given as follows:

Theorem 6.9. *Representation Theorem for equivalence classes*

For any finite dimensional canonical pseudo linear system, there exists a uniquely determined isomorphic real time standard system.

[proof] Note that F_s in the real time standard system is the quasi-reachable standard form.

Let $\sigma = ((\boldsymbol{R}^n, F), g, h, h^0)$ be any finite dimensional canonical pseudo linear system. For the real time standard form $((\boldsymbol{R}^n, F_s), g_s, h_s, h^0)$ and a linear operator $T : \boldsymbol{R}^n \to \boldsymbol{R}^n$ such that $TF = F_sT$ and $Tg = g_s$ hold , let $h_s := h \cdot T^{-1}$. Then T is a pseudo linear system morphism $: \sigma = ((\boldsymbol{R}^n, F), g, h, h^0) \to \sigma_s = ((\boldsymbol{R}^n, F_s), g_s, h_s, h^0)$. T is bijective and σ_s is the only real time standard system. By Corollary (6.6), the behaviors of σ and σ_s are the same.

$$F_s = \left[\begin{array}{llllllllll}
0 \cdots & 0\ \alpha_{11} & 0 \cdots\cdots 0 & \alpha_{21} & 0 \cdots 0 & & 0\ \alpha_{m1} \\
1 \ddots & \vdots\ \alpha_{12} & 0 \cdots\ 0 & \alpha_{22} & \vdots & & \alpha_{m2} \\
0 \ddots & \vdots\ \ \vdots & \vdots\ \ \vdots & \vdots & & & \vdots \\
\vdots \ddots 1\ 0 & \vdots & \vdots\ \ \vdots & \vdots & & & \vdots \\
0 \ddots 0\ 1 & \alpha_{1m_1} & \vdots\ \ \vdots & \alpha_{2m_1} & \vdots & & \\
0 \cdots 0\ 0 & 0 & 0 \cdots\cdots 0 & \alpha_{2m_1+1} & 0 \cdots 0 & & \\
0 \ \cdots & \vdots & 1 \ddots\ \vdots & \vdots & \vdots & & \\
0 \ \cdots & \vdots & 0 \ddots \ddots \vdots & \vdots & & & \\
0 \ \cdots & \vdots & \vdots \ddots 1\ 0 & \vdots & \vdots \cdots & & \\
0 \ \cdots & & 0 \cdots 0\ 1 & \alpha_{2m_1+m_2} & 0 \ddots & & \\
0 \ \cdots & & \cdots\cdots 0 & 0 & 0 \ddots & & \\
0 \ \cdots & & \cdots\cdots & \vdots & \ddots \ddots \vdots & \vdots & \\
0 \ \cdots & & \cdots\cdots & \vdots & \ddots \ddots 0 \cdots\cdots 0 & & \vdots \\
0 \ \cdots & & \cdots\cdots & \vdots & \ddots\ 0\ 1 & \vdots & \vdots \\
0 \ \cdots & & \cdots\cdots & \vdots & \ddots \ddots 0 \ddots\ 0 & \vdots & \vdots \\
0 \ \cdots & & \cdots\cdots & \vdots & \ddots \ddots \vdots \ddots & 1\ 0 & \alpha_{mn-1} \\
0 \ \cdots & & \cdots\cdots & \cdots & \cdots\cdots 0 \cdots & 0\ 1 & \alpha_{mn}
\end{array}\right].$$

Definition 6.10. For any time-invariant input response map $a \in F(U^*, Y)$, the corresponding linear input/output map $A : ((A(N \times U, \boldsymbol{R}), S_r) \to (F(N, Y), S_l)$ satisfies $A(e_{(s,u)})(t) = a(u^{s+t+1}) - a(u^{s+t})$.

Therefore, the A can be represented by the next infinite matrix $(I/O)_a$. This $(I/O)_a$ is said to be an Input/output matrix of a.

$$\begin{array}{c} (I/O)_a = \\ t \end{array} \left(\begin{array}{ccc} & & (s,u) \\ & & \vdots \\ & & \vdots \\ & & \vdots \\ \cdots & \cdots & a(u^{s+t+1}) - a(u^{s+t}) \\ & & \end{array} \right)$$

Since $S_l^s(\chi(u))(t) = (\chi(u))(t+s) = a(\omega|u) - a(\omega), \omega \in U^*, |\omega| = t+s$ holds, the column vectors of Input/output matrix of $(I/O)_a$ may be expressed by $S_l^s(\chi(u)) = S_l^s I(u)$.

Theorem 6.11. *Theorem for existence criterion*

For a time-invariant input response map $a \in F(U^*, Y)$, the following conditions are equivalent:

1) The time-invariant input response map $a \in F(U^*, Y)$ has the behavior of a n-dimensional canonical pseudo linear system.

2) There exist n linearly independent vectors and no more than n linearly independent vectors in a set $\{S_l^i(\chi(u)); u \in U, i \in N, 1 \leq i \leq n\}$.
3) The rank of the Input/output matrix $(I/O)_a$ *of a is n.*

Theorem 6.12. *Theorem for a realization procedure*

Let a time-invariant input response map $a \in F(U^*, Y)$ *satisfy the condition of Theorem (6.11). Then the real time standard system* $\sigma_s = ((\boldsymbol{R}^n, F_s), g_s, h_s, h^0)$ *which realizes a can be obtained by the following procedure:*

1) Select the linearly independent vectors $\{S_l^j\chi(u_i));\ 1 \leq j \leq m_i,\ 1 \leq i \leq m\}$ *in order of the set* $\{\chi(u_1),\ S_l\chi(u_1), \cdots, S_l^{m_1-1}chi(u_1), \chi(u_2),\ S_l\chi(u_2), \cdots, S_l^{m_2-1}\chi(u_2), \cdots, \chi(u_m),\ S_l\chi(u_m), \cdots, S_l^{m_m-1}\chi(u_m)\}$. *Let* $n := \mathrm{rank}\ I/O_a = m_1 + m_2 + \cdots + m_m$.
2) Let the state space be $\boldsymbol{R}^n$. *Let the map* $g_s : U \to \boldsymbol{R}^n$ *be* $g_s(u_i) := \mathbf{e}_{m_1+\cdots+m_{i-1}+1}$ *for* $u_i \in U$ *and* $1 \leq i \leq m$ *and* $F_s^j g_s(u_i) := \mathbf{e}_{m_1+\cdots+m_{i-1}+1+j}$ *for* $1 \leq j \leq m_i - 1$. *And let* $F_s^{m_i} g_s(u_i) := \sum_{j=1}^{m_1+\cdots+m_i} \alpha_{i,j}\mathbf{e}_j$ *for* $u_i \in U$ *and* $S_l^{m_i}\chi(u_i) := \sum_{j=1}^{m_1+\cdots+m_i} \alpha_{i,j}\chi(u_j)$.
3) Let the output map $h_s = [a(u_1) - a(1), a(u_1^2) - a(u_1), \cdots, a(u_1^{m_1}) - a(u_1^{m_1-1}), , \cdots, a(u_m) - a(1),\ a(u_m^2) - a(u_m), \cdots, a(u_m^{m_m}) - a(u_m^{m_m-1})]$
4) Let $F_s \in \boldsymbol{R}^{n\times n}$ *be the* F_s *in Definition (6.8).*

[proof] Let $R(\chi) = \{S_l^i(\chi(u)); u \in U, i \in N\}$. By Theorem (6.3), $((\ll S_l^N(\chi(U)) \gg, S_l), \chi, 0, a(1))$ is a canonical pseudo linear system that realizes a time-invariant input response map $a \in F(U^*, Y)$. The linearly independent vectors $\{S_l^j(\chi(u_i)) \in \gg = \ll R(\chi) \gg; u_i \in U, 1 \leq i \leq m,\ 0 \leq j \leq m_i - 1\}$ are in order of the numerical value. Let a linear map $T : \ll R(\chi) \gg \to \boldsymbol{R}^n$ be $T(\chi(u_i)) = \mathbf{e}_{m_1+\cdots+m_{i-1}+1}$ for any $i(1 \leq i \leq n)$ and $T(S_l^j\chi(u_i)) := \mathbf{e}_{m_1+\cdots+m_{i-1}+1+j}$ for $1 \leq j \leq m_i - 1$. Then, by step 2), $T\chi = g_s$ holds and by step 3), $h_s \cdot T = 0$ holds, and by step 4), $F_s \cdot T = T \cdot F_s$ holds. Consequently, T is bijective and a pseudo linear system morphism $: ((\ll S_l^N(\chi(U)) \gg, S_l), \chi, 0, a(1)) \to \sigma_s = ((\boldsymbol{R}^n, F_s), g_s, h_s, a(1))$.

By Corollary (6.6), the behavior of σ_s is a. It follows that the choice of $\{S_l^i(\chi(u_i)); u_i \in U, i \in N\})$ for $i(1 \leq i \leq m\}$ are in order of the numerical value and the determination of map T imply that σ_s is the real time standard system.

6.3 Partial Realization of Pseudo Linear Systems

Here we consider a partial realization problem by multi-experiment. Let $\underline{a}$ be an $\underline{N}$ sized time-invariant input response map ($\in F(U_{\underline{N}}^*, Y)$, where $\underline{N} \in N$ and $U_{\underline{N}}^* := \{\omega \in U^*; |\omega| \leq \underline{N}\}$. The $\underline{a}$ is said to be a partial time-invariant input response map.

A finite dimensional pseudo linear system $\sigma = ((X, F), g, h.x^0)$ is said to be a partial realization of $\underline{a}$ if $h^0 + h(\sum_{j=1}^{|\omega|} F^{|\omega|-j} g(\omega(j))) = \underline{a}(\omega)$ holds for any $\omega \in U_{\underline{N}}^*$.

A partial realization problem of pseudo linear systems can be stated as follows:

< For any given partial time-invariant input response $\underline{a} \in F(U_{\underline{N}}^*, Y)$, find a partial realization σ of $\underline{a}$ such that the dimensions of state space X of σ is minimum, where the σ is said to be a minimal partial realization of $\underline{a}$.

In section 6.1, we stated a representation theorem for the time-invariant input response maps. The theorem says that any time-invariant input response map can be characterized by the modified impulse response. Note that the modified impulse response $I : U \to F(N, Y)$ can be represented by $(I(u)(t)) = a(u^t) - a(u^{t-1})$ for $u \in U, t \in N$ and the time-invariant input response map $a \in F(U^*, Y)$.

For any given partial time-invariant input response $\underline{a} \in F(U_{\underline{N}}^*, Y)$, this correspondence can determine a partial modified impulse response $\underline{I} : U \to F(N_{\underline{N}-1}, Y)$, where $N_{\underline{N}-1} := \{1, 2, , \underline{N} - 1;$ for some $\underline{N} \in N\}$.

$$\begin{array}{c}(I/O)_{\underline{a}\ (p,\underline{N}-p)} \\ t\end{array} = \begin{pmatrix} & & (s,u) \\ & & \vdots \\ & & \vdots \\ & & \vdots \\ \cdots & \cdots & a(u^{s+t+1}) - a(u^{s+t}) \\ & & \end{pmatrix},$$

where $0 \leq s \leq p, 0 \leq t \leq \underline{N} - p$ and $u \in U$.

When we actually treat algebraically approximate and noisy realization problems, we will use a notation $H_{\underline{a}\ (n_1+n_2, \underline{N}-n_1-n_2)}(n_1, n_2)$ expressed as follows:

$$H_{\underline{a}\ (n_1+n_2, \underline{N}-n_1-n_2)}(n_1, n_2) = [I_{\underline{a}}(0), \cdots, S_l^{n_1-1} I_{\underline{a}}(0), I_{\underline{a}}(1), \cdots, S_l^{n_2-1} I_{\underline{a}}(1)].$$

Theorem 6.13. *Let $(I/O)_{\underline{a}\ (p,\underline{N}-p)}$ be the finite-sized Input/output matrix of $\underline{a} \in F(U_{\underline{N}}^*, Y)$. Then there exists a natural partial realization of $\underline{a}$ if and only if the following conditions hold:*
rank $(I/O)_{\underline{a}\ (p,\underline{N}-p)} =$ rank $(I/O)_{\underline{a}\ (p,\underline{N}-p-1)} =$ rank $(I/O)_{\underline{a}\ (p+1,\underline{N}-p)}$ *for some $p \in N$.*

Theorem 6.14. *Let a partial time-invariant input response $\underline{a} \in F(U_{\underline{N}}^*, Y)$ satisfy the condition of Theorem (6.13), then the real time standard system $\sigma_s = ((\boldsymbol{R}^n, F_s), g_s, h_s, h^0)$ that realizes $\underline{a}$ can be obtained by the following algorithm.*

Set $n :=$ rank $(I/O)_{\underline{a}\ (p,\underline{N}-p)}$, where $(I/O)_{\underline{a}\ (p,\underline{N}-p)}$ is the finite Input/output matrix of $\underline{a} \in F(U_{\underline{N}}^*, Y)$.

1) Select the linearly independent vectors $\{S_l^j(\chi(u_i)); 1 \le i \le m,\ 0 \le j \le m_i - 1\}$ *from* $(I/O)_{\underline{a}\ (p,\underline{N}-p)}$ *in order of the numerical value.*
2) Let the state space be $\boldsymbol{R}^n$. *Let the map* $g_s : U \to \boldsymbol{R}^n$ *be* $g_s(u_i) := \mathbf{e}_{m_1+\cdots+m_{i-1}+1}$ *for* $u_i \in U$ *such that* $1 \le i \le m$ *and* $F_s^j g_s(u_i) := \mathbf{e}_{m_1+\cdots+m_{i-1}+1+j}$ *for* $1 \le j \le m_i - 1$. *And let* $S_l^{m_i} g_s(u_i) := \sum_{j=1}^{m_1+\cdots+m_i} \alpha_{i,j}\mathbf{e}_j$ *for* $u_i \in U$.
3) Let the output map $h_s = [a(u_1) - a(1), a(u_1^2) - a(u_1), \cdots, a(u_1^{m_1}) - a(u_1^{m_1-1}), , \cdots, a(u_m) - a(1), a(u_m^2) - a(u_m), \cdots, a(u_m^{m_m}) - a(u_m^{m_m-1})]$.
4) Let F_s *be the* F_s *in Fig. 2,*
where $S_l^{m_i}\chi(u_i) := \sum_{j=1}^{m_1+\cdots+m_i} \alpha_{i,j}\chi(u_j)$, $\alpha_{i,j} \in \boldsymbol{R}$ *holds in the sense of* $F(N_{\underline{N}-p}, Y)$ *and* $\underline{S_l} : F(N_p, Y) \to F(N_{p-1}, Y); a \mapsto \underline{S_l a}[; t \mapsto \underline{a}(t+1)$ *for some* $p \in N$.

6.4 Real-Time Partial Realization of Pseudo Linear Systems

In general, it is well known that non-linear systems can only be determined by multi-experiments. However, for pseudo linear systems, special single-experiments to mimic multi-experiments were given in the reference [Matsuo and Hasegawa, 2003].

In this section, the results are introduced as previously described.

Problem 6.15. Real time partial realization problem

Let a physical object (equivalently, $a \in F(U^*, Y)$) be a finite dimensional pseudo linear system. Then for any given finite data $\{\underline{a}(\underline{\omega})$; an input $\underline{\omega}$ is finite length $\}$, find a pseudo linear system $\sigma = ((\boldsymbol{R}^n, F), g, h, h^0)$ and an input $\underline{\omega}$ such that $a_\sigma(\omega) = a(\omega)$ for any $\omega \in U^*$.

Definition 6.16. For a finite dimensional pseudo linear system, if there exists a solution of a real time partial realization problem, then an input $\underline{\omega} \in U^*$ of the solution is said to be a (real time partial) realization signal.

Lemma 6.17. *Let a given time invariant input response map* $a \in F(U^*, Y)$ *have the behavior of a pseudo linear system whose state space is less than L dimensional. Then there exists an input of finite length* $\underline{\omega} \in U^*$ *such that the following algorithm provides a finite Input/output matrix, where* $p := max\{L_1, L_2, \cdots, L_m\}$.

1) Find an integer L_1 *such that row vectors* $\{\underline{S_l}^i(\chi(u_1)) \in \boldsymbol{R}^{L-1}; 0 \le i \le L_1 - 1\}$ *are linearly independent and* $\{\underline{S_l}^i(\chi(u_1)) \in \boldsymbol{R}^{L-1}; 0 \le i \le L_1\}$ *are linearly dependent. Namely, feed an input* $\omega_1 := u_1^{L_1+L}$ *into the plant.*
2) Find an integer L_2 *such that row vectors* $\{\underline{S_l}^i(\chi(u_j)) \in \boldsymbol{R}^{L-1}; 0 \le i \le L_j - 1, 1 \le j \le 2\}$ *are linearly independent and* $\{\underline{S_l}^i(\chi(uj)) \in \boldsymbol{R}^{L-1}$, $\underline{S_l}^{L_2}(\chi(u_2)) \in \boldsymbol{R}^{L-1}; 0 \le i \le L_j - 1, 1 \le j \le 2\}$ *are linearly dependent.*

Namely, feed a further input $\omega_2 := u_1^{L_1+L-1}|u_2$ *into the plant.*
3) Find an integer L_3 *such that row vectors* $\{\underline{S_l}^i(\chi(u_j)) \in \boldsymbol{R}^{L-1}; 0 \leq i \leq L_j - 1, 1 \leq j \leq 3\}$ *are linearly independent and* $\{\underline{S_l}^i(\chi(u_j)) \in \boldsymbol{R}\ L-1,$ $\underline{S_l}^{L_3}(\chi(u_3)) \in \boldsymbol{R}^{L-1}; 0 \leq i \leq L_j - 1, 1 \leq j \leq 3\}$ *are linearly dependent. Namely, feed a further input* $\omega_3 := u_1^{L_3+L-1}|u_3$ *into the plant.*

$\vdots$

$\vdots$

m) Find an integer L_m *such that row vectors* $\{\underline{S_l}^i(\chi(u_j)) \in \boldsymbol{R}^{L-1}; 0 \leq i \leq L_j - 1, 1 \leq j \leq m\}$ *are linearly independent and* $\{\underline{S_l}^i(\chi(u_j)) \in \boldsymbol{R}^{L-1},$ $\underline{S_l}^{L_m}(\chi(u_m)) \in \boldsymbol{R}^{L-1}; 0 \leq i \leq L_j - 1, 1 \leq j \leq m\}$ *are linearly dependent. Namely, feed a further input* $\omega_m := u_1^{Lm+L-1}|u_m$ *into the plant.*
Let $\omega = \omega_m|\omega_{m-1}|\cdots|\omega_2|\omega_1$.

Making row vectors of a matrix from the row vectors $\{\underline{S_l}^i(\chi(uj)) \in \boldsymbol{R}^{L-1}; 0 \leq i \leq L_j - 1, 1 \leq j \leq m\}$ *obtained by the above iterations, we will obtain a finite Input/output matrix* $H_{\underline{a}\ (L-1,p)}$.

Theorem 6.18. *Let a given time-invariant input response map* $a \in F(U^*, Y)$ *have the behavior of a pseudo linear system whose state space is less than L-dimensional. Then there exists a realization signal such that the quasi-reachable standard system* $\sigma_s = ((\boldsymbol{R}^n, F_s), g_s, h_s, h^0)$ *that realizes a can be obtained by the following algorithm:*

1) Find a finite Input/output matrix $(I/O)_{\underline{a}\ (L-1,p)}$ *based upon the algorithm given in Lemma (6.17).*
2) Apply the algorithm given in Theorem (6.14) to the above finite Input/output matrix $(I/O)_{\underline{a}\ (L-1,p)}$.

Theorem 6.19. *Let the modified impulse response* $I_a(u) \in F(\mathrm{M}, Y)$ *satisfy the conditions of Theorem (6.13). Then the pseudo linear system* $\sigma = ((X, F_s), g_s, h_s, h^0)$ *which realizes a can be obtained by the following algorithm:*

1) Select n_1 *independent vectors on the vectors* $\{S_l^s I_a(0) : 0 \leq s \leq p\}$. *And select* n_2 *independent vectors in* $\{S_l^s I_a(1) : 0 \leq s \leq p\}$.
2) Let the state space be $\boldsymbol{R}^n$. *And let* g_s^0 *and* g_s *be as follows:* $g_s^0 = \mathbf{e}_1$, $g_s = \mathbf{e}_{n_1+1} - \mathbf{e}_1$, *where* Ag_s *is given by* $g_s = 0$ *if* $S_l I_a(0) = 0$ *holds. Moreover,* $n = n_1 + n_2$ *and* $\mathbf{e}_i = [0, \cdots, 0, \overset{i}{1}, 0, \cdots, 0]^T$ *hold.*
3) $F_s \in \boldsymbol{R}^{n \times n}$ *is the same as* F_s *in Definition (6.8).*
$S_l^{n_1} I_a(0) = \sum_{i=1}^{n_1} \alpha_{1i} S_l^{i-1} I_a(0)$.
$S_l^{n_2} I_a(1)$
$= \sum_{i=1}^{n_1} \alpha_{2i} S_l^{i-1} I_a(0) + \sum_{i=1}^{n_2} \alpha_{2n_1+i} S_l^{i-1} I_a(1)$.
4) Let h_s *be* $h_s = [a(0) - a(1), a(0^2) - a(0), \cdots, a(0^{n_1}) - a(0^{n_1-1}), a(1) - a(1), a(0|1) - a(0), \cdots, a(0^{n_1-1}|1) - a(0^{n_1-1})]$.
5) Let h^0 *be* $h^0 = a(1)D$

6.5 Algebraically Approximate Realization of Pseudo Linear Systems

In this section, we discuss the algebraically approximate realization problems of pseudo linear systems.

We will discuss the algebraically approximate realization problem under the assumption that the set U of input's values is a finite set $U = \{u_j : 1 \leq j \leq m\}$ for an finite integer $m \in N$. In the reference [Matsuo and Hasegawa, 2003], we showed that this assumption is not so special. However, for simplicity of our discussion, we assume that the set U of input's values is $U = \{u_1,\ u_2,\ u_3\}$ in our examples.

Roughly speaking, the algebraically approximate realization of pseudo linear systems can be stated as follows:

< For any given partial data of a pseudo linear system, find, using only algebraic calculations, a pseudo linear system which approximates the given data. >

In order to make our discussion simple, we assume that the set Y of output's value is the set $\boldsymbol{R}$ of real numbers, namely 1-output.

Theorem 6.20. *Algebraic algorithm for approximate realization*

Let an input response map $\underline{a}$ be a considered object which is a pseudo linear system. Then an approximate realization $\sigma = ((\boldsymbol{R}^n, F_s), g_s, h_s, h^0)$ of $\underline{a}$ is given by the following algorithm:

1) Based on the ratio of the square root of eigenvalues for a matrix $H_{\underline{a}\ (p,\bar{p})}(p,0,0)H_{\underline{a}\ (p,\bar{p})}(p,0,0)^T$, determine the value n_1 of rank for the matrix $H_{\underline{a}\ (p,\bar{p})}(p,0,0)$, where $n_1 \leq p$.
Namely, determine the value n_1 of rank for the matrix $H_{\underline{a}\ (p,\bar{p})}(p,0,0)$ such that the ratio of the square root of eigenvalues for the covariance matrix becomes very small. The small ratio indicates the nearness of approximation degree.

2) The algebraic CLS method is used as follows:
① Based on Proposition (2.14), determine coefficients $\{\alpha_{1i} : 1 \leq i \leq n_1\}$. The Q in Proposition (2.14) can be considered as the matrix composed from the eigenvectors of $H_{\underline{a}\ (n_1+1,L)}(n_1+1,0)H^T_{\underline{a}\ (n_1+1,L)}(n_1+1,0)$.
Let a matrix $A_1 \in \boldsymbol{R}^{1\times(n_1+1)}$ be $A_1 = [\alpha_{11}, \alpha_{12}, \cdots, \alpha_{1n_1}, -1]$.
② Determine the error vectors $\{\underline{S_l^i}\bar{I}_{\underline{a}}(u_1) \in \boldsymbol{R}^{L\times 1} :\ 0 \leq i \leq n_1\}$ by using the equation $[\bar{I}_{\underline{a}}(u_1), \underline{S_l}\bar{I}_{\underline{a}}(u_1), \cdots, \underline{S_l^{n_1}}\bar{I}_{\underline{a}}(u_1)]^T :=$ $A_1^T[A_1A_1^T]^{-1}A_1H^T_{\underline{a}\ (n_1+1,L)}(n_1+1,0)$ and $H^T_{\underline{a}\ (n_1,L)}(n_1+1,0) :=$ $[I_{\underline{a}}(u_1), \cdots, S_l^{n_1-1}I_{\underline{a}}(u_1), S_l^{n_1}I_{\underline{a}}(u_1)]$.
③ Let $h_{1s} \in \boldsymbol{R}^{1\times n_1}$ be h_{1s} $= [(I_{\underline{a}}(u_1))(0) - (\bar{I}_{\underline{a}}(u_1))(0), (S_lI_{\underline{a}}(u_1))(0) - (S_l\bar{I}_{\underline{a}}(u_1))(0), \cdots,$ $(S_l^{n_1-1}I_{\underline{a}}(u_1))(0) - (S_l^{n_1-1}\bar{I}_{\underline{a}}(u_1))(0)]$.

3) Based on the ratio of the square root of eigenvalues for a matrix $H_{\underline{a}\ (n_1+p,\bar{p})}(n_1,p,0)H_{\underline{a}\ (n_1+p,\bar{p})}(n_1,p,0)^T$*, determine the value* n_2 *of rank for the matrix* $H_{\underline{a}\ (n_1+p,\bar{p})}(n_1,p,0)$*, where* $n_2 \le p$*.*
Namely, determine the value n_2 *of rank for the matrix* $H_{\underline{a}\ (p,\bar{p})}(n_1,p,0)$ *such that the ratio of the square root of eigenvalues for the covariance matrix becomes very small. The small ratio indicates the nearness of approximation degree.*

4) The algebraic CLS method is used as follows:
① Based on Proposition (2.14), determine coefficients $\{\alpha_{2i} : 1 \le i \le n_1+n_2\}$.
The Q in Proposition (2.14) can be considered as the matrix composed from the eigenvectors of $H^T_{\underline{a}\ (n_1+n_2+1,L)}(n_1,n_2+1,0)H^T_{\underline{a}\ (n_1+n_2+1,L)}(n_1,n_2+1,0)$.
Let a matrix $A_2 \in \boldsymbol{R}^{1\times(n_1+n_2+1)}$ *be* $A_2 = [\alpha_{21},\alpha_{22},\cdots,\alpha_{2n_1+n_2},-1]$.
② Determine the error vectors $\{\underline{S}_l^{i_j}\bar{I}_{\underline{a}}(u_j) \in \boldsymbol{R}^{L\times 1} : 0 \le i_1 \le n_1-1,\ 0 \le i_2 \le n_2, 1 \le j \le 2\}$ *by using the equation* $[\bar{I}_{\underline{a}}(u_1),\cdots,\underline{S}_l^{n_1-1}\bar{I}_{\underline{a}}(u_1),\bar{I}_{\underline{a}}(u_2),\cdots,\underline{S}_l^{n_2-1}\bar{I}_{\underline{a}}(u_2),\underline{S}_l^{n_2}\bar{I}_{\underline{a}}(u_2)]^T :=$ $A_2^T[A_2A_2^T]^{-1}A_2H^T_{\underline{a}\ (n_1+n_2+1,L)}(n_1,n_2+1,0)$ *and* $H^T_{\underline{a}\ (n_1+n_2;1,L)}(n_1,n_2+1,0) :=$ $[I_{\underline{a}}(u_1),\cdots,S_l^{n_1-1}I_{\underline{a}}(u_1),I_{\underline{a}}(u_2),\cdots,S_l^{n_2-1}I_{\underline{a}}(u_2),S_l^{n_2}I_{\underline{a}}(u_2)]$.
③ Let $h_{2s} \in \boldsymbol{R}^{1\times n_2}$ *be* h_{2s}
$= [(I_{\underline{a}}(u_2))(0) - (\bar{I}_{\underline{a}}(u_2))(0), (S_l I_{\underline{a}}(u_2))(0) - (S_l\bar{I}_{\underline{a}}(u_2))(0),$
$\cdots, (S_l^{n_2-1}I_{\underline{a}}(u_2))(0) - (S_l^{n_2-1}\bar{I}_{\underline{a}}(u_2))(0)]$.

5) Based on the ratio of the square root of eigenvalues for a matrix $H_{\underline{a}\ (n_1+p,\bar{p})}(n_1,n_2,q)H_{\underline{a}\ (n_1+n_2+q,\bar{q})}(n_1,n_2,q)^T$*, determine the value* n_3 *of rank for the matrix* $H_{\underline{a}\ (n_1+n_2+q,\bar{q})}(n_1,n_2,q)$*, where* $n_3 \le q$*.*
Namely, determine the value n_3 *of rank for the matrix* $H_{\underline{a}\ (n_1+n_2+q,\bar{q})}(n_1,n_2,q)$ *such that the ratio of the square root of eigenvalues for the covariance matrix becomes very small. The small ratio indicates the nearness of approximation degree.*

6) The algebraic CLS method is used as follows:
① Based on Proposition (2.14), determine coefficients $\{\alpha_{3i} : 1 \le i \le n_1+n_2+n_3\}$.
The Q in Proposition (2.14) can be considered as the matrix composed from the eigenvectors of $H_{\underline{a}\ (n_1+n_2+n_3+1,L)}(n_1,n_2,n_3+1)$ $\times H^T_{\underline{a}\ (n_1+n_2+n_3+1,L)}(n_1,n_2,n_3+1)$.
Let a matrix $A_3 \in \boldsymbol{R}^{1\times(n_1+n_2+n_3+1)}$ *be* $A_3 = [\alpha_{31},\alpha_{32},\cdots,\alpha_{3n_1+n_2+n_3},-1]$.
② Determine the error vectors $\{\underline{S}_l^{i_j}\bar{I}_{\underline{a}}(u_j) \in \boldsymbol{R}^{L\times 1} : 0 \le i_j \le n_j-1$ *for* $1 \le j \le 2,\ 0 \le i_3 \le n_3\}$ *by using the equation* $[\bar{I}_{\underline{a}}(u_1),\cdots,\underline{S}_l^{n_1-1}\bar{I}_{\underline{a}}(u_1),\bar{I}_{\underline{a}}(u_2),\cdots,\underline{S}_l^{n_2-1}\bar{I}_{\underline{a}}(u_2),\underline{S}_l\bar{I}_{\underline{a}}(u_3),\cdots,$ $\underline{S}_l^{n_3}\bar{I}_{\underline{a}}(u_3)]^T := A_3^T[A_3A_3^T]^{-1}A_3H^T_{\underline{a}\ (n_1+n_2+n_3+1,L)}(n_1,n_2,n_3+1)$ *and* $H^T_{\underline{a}\ (n_1+n_2+n_3+1,L)}(n_1,n_2,n_3+1) := [I_{\underline{a}}(u_1),\cdots,S_l^{n_1-1}I_{\underline{a}}(u_1),I_{\underline{a}}(u_2),\cdots,$ $S_l^{n_2-1}I_{\underline{a}}(u_2),S_l^{n_2}I_{\underline{a}}(u_2),I_{\underline{a}}(u_3),\cdots,S_l^{n_3-1}I_{\underline{a}}(u_3),S_l^{n_3}I_{\underline{a}}(u_3),]$.

③ *Let* $h_{3s} \in \boldsymbol{R}^{1\times n_3}$ *be* $h_{3s} =$
$[(I_{\underline{a}}(u_3))(0) - (\bar{I}_{\underline{a}}(u_3))(0), (S_l I_{\underline{a}}(u_3))(0) - (S_l \bar{I}_{\underline{a}}(u_3))(0),$
$\cdots, (S_l^{n_3-1} I_{\underline{a}}(u_3))(0) - (S_l^{n_3-1} \bar{I}_{\underline{a}}(u_3))(0)].$

⋮

*2*m-1) Based on the ratio of the square root of eigenvalues for a matrix* $H_{\underline{a}\ (n_1+\cdots+n_m+p,\bar{p})}(n_1,\cdots,n_m,q) H_{\underline{a}\ (n_1+\cdots+n_m+q,\bar{q})}(n_1,\cdots,n_m,q)^T$*, determine the value* n_m *of rank for the matrix* $H_{\underline{a}\ (n_1+\cdots+n_m+q,\bar{q})}(n_1,\cdots,n_m,q)$*, where* $n_m \le q$.
Namely, determine the value n_m *of rank for the matrix* $H_{\underline{a}\ (n_1+\cdots+n_m+q,\bar{q})}(n_1,\cdots,n_m,q)$ *such that the ratio of the square root of eigenvalues for the covariance matrix becomes very small. The small ratio indicates the nearness of approximation degree.*

*2*m) The algebraic CLS method is used as follows:*
① *Based on Proposition (2.14), determine coefficients* $\{\alpha_{mi} : 1 \le i \le n_1 + \cdots + n_m\}$.
The Q in Proposition (2.14) can be considered as the matrix composed from the eigenvectors of $H^T_{\underline{a}\ (n_1+\cdots+n_m+1,L)}(n_1,\cdots,n_m+1)$ $\times H^T_{\underline{a}\ (n_1+\cdots+n_m+1,L)}(n_1,\cdots,n_m+1)$.
Let a matrix $A_m \in \boldsymbol{R}^{1\times(n_1+\cdots+n_m+1)}$ *be*
$A_m = [\alpha_{m1}, \alpha_{m2}, \cdots, \alpha_{mn_1+\cdots+n_m}, -1]$.
② *Determine the error vectors* $\{\underline{S}_l^{i_j} \bar{I}_{\underline{a}}(u_j) \in \boldsymbol{R}^{L\times 1} : 0 \le i_j \le n_j - 1 \text{ for } 1 \le j \le m-1,\ 0 \le i_m \le n_m\}$ *by using the equation*
$[\bar{I}_{\underline{a}}(u_1), \cdots, \underline{S}_l^{n_1-1} \bar{I}_{\underline{a}}(u_1), \bar{I}_{\underline{a}}(u_2), \cdots, \underline{S}_l \bar{I}_{\underline{a}}(u_m), \cdots, \underline{S}_l^{n_m} \bar{I}_{\underline{a}}(u_m)]^T :=$
$A_m^T [A_m A_m^T]^{-1} A_m H^T_{\underline{a}\ (n_1+\cdots+n_m+1,L)}(n_1,\cdots,n_m+1)$ *and*
$H^T_{\underline{a}\ (n_1+\cdots+n_m+1,L)}(n_1,\cdots,n_m+1) := [I_{\underline{a}}(u_1), \cdots, S_l^{n_1-1} I_{\underline{a}}(u_1), I_{\underline{a}}(u_2), \cdots,$
$S_l^{n_2-1} I_{\underline{a}}(u_2), \cdots, I_{\underline{a}}(u_m), \cdots, S_l^{n_m-1} I_{\underline{a}}(u_m), S_l^{n_m} I_{\underline{a}}(u_m),]$.
③ *Let* $h_{ms} \in \boldsymbol{R}^{1\times n_m}$ *be* $h_{ms} =$
$[(I_{\underline{a}}(u_m))(0) - (\bar{I}_{\underline{a}}(u_m))(0), (S_l I_{\underline{a}}(u_m))(0) - (S_l \bar{I}_{\underline{a}}(u_m))(0),$
$\cdots, (S_l^{n_m-1} I_{\underline{a}}(u_m))(0) - (S_l^{n_m-1} \bar{I}_{\underline{a}}(u_m))(0)]$.

*2*m+1) Let* $g_s \in F(U, \boldsymbol{R}$ *be* $g_s(u_1) := \mathbf{e}_1,\ g_s(u_2) := \mathbf{e}_{n_1+1},\ \cdots,$
$g_s(u_m) := \mathbf{e}_{n_1+\cdots+n_{m-1}+1}$.
Let $F_s \in \boldsymbol{R}^{n\times n}$ *be the same as in Theorem (6.19).*
Let $h_s \in \boldsymbol{R}^{1\times n}$ *be* $h_s := [h_{1s}, h_{2s}, \cdots, h_{ms}]$,
where $n := n_1 + n_2 + \cdots + n_m$.

[proof] By 1) and 3), the reduction part in the data can be excluded in the sense of the number of dimensions by using the ratio of the matrix norm, which produces a degree of information loss. The matrices A_1 in 2), A_2 in 4), A_3 in 6), $\cdots$ and A_m in 2*m) correspond to the matrix A in Lemma (2.17). Hence, the reduced part of the given finite-sized Input/output matrix were obtained. Therefore, applying Theorem (6.19), we can obtain g_s, F_s and h_s by 2*m+1).

For the real time standard system $\sigma_s = ((\boldsymbol{R}^n, F_s), g_s^0, g_s, h_s, h_s^0)$, its modified impulse responses $I(u_1)(i) := h_s F_s^i g_s(u_1)$, $I(u_2)(i) := h_s F_s^i g_s(u_2)$ and $I(u_3)(i) := h_s F_s^i g_s(u_3)$ are written by $I(1)\_(n_1, n_2, n_3)$ and $I(2)\_(n_1, n_2, n_3)$ and $I(3)\_(n_1, n_2, n_3)$ respectively.

Example 6.21. Let the signals be the modified impulse responses of the following 3-dimensional pseudo linear system: $\sigma = ((\boldsymbol{R}^3, F), g, h, h^0)$, where

$$F = \begin{bmatrix} 0.9 & 0.3 & 0.6 \\ 0 & -0.9 & 0 \\ 0 & 0 & -0.3 \end{bmatrix}, \ g(u_1) = \mathbf{e}_1,\ g(u_2) = \mathbf{e}_2,\ g(u_3) = \mathbf{e}_3,$$

$h = [17,\ -4,\ 7]$, $h^0 = 1$.

Then the algebraically approximate realization problem is solved as follows:

covariance matrix	eigenvalues			
	1	2	3	4
$H^T_{\underline{a}\ (3,50)}(3,0,0)H_{\underline{a}\ (3,50)}(3,0,0)$	3751	0	0	
$H^T_{\underline{a}\ (3,50)}(1,2,0)H_{\underline{a}\ (3,50)}(1,2,0)$	1599	439	0	
$H^T_{\underline{a}\ (4,50)}(1,1,2)H_{\underline{a}\ (4,50)}(1,1,2)$	2222	240	1.5	0
covariance matrix	square root of eigenvalues			
$H^T_{\underline{a}\ (3,50)}(3,0,0)H_{\underline{a}\ (3,50)}(3,0,0)$	61.2	0	0	
$H^T_{\underline{a}\ (3,50)}(1,2,0)H_{\underline{a}\ (3,50)}(1,2,0)$	40	21	0	
$H^T_{\underline{a}\ (4,50)}(1,1,2)H_{\underline{a}\ (4,50)}(1,1,2)$	47.1	15.5	1.2	0

1) Since the ratio $\frac{1.2}{47.1} = 0.03$ obtained by the square root of $H^T_{\underline{a}\ (4,50)}(1,1,2)H_{\underline{a}\ (4,50)}(1,1,2)$ is not so large, the approximate pseudo linear system obtained by the algebraic CLS method may not be so good.
2) After determining the numbers n_1, n_2 and n_3 of dimensions which are 1, 1 and 0, we will continue the algebraically approximate realization algorithm.

Therefore, the modified impulse responses $I(1)$, $I(2)$ and $I(3)$ of an approximate pseudo linear system obtained by the algebraic CLS method is constructed for a 2-dimensional space.

The 2-dimensional pseudo linear system $\sigma_2 = ((\boldsymbol{R}^2,\ F_2),\ g_2,\ h_2)$ obtained by the algebraic CLS method can be expressed as follows:

$$F_2 = \begin{bmatrix} 0.9 & 0.3 \\ 0 & -0.9 \end{bmatrix}, \ h_2 = [17,\ -4],\ g_2(u_1) = [1,\ 0]^T, g_2(u_2) = [0,\ 1]^T,$$

$g_2(u_3) = [0.48,\ 0.05]^T$, $h^0 = 1$.

The 3-dimensional pseudo linear system $\sigma_3 = ((\boldsymbol{R}^3,\ F_3),\ g_3,\ h_3)$ obtained by the algebraic CLS method can be expressed as follows:

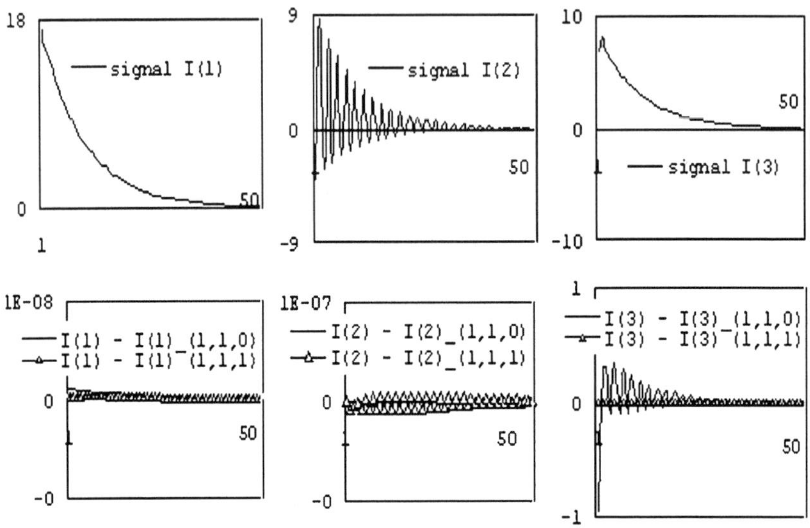

Fig. 6.1 In Example (6.21), the left are the original signal $I(1)$ and the difference between it and the approximate signal $I(1)\_(1,1,0)$ or $I(1)\_(1,1,1)$. The middle are the original signal $I(2)$ and the difference between it and the approximate signal $I(2)\_(1,1,0)$ or $I(2)\_(1,1,1)$. The right are the original signal $I(3)$ and the difference between it and the approximate signal $I(3)\_(1,1,0)$ or $I(3)\_(1,1,1)$.

$$F_3 = \begin{bmatrix} 0.9 & 0.3 & 0.6 \\ 0 & -0.9 & 0 \\ 0 & 0 & -0.3 \end{bmatrix}, \; h_3 = [17,\ -4,\ 7],\; g_3(u_1) = [1,\ 0,\ 0]^T,$$

$$g_3(u_2) = [0,\ 1,\ 0]^T,\; g_3(u_3) = [0,\ 0,\ 1]^T,\; h^0 = 1.$$

We can show that the algorithm for approximate realization given by the analytic CLS method in the reference [Hasegawa, 2008] produces the same systems as the above ones in the sense of the numerical calculation.

In this example, the original signals are considered as the modified impulse responses of a 3-dimensional pseudo linear system and the desirable modified impulse responses are obtained by the algebraic CLS method within our expectations. The model obtained by the algebraic CLS method is a 2-dimensional pseudo linear system.

For reference, a 3-dimensional pseudo linear system is also given by the algebraic CLS method. The system completely reconstructs the original system.

Just as we thought, the following table and Fig. 6.1 truly indicate that the 2-dimensional pseudo linear system obtained by the algebraic CLS method is a somewhat good approximation. For reference, the modified impulse responses of the same dimensional pseudo linear system as the original system are also shown.

dimen-sion	ratio of matrices	mean values of square root for sum of signal	signal by CLS	error	cosine ① and ②	error ratio
		①	②	③	$\cos\theta$	③/①
I(1)_(1,1,0)	0	0.78	0.78	0	1	0
I(2)_(1,1,0)	0	0.3265	0.3265	0	1	0
I(3)_(1,1,0)	0.03	0.381	0.38	0.02	0.998	0.06
I(1)_(1,1,1)	0	0.78	0.78	0	1	0
I(2)_(1,1,1)	0	0.3265	0.3265	0	1	0
I(3)_(1,1,1)	0	0.381	0.381	0	1	0

Example 6.22. Let the signals be the modified impulse responses of the following 4-dimensional pseudo linear system: $\sigma = ((\boldsymbol{R}^4, F), g, h, h^0)$, where

$$F = \begin{bmatrix} 0.8 & 0.2 & 0 & 0 \\ 0 & 0.6 & 0 & 0.3 \\ 0 & 0 & -0.7 & 0.2 \\ 0 & 0 & 0 & -0.8 \end{bmatrix}, \; g(u_1) = \mathbf{e}_1, \;\; g(u_2) = \mathbf{e}_3, \;\; g(u_3) = \mathbf{e}_4,$$

$h = [16, \; 5, \; -15, \; 2], \; h^0 = 1.$

Then the algebraically approximate realization problem is solved as follows:

covariance matrix	eigenvalues				
	1	2	3	4	5
$H^T_{\underline{a}\,(2,50)}(2,0,0)H_{\underline{a}\,(2,50)}(2,0,0)$	1166	0			
$H^T_{\underline{a}\,(3,50)}(1,2,0)H_{\underline{a}\,(3,50)}(1,2,0)$	874	495	0		
$H^T_{\underline{a}\,(4,50)}(1,1,2)H_{\underline{a}\,(4,50)}(1,1,2)$	852	615	131	1.1	
$H^T_{\underline{a}\,(5,50)}(1,1,3)H_{\underline{a}\,(5,50)}(1,1,3)$	1021	654	135	1.1	0
covariance matrix	square root of eigenvalues				
$H^T_{\underline{a}\,(4,50)}(1,1,2)H_{\underline{a}\,(4,50)}(1,1,2)$	29.2	24.8	11.4	1.04	
$H^T_{\underline{a}\,(5,50)}(1,1,3)H_{\underline{a}\,(5,50)}(1,1,3)$	32	25.6	11.6	1.04	0

1) Since the ratio $\frac{1.04}{29.2} = 0.04$ obtained by the square root of $H^T_{\underline{a}\,(4,50)}(1,1,2)H_{\underline{a}\,(4,50)}(1,1,2)$ is not so large, the approximate pseudo linear system obtained by the algebraic CLS method may not be so good. The reason is likely to be caused by rapid damping in Fig. 6.2.
2) After determining the numbers n_1, n_2 and n_3 of dimensions which are 1, 1 and 1, we will continue the algebraically approximate realization algorithm.

Therefore, the modified impulse responses $I(1)$, $I(2)$ and $I(3)$ of an approximate pseudo linear system obtained by the algebraic CLS method is constructed for a 3-dimensional space.

The 3-dimensional pseudo linear system $\sigma_2 = ((\boldsymbol{R}^3, \; F_2), \; g_2, \; h_2, \; h^0)$ obtained by the algebraic CLS method can be expressed as follows:

$$F_2 = \begin{bmatrix} 0.8 & 0 & 0.16 \\ 0 & -0.7 & 0.25 \\ 0 & 0 & -0.76 \end{bmatrix}, \; h_2 = [16, \; -15, \; 2.2], \; g_2(u_1) = [1, \; 0, \; 0]^T,$$

$g_2(u_2) = [0, \; 1, \; 0]^T, \; g_2(u_3) = [0, \; 0, \; 1]^T, \; h^0 = 1.$

For reference, a 4-dimensional pseudo linear system $\sigma_3 = ((\boldsymbol{R}^4, F_3), g_3, h_3, h^0)$ obtained by the algebraic CLS method can be expressed as follows:

$$F_3 = \begin{bmatrix} 0.8 & 0 & 0 & 0.06 \\ 0 & -0.7 & 0 & -0.26 \\ 0 & 0 & 0 & 0.48 \\ 0 & 0 & 1 & -0.2 \end{bmatrix},\ h_3 = [16,\ -15,\ 2,\ -3.1],\ g_3(u_1) = [1,\ 0,\ 0,\ 0]^T,$$

$$g_3(u_2) = [0,\ 1,\ 0,\ 0]^T,\ g_3(u_3) = [0,\ 0,\ 1,\ 0]^T,\ h^0 = 1.$$

We can show that the algorithm for approximate realization given by the analytic CLS method in the reference [Hasegawa, 2008] produces the same systems as the above ones in the sense of the numerical calculation.

In this example, the original signals are considered as the modified impulse responses of a 4-dimensional pseudo linear system and the desirable modified impulse responses are obtained by the algebraic CLS method within our expectations. The model obtained by the algebraic CLS method is a 3-dimensional pseudo linear system.

For reference, a 4-dimensional pseudo linear system is also given by the algebraic CLS method. The system completely reconstructs the original system.

Just as we thought, the following table and Fig. 6.2 truly indicate that the 3-dimensional pseudo linear system obtained by the algebraic CLS method is

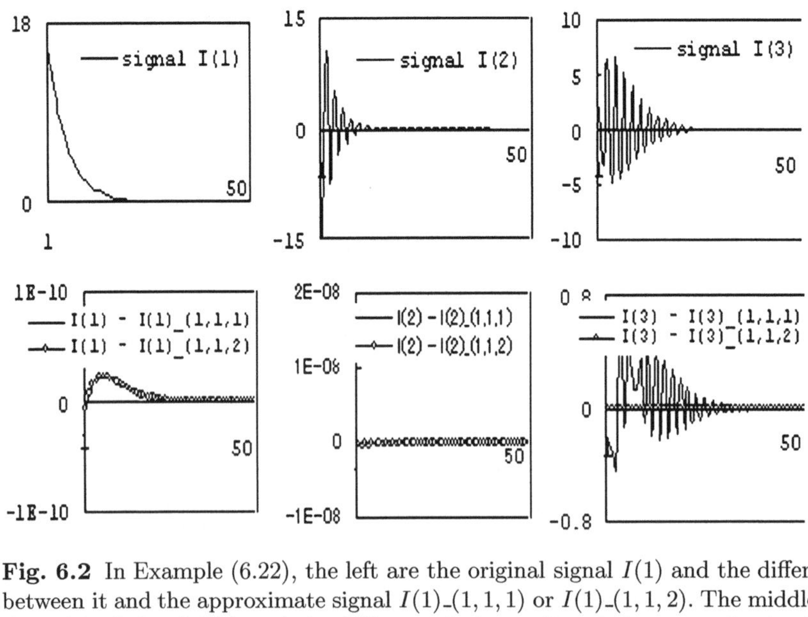

Fig. 6.2 In Example (6.22), the left are the original signal $I(1)$ and the difference between it and the approximate signal $I(1)\_(1,1,1)$ or $I(1)\_(1,1,2)$. The middle are the original signal $I(2)$ and the difference between it and the approximate signal $I(2)\_(1,1,1)$ or $I(2)\_(1,1,2)$. The right are the original signal $I(3)$ and the difference between it and the approximate signal $I(3)\_(1,1,1)$ or $I(3)\_(1,1,2)$.

not such a good approximation. For reference, the modified impulse responses of the same dimensional pseudo linear system as the original system are also shown.

dimen-sion	ratio of matrices	mean values of square root for sum of signal ①	signal by CLS ②	error ③	cosine ① and ② $\cos\theta$	error ratio ③/①
I(1)_(1,1,1)	0	0.533	0.533	0	1	0
I(2)_(1,1,1)	0	0.42	0.42	0	1	0
I(3)_(1,1,1)	0.05	0.30	0.304	0.027	0.996	0.09
I(1)_(1,1,2)	0	0.533	0.533	0	1	0
I(2)_(1,1,2)	0	0.42	0.42	0	1	0
I(3)_(1,1,2)	0	0.30	0.30	0	1	0

Example 6.23. Let the signals be the modified impulse responses of the following 5-dimensional pseudo linear system: $\sigma = ((\boldsymbol{R}^5, F), g, h, h^0)$, where

$$F = \begin{bmatrix} 0 & 0.4 & 0 & 0 & 0 \\ 0.9 & 0.6 & 0 & 0 & 0.2 \\ 0 & 0 & 0 & -0.7 & 0.1 \\ 0 & 0 & 1 & -0.7 & 0.7 \\ 0 & 0 & 0 & 0 & 0.8 \end{bmatrix}, \ g(u_1) = \mathbf{e}_1, \ g(u_2) = \mathbf{e}_3, \ g(u_3) = \mathbf{e}_5,$$

$h = [16, \ -1, \ -15, \ 5, \ -1]$, $h^0 = 1$.

Then the algebraically approximate realization problem is solved as follows:

covariance matrix	— eigenvalues 1	2	3	4	5	6
$H^T_{\underline{a}\ (3,50)}(3,0,0)H_{\underline{a}\ (3,50)}(3,0,0)$	774	136	0			
$H^T_{\underline{a}\ (5,50)}(2,3,0)H_{\underline{a}\ (5,50)}(2,3,0)$	806	409	271	58	0	
$H^T_{\underline{a}\ (6,50)}(2,2,2)H_{\underline{a}\ (6,50)}(2,2,2)$	1061	638	246	58	3.2	0
covariance matrix	square root of eigenvalues					
$H^T_{\underline{a}\ (6,50)}(2,2,2)H_{\underline{a}\ (6,50)}(2,2,2)$	32.6	25.3	15.7	7.6	1.8	0

1) Since the ratio $\frac{1.8}{32.6} = 0.06$ obtained by the square root of $H^T_{\underline{a}\ (6,50)}(2,2,2)H_{\underline{a}\ (6,50)}(2,2,2)$ is not so small, the approximate pseudo linear system obtained by the algebraic CLS method may not be good.
2) After determining the numbers n_1, n_2 and n_3 of dimensions which are 2, 2 and 0, we will continue the algebraically approximate realization algorithm.

Therefore, the modified impulse responses $I(1)$, $I(2)$ and $I(3)$ of an approximate pseudo linear system obtained by the algebraic CLS method is constructed for a 4-dimensional space.

The 4-dimensional pseudo linear system $\sigma_2 = ((\boldsymbol{R}^4, \ F_2), \ g_2, \ h_2, h^0)$ obtained by the algebraic CLS method can be expressed as follows:

$$F_2 = \begin{bmatrix} 0 & 0.36 & 0 & 0 \\ 1 & 0.6 & 0 & 0 \\ 0 & 0 & 0 & -0.7 \\ 0 & 0 & 1 & -0.7 \end{bmatrix},\ h_2 = [16,\ -1,\ -15,\ 5.05],\ g_2(u_1) = [1,\ 0,\ 0,\ 0]^T,$$

$$g_2(u_2) = [0,\ 0,\ 1,\ 0]^T,\ g_2(u_3) = [0.26,\ 0.9,\ 0.13,\ -0.42]^T,\ h^0 = 1.$$

For reference, a 5-dimensional pseudo linear system σ_3=$((\boldsymbol{R}^5,\ F_3),\ g_3,\ h_3, h^0)$ obtained by the algebraic CLS method can be expressed as follows:

$$F_3 = \begin{bmatrix} 0 & 0.36 & 0 & 0 & 0 \\ 1 & 0.6 & 0 & 0 & 0.2 \\ 0 & 0 & 0 & -0.7 & 0.1 \\ 0 & 0 & 1 & -0.7 & 0.7 \\ 0 & 0 & 0 & 0 & 0.8 \end{bmatrix},\ h_3 = [16, -0.9, -15, 5, -1],$$

$$g_3(u_1) = [1, 0, 0, 0, 0]^T, g_3(u_2) = [0, 0, 1, 0, 0]^T, g_3(u_3) = [0, 0, 0, 0, 1]^T, h^0 = 1.$$

We can show that the algorithm for approximate realization given by the analytic CLS method in the reference [Hasegawa, 2008] produces the same systems as the above ones in the sense of the numerical calculation.

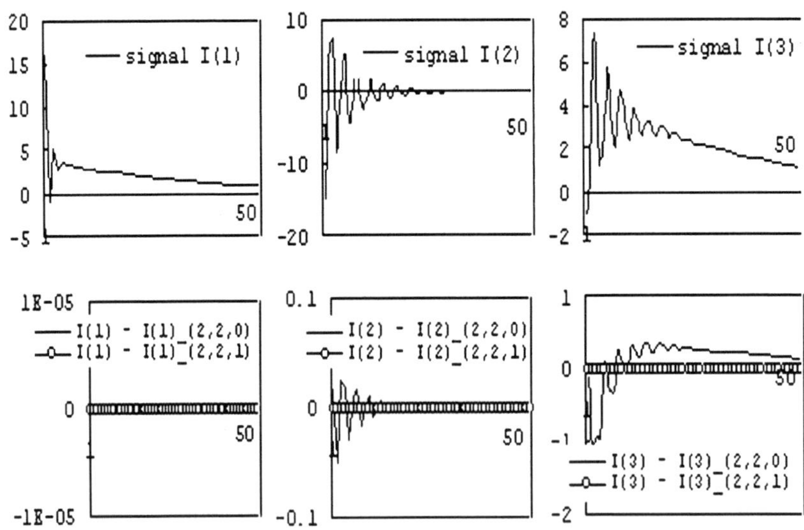

Fig. 6.3 In Example (6.23), the left are the original signal $I(1)$ and the difference between it and the approximate signal $I(1)\_(2, 2, 0)$ or $I(1)\_(2, 2, 1)$. The middle are the original signal $I(2)$ and the difference between it and the approximate signal $I(2)\_(2, 2, 0)$ or $I(2)\_(2, 2, 1)$. The right are the original signal $I(3)$ and the difference between it and the approximate signal $I(3)\_(2, 2, 0)$ or $I(3)\_(2, 2, 1)$.

In this example, the original signals are considered as the modified impulse responses of a 5-dimensional pseudo linear system and the desirable modified impulse responses are obtained by the algebraic CLS method within our expectations. The model obtained by the algebraic CLS method is a 4-dimensional pseudo linear system.

For reference, a 5-dimensional pseudo linear system is also given by the algebraic CLS method. The system completely reconstructs the original system.

Just as we expected, the following table and Fig. 6.3 truly indicate that the 4-dimensional pseudo linear system obtained by the algebraic CLS method is not a good approximation. For reference, the modified impulse responses of the same dimensional pseudo linear system as the original system are also shown.

dimension	ratio of matrices	mean values of square root for sum of signal	signal by CLS	error	cosine ① and ②	error ratio
		①	②	③	$\cos\theta$	③/①
I(1)_(2,2,0)	0	0.435	0.430	0.007	0.9999	0.02
I(2)_(2,2,0)	0	0.420	0.421	0.001	0.9999	0.002
I(3)_(2,2,0)	0.0002	0.362	0.354	0.047	0.99	0.12
I(1)_(2,2,1)	0	0.435	0.435	0	1	0
I(2)_(2,2,1)	0	0.420	0.420	0	1	0
I(3)_(2,2,1)	0	0.362	0.362	0	1	0

Example 6.24. Let the signals be the modified impulse responses of the following 6-dimensional pseudo linear system $\sigma = ((\boldsymbol{R}^6, F), g, h, h^0)$, where

$$F = \begin{bmatrix} 0 & 0.2 & 0 & 0.2 & 0 & 0 \\ 1 & 0.7 & 0 & 0.3 & 0 & -0.3 \\ 0 & 0 & 0 & 0.1 & 0 & -0.1 \\ 0 & 0 & 1 & -0.5 & 0 & 0.4 \\ 0 & 0 & 0 & 0 & 0 & 0.3 \\ 0 & 0 & 0 & 0 & 1 & -0.6 \end{bmatrix}, \; g(u_1) = \mathbf{e}_1, \; g(u_2) = \mathbf{e}_3, \; g(u_3) = \mathbf{e}_5,$$

$$h = [16, \; -1, \; -14, \; 4, \; -8, \; 2], \; h^0 = 1.$$

Then the algebraically approximate realization problem is solved as follows:

covariance matrix	eigenvalues						
	1	2	3	4	5	6	7
$H_{\underline{a}\ (3,50)}(3,0,0)H^T_{\underline{a}\ (3,50)}(3,0,0)$	293	36.7	0				
$H_{\underline{a}\ (5,50)}(2,3,0)H^T_{\underline{a}\ (5,50)}(2,3,0)$	492	54	0.6	0.06	0		
$H_{\underline{a}\ (6,50)}(2,2,2)H^T_{\underline{a}\ (6,50)}(2,2,2)$	563	51	9.1	0.7	0.15	0.001	
$H_{\underline{a}\ (7,50)}(2,2,3)H^T_{\underline{a}\ (7,50)}(2,2,3)$	563	52	13.7	0.8	0.2	0.001	0
covariance matrix	square root of eigenvalues						
$H^T_{\underline{a}\ (6,50)}(2,2,2)H_{\underline{a}\ (6,50)}(2,2,2)$	23.7	7.1	3	0.8	0.4	0.03	
$H^T_{\underline{a}\ (7,50)}(2,2,3)H_{\underline{a}\ (7,50)}(2,2,3)$	23.7	7.2	3.7	0.9	0.4	0.03	0

1) Since the ratios $\frac{0.4}{23.7} = 0.02$ and $\frac{0.03}{23.7} = 0.001$ obtained by the square root of $H^T_{\underline{a}\ (7,50)}(2,2,3)H_{\underline{a}\ (7,50)}(2,2,3)$ are small , the approximate pseudo linear system obtained by the algebraic CLS method may be good.

However, the ratio $\frac{0.4}{23.7} = 0.02$ of $H^T_{\underline{a}\ (7,50)}(2,2,3)H_{\underline{a}\ (7,50)}(2,2,3)$ comes from $H^T_{\underline{a}\ (5,50)}(2,3,0)H_{\underline{a}\ (5,50)}(2,3,0)$, hence the reduction from it is not good.

2) After determining the numbers n_1, n_2 and n_3 of dimensions which are 2, 2 and 1, we will continue the algebraically approximate realization algorithm.

Therefore, the modified impulse responses $I(1)$, $I(2)$ and $I(3)$ of an approximate pseudo linear system obtained by the algebraic CLS method is constructed for a 5-dimensional space.

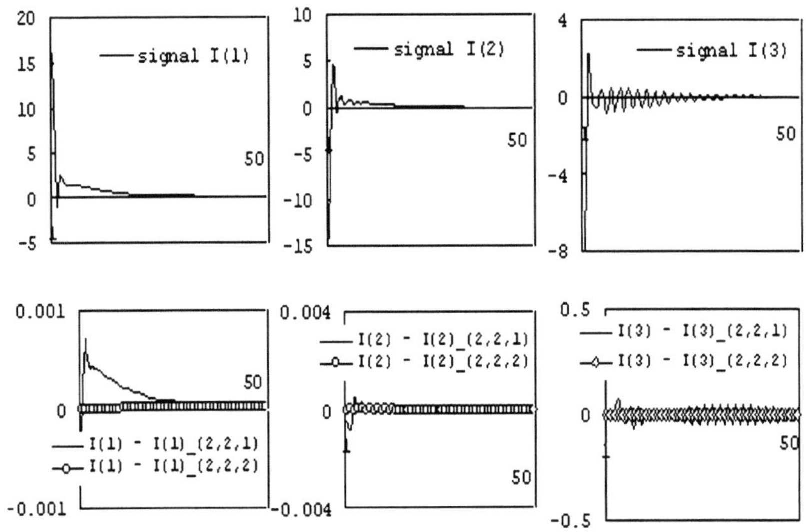

Fig. 6.4 In Example (6.24), the left are the original signal $I(1)$ and the difference between it and the approximate signal $I(1)\_(2,2,1)$ or $I(1)\_(2,2,2)$. The middle are the original signal $I(2)$ and the difference between it and the approximate signal $I(2)\_(2,2,1)$ or $I(2)\_(2,2,2)$. The right are the original signal $I(3)$ and the difference between it and the approximate signal $I(3)\_(2,2,1)$ or $I(3)\_(2,2,2)$.

The 5-dimensional pseudo linear system $\sigma_3 = ((\boldsymbol{R}^5,\ F_3),\ g_3,\ h_3, h^0)$ obtained by the algebraic CLS method can be expressed as follows:

$$F_3 = \begin{bmatrix} 0 & 0.2 & 0 & 0.2 & -0.69 \\ 1 & 0.7 & 0 & 0.3 & 1.68 \\ 0 & 0 & 0 & 0.1 & -1.07 \\ 0 & 0 & 1 & -0.5 & -1.95 \\ 0 & 0 & 0 & 0 & -0.94 \end{bmatrix},\ h_3 = [16, -1, -14, 4, -8],$$

$$g_3(u_1) = [1,0,0,0,0]^T, g_3(u_2) = [0,0,1,0,0]^T, g_3(u_3) = [0,0,0,0,1]^T,\ h^0 = 1.$$

For reference, a 6-dimensional pseudo linear system $\sigma_4 = ((\boldsymbol{R}^6, F_4), g_4, h_4, h^0)$ obtained by the algebraic CLS method can be expressed as follows:

$$F_4 = \begin{bmatrix} 0 & 0.2 & 0 & 0.2 & 0 & 0 \\ 1 & 0.7 & 0 & 0.3 & 0 & -0.3 \\ 0 & 0 & 0 & 0.1 & 0 & -0.1 \\ 0 & 0 & 1 & -0.5 & 0 & 0.4 \\ 0 & 0 & 0 & 0 & 0 & 0.3 \\ 0 & 0 & 0 & 0 & 1 & -0.6 \end{bmatrix}, \; h_4 = [16,\ -1,\ -14,\ 4,\ -8,\ 2],$$

$$g_4(u_1) = [1,\ 0,\ 0,\ 0,\ 0,\ 0]^T, g_4(u_2) = [0,\ 0,\ 1,\ 0,\ 0,\ 0]^T,$$
$$g_4(u_3) = [0,\ 0,\ 0,\ 0,\ 1,\ 0]^T,\ h^0 = 1.$$

We can show that the algorithm for approximate realization given by the analytic CLS method in the reference [Hasegawa, 2008] produces the same systems as the above ones in the sense of the numerical calculation.

In this example, the original signals are considered as the modified impulse responses of a 6-dimensional pseudo linear system and the desirable modified impulse responses are obtained by the algebraic CLS method within our expectations. The model obtained by the algebraic CLS method is a 5-dimensional pseudo linear system.

For reference, a 6-dimensional pseudo linear system is also given by the algebraic CLS method. The system completely reconstructs the original system.

Just as we expected, the following table and Fig. 6.4 truly indicate that the 5-dimensional pseudo linear system obtained by the algebraic CLS method is a good approximation. For reference, the modified impulse responses of the same dimensional pseudo linear system as the original system are also shown. Hence, there exists a good approximation for the given system.

dimen-sion	ratio of matrices	mean values of square root for sum of signal	signal by CLS	error	cosine ① and ②	error ratio
		①	②	③	$\cos\theta$	③/①
I(1)_(2,2,1)	0	0.3353	0.3353	0	1	0
I(2)_(2,2,1)	0	0.29393	0.29393	0	1	0
I(3)_(2,2,1)	0.003	0.171	0.1711	0.006	0.999	0.02
I(1)_(2,2,2)	0	0.3353	0.3353	0	1	0
I(2)_(2,2,2)	0	0.29393	0.29393	0	1	0
I(3)_(2,2,2)	0	0.171	0.171	0	1	0

6.6 Algebraically Noisy Realization of Pseudo Linear Systems

In this section, we discuss the algebraically noisy realization problem of pseudo linear systems.

For noise $\{\bar{\gamma}(t) : t \in N\}$ added to the unknown pseudo linear system a, we will obtain the observed data $\{\hat{\gamma}(|\omega|) + \bar{\gamma}(|\omega|) : \omega \in U^*\}$.

For any given $\{\hat{\gamma}(|\omega|)+\bar{\gamma}(|\omega|) : \omega \in U^*\}$, σ which satisfies $a_\sigma(\omega) \approx \hat{\gamma}(|\omega|) : \omega \in U^*$ is called a noisy realization of a.

Roughly speaking, we can propose the following algebraically noisy realization problem:

For any given $\{\hat{\gamma}(|\omega|) + \bar{\gamma}(|\omega|) : \omega \in U^*\}$, find, using only algebraic calculations, a pseudo linear system σ which satisfies $a_\sigma(\omega) \approx \hat{\gamma}(|\omega|)$ for any $\omega \in U^*$. In order to make our discussion simple, we assume that the set Y of output is the set $\boldsymbol{R}$ of real numbers, namely 1-output.

A situation for algebraically noisy realization problem 6.25

Let the observed object be a pseudo linear system and noise be added to output. Then we will obtain the data $\{\gamma(t) = \hat{\gamma}(t) + \bar{\gamma}(t) : 0 \leq t \leq \underline{N}\}$ for some integer $\underline{N} \in N$, where $\hat{\gamma}(t)$ is the exact signal which comes from the observed pseudo linear system and $\bar{\gamma}(t)$ is the noise added at the time of observation.

Problem 6.26. Problem statement of an algebraically noisy realization for Pseudo Linear Systems.

Let $H_{\underline{a}\ (p,\bar{p})}$ be the measured finite-sized Input/output matrix. Then find, using only algebraic calculations, the cleaned-up Input/output matrix $\hat{H}_{\underline{a}\ (p,\bar{p})}$ such that $H_{\underline{a}\ (p,\bar{p})} = \hat{H}_{\underline{a}\ (p,\bar{p})} + \bar{H}_{\underline{a}\ (p,\bar{p})}$ holds.

Namely, find, using only algebraic calculations, a minimal dimensional pseudo linear system $\sigma = ((\boldsymbol{R}^n, F_r), g_r, h_r, h^0))$ which realizes $\hat{H}_{\underline{a}\ (p,\bar{p})}$.

Theorem 6.27. *Algebraic algorithm for noisy realization*

Let a partial input response map $\underline{a}$ be a considered object which is a pseudo linear system. Then a noisy realization $\sigma = ((\boldsymbol{R}^n, F_s), g_s, h_s, h^0)$ of $\underline{a}$ is given by the following algorithm:

1) *Based on the square root of eigenvalues for a matrix $H_{\underline{a}\ (p,\bar{p})}(p,0,0)H_{\underline{a}\ (p,\bar{p})}(p,0,0)^T$, determine the value n_1 of rank for the matrix $H_{\underline{a}\ (p,\bar{p})}(p,0,0)$, where $n_1 \leq p$.*
Namely, determine the value n_1 of rank for the matrix $H_{\underline{a}\ (p,\bar{p})}(p,0,0)$ such that a set of the square root of eigenvalues for the covariance matrix composed of relatively small and equally-sized numbers is excluded, where the signal part effected by the set may be the noisy part.
2) *The algebraic CLS method is used as follows:*
① Based on Proposition (2.14), determine coefficients $\{\alpha_{1i} : 1 \leq i \leq n_1\}$. The Q in Proposition (2.14) can be considered as the matrix composed from the eigenvectors of $H_{\underline{a}\ (n_1+1,L)}(n_1+1,0)H^T_{\underline{a}\ (n_1+1,L)}(n_1+1,0)$.
Let a matrix $A_1 \in \boldsymbol{R}^{1\times(n_1+1)}$ be $A_1 = [\alpha_{11}, \alpha_{12}, \cdots, \alpha_{1n_1}, -1]$.
② Determine the error vectors $\{\underline{S_l^i}\bar{I}_{\underline{a}}(u_1) \in \boldsymbol{R}^{L\times 1} : 0 \leq i \leq n_1\}$ by using the equation $[\bar{I}_{\underline{a}}(u_1), \underline{S_l}\bar{I}_{\underline{a}}(u_1), \cdots, \underline{S_l^{n_1}}\bar{I}_{\underline{a}}(u_1)]^T :=$ $A_1^T[A_1A_1^T]^{-1}A_1H^T_{\underline{a}\ (n_1+1,L)}(n_1+1,0)$ and $H^T_{\underline{a}\ (n_1,L)}(n_1,0,0) :=$

$[I_{\underline{a}}(u_1), \cdots, S_l^{n_1-1} I_{\underline{a}}(u_1), S_l^{n_1} I_{\underline{a}}(u_1)]$.

③ *Let* h_{1s} *be* $h_{1s} = [I_{\underline{a}}(1) - \bar{I}_{\underline{a}}(1), I_{\underline{a}}(2) - \bar{I}_{\underline{a}}(2), \cdots, I_{\underline{a}}(n_1) - \bar{I}_{\underline{a}}(n_1)]$.

3) Based on the square root of eigenvalues for a matrix $H_{\underline{a}\ (n_1+p,\bar{p})}(n_1, p, 0) H_{\underline{a}\ (n_1+p,\bar{p})}(n_1, p, 0)^T$*, determine the value* n_2 *of rank for the matrix* $H_{\underline{a}\ (n_1+p,\bar{p})}(n_1, p, 0)$*, where* $n_2 \leq p$.
Namely, determine the value n_2 *of rank for the matrix* $H_{\underline{a}\ (p,\bar{p})}(n_1, p, 0)$ *such that a set of the square root of eigenvalues for the covariance matrix composed of relatively small and equally-sized numbers is excluded, where the signal part effected by the set may be the noisy part.*

4) The algebraic CLS method is used as follows:
① *Based on Proposition (2.14), determine coefficients* $\{\alpha_{2i} : 1 \leq i \leq n_1 + n_2\}$.
The Q in Proposition (2.14) can be considered as the matrix composed from the eigenvectors of $H^T_{\underline{a}\ (n_1+n_2+1,L)}(n_1, n_2+1, 0) H^T_{\underline{a}\ (n_1+n_2+1,L)}(n_1, n_2+1, 0)$.
Let a matrix $A_2 \in \boldsymbol{R}^{1\times(n_1+n_2+1)}$ *be* $A_2 = [\alpha_{21}, \alpha_{22}, \cdots, \alpha_{2n_1+n_2}, -1]$.
② *Determine the error vectors* $\{\underline{S}_l^{i_j} \bar{I}_{\underline{a}}(u_j) \in \boldsymbol{R}^{L\times 1} : 0 \leq i_1 \leq n_1 - 1,\ 0 \leq i_2 \leq n_2, 1 \leq j \leq 2\}$ *by using the equation*
$[\bar{I}_{\underline{a}}(u_1), \cdots, \underline{S}_l^{n_1-1} \bar{I}_{\underline{a}}(u_1), \bar{I}_{\underline{a}}(u_2), \cdots, \underline{S}_l^{n_2-1} \bar{I}_{\underline{a}}(u_2), \underline{S}_l^{n_2} \bar{I}_{\underline{a}}(u_2)]^T := A_2^T [A_2 A_2^T]^{-1} A_2 H^T_{\underline{a}\ (n_1+n_2+1,L)}(n_1, n_2+1, 0)$ *and*
$H^T_{\underline{a}\ (n_1+n_2;1,L)}(n_1, n_2+1, 0) :=$
$[I_{\underline{a}}(u_1), \cdots, S_l^{n_1-1} I_{\underline{a}}(u_1), I_{\underline{a}}(u_2), \cdots, S_l^{n_2-1} I_{\underline{a}}(u_2), S_l^{n_2} I_{\underline{a}}(u_2)]$.
③ *Let* h_{2s} *be* $h_{2s} = [(I_{\underline{a}}(u_2))(0) - (\bar{I}_{\underline{a}}(u_2))(0), (S_l I_{\underline{a}}(u_2))(0) - (S_l \bar{I}_{\underline{a}}(u_2))(0), \cdots, (S_l^{n_2-1} I_{\underline{a}}(u_2))(0) - (S_l^{n_2-1} \bar{I}_{\underline{a}}(u_2))(0)]$.

5) Based on the square root of eigenvalues for a matrix $H_{\underline{a}\ (n_1+n_2+p,\bar{p})}(n_1, n_2, q) H_{\underline{a}\ (n_1+n_2+q,\bar{q})}(n_1, n_2, q)^T$*, determine the value* n_3 *of rank for the matrix* $H_{\underline{a}\ (n_1+n_2+q,\bar{q})}(n_1, n_2, q)$*, where* $n_3 \leq q$.
Namely, determine the value n_3 *of rank for the matrix* $H_{\underline{a}\ (n_1+n_2+q,\bar{q})}(n_1, n_2, q)$ *such that a set of the square root of eigenvalues for the covariance matrix composed of relatively small and equally-sized numbers is excluded, where the signal part effected by the set may be the noisy part.*

6) The algebraic CLS method is used as follows:
① *Based on Proposition (2.14), determine coefficients* $\{\alpha_{3i} : 1 \leq i \leq n_1 + n_2 + n_3\}$.
The Q in Proposition (2.14) can be considered as the matrix composed from the eigenvectors of $H_{\underline{a}\ (n_1+n_2+n_3+1,L)}(n_1, n_2, n_3+1) \times H^T_{\underline{a}\ (n_1+n_2+n_3+1,L)}(n_1, n_2, n_3+1)$.
Let a matrix $A_3 \in \boldsymbol{R}^{1\times(n_1+n_2+n_3+1)}$ *be* $A_3 = [\alpha_{31}, \alpha_{32}, \cdots, \alpha_{3n_1+n_2+n_3}, -1]$.
② *Determine the error vectors* $\{\underline{S}_l^{i_j} \bar{I}_{\underline{a}}(u_j) \in \boldsymbol{R}^{L\times 1} : 0 \leq i_j \leq n_j - 1$ *for* $1 \leq j \leq 2,\ 0 \leq i_3 \leq n_3\}$ *by using the equation*
$[\bar{I}_{\underline{a}}(u_1), \cdots, \underline{S}_l^{n_1-1} \bar{I}_{\underline{a}}(u_1), \bar{I}_{\underline{a}}(u_2), \cdots, \underline{S}_l^{n_2-1} \bar{I}_{\underline{a}}(u_2), \underline{S}_l \bar{I}_{\underline{a}}(u_3), \cdots, \underline{S}_l^{n_3} \bar{I}_{\underline{a}}(u_3)]^T := A_3^T [A_3 A_3^T]^{-1} A_3 H^T_{\underline{a}\ (n_1+n_2+n_3+1,L)}(n_1, n_2, n_3+1)$ *and*
$H^T_{\underline{a}\ (n_1+n_2+n_3+1,L)}(n_1, n_2, n_3+1) := [I_{\underline{a}}(u_1), \cdots, S_l^{n_1-1} I_{\underline{a}}(u_1), I_{\underline{a}}(u_2), \cdots,$

$S_l^{n_2-1}I_{\underline{a}}(u_2), S_l^{n_2}I_{\underline{a}}(u_2), I_{\underline{a}}(u_3), \cdots, S_l^{n_3-1}I_{\underline{a}}(u_3), S_l^{n_3}I_{\underline{a}}(u_3),]$.

③ *Let* $h_{3s} \in \boldsymbol{R}^{1\times n_3}$ *be* $h_{3s} =$
$[(I_{\underline{a}}(u_3))(0) - (\bar{I}_{\underline{a}}(u_3))(0), (S_l I_{\underline{a}}(u_3))(0) - (S_l \bar{I}_{\underline{a}}(u_3))(0),$
$\cdots, (S_l^{n_3-1}I_{\underline{a}}(u_3))(0) - (S_l^{n_3-1}\bar{I}_{\underline{a}}(u_3))(0)]$.

⋮

*2*m-1) Based on the square root of eigenvalues for a matrix*
$H_{\underline{a}\ (n_1+\cdots+n_{m-1}+p,\bar{p})}(n_1,\cdots,n_{m-1},q)H_{\underline{a}\ (n_1+\cdots+n_{m-1}+q,\bar{q})}(n_1,\cdots,n_{m-1},q)^T$,
determine the value n_m *of rank for the matrix*
$H_{\underline{a}\ (n_1+\cdots+n_{m-1}+p,q)}(n_1,\cdots,n_{m-1},q)$, *where* $n_m \le q$.
Namely, determine the value n_3 *of rank for the matrix*
$H_{\underline{a}\ (n_1+\cdots,n_{m-1}+q,\bar{q})}(n_1,\cdots,n_{m-1},q)$ *such that a set of the square root of eigenvalues for the covariance matrix*
composed of relatively small and equally-sized numbers is excluded, where the signal part effected by the set may be the noisy part.

*2*m) The algebraic CLS method is used as follows:*

① *Based on Proposition (2.14), determine coefficients* $\{\alpha_{mi} : 1 \le i \le n_1 + \cdots + n_m\}$.
The Q in Proposition (2.14) can be considered as the matrix composed from the eigenvectors of $H^T_{\underline{a}\ (n_1+\cdots+n_m+1,L)}(n_1,\cdots,n_m+1)$
$\times H^T_{\underline{a}\ (n_1+\cdots+n_m+1,L)}(n_1,\cdots,n_m+1)$.
Let a matrix $A_m \in \boldsymbol{R}^{1\times(n_1+\cdots+n_m+1)}$ *be*
$A_m = [\alpha_{m1}, \alpha_{m2}, \cdots, \alpha_{mn_1+\cdots+n_m}, -1]$.

② *Determine the error vectors* $\{\underline{S}_l^{i_j}\bar{I}_{\underline{a}}(u_j) \in \boldsymbol{R}^{L\times 1} :$
$0 \le i_j \le n_j - 1$ *for* $1 \le j \le m-1,\ 0 \le i_m \le n_m\}$ *by using the equation*
$[\bar{I}_{\underline{a}}(u_1), \cdots, \underline{S}_l^{n_1-1}\bar{I}_{\underline{a}}(u_1), \bar{I}_{\underline{a}}(u_2), \cdots, \underline{S}_l\bar{I}_{\underline{a}}(u_m), \cdots, \underline{S}_l^{n_m}\bar{I}_{\underline{a}}(u_m)]^T :=$
$A_m^T[A_m A_m^T]^{-1}A_m H^T_{\underline{a}\ (n_1+\cdots+n_m+1,L)}(n_1,\cdots,n_m+1)$ *and*
$H^T_{\underline{a}\ (n_1+\cdots+n_m+1,L)}(n_1,\cdots,n_m+1) := [I_{\underline{a}}(u_1), \cdots, S_l^{n_1-1}I_{\underline{a}}(u_1), I_{\underline{a}}(u_2), \cdots,$
$S_l^{n_2-1}I_{\underline{a}}(u_2), \cdots, I_{\underline{a}}(u_m), \cdots, S_l^{n_m-1}I_{\underline{a}}(u_m), S_l^{n_m}I_{\underline{a}}(u_m),]$.

③ *Let* h_{ms} *be* $h_{ms} = [(I_{\underline{a}}(u_m))(0) - (\bar{I}_{\underline{a}}(u_m))(0), (S_l I_{\underline{a}}(u_m))(0) -$
$(S_l\bar{I}_{\underline{a}}(u_m))(0), \cdots, (S_l^{n_m-1}I_{\underline{a}}(u_m))(0) - (S_l^{n_m-1}\bar{I}_{\underline{a}}(u_m))(0)]$.

*2*m+1) Let* $g_s \in F(U, \boldsymbol{R})$ *be* $g_s(u_1) := \mathbf{e}_1$, $g_s(u_2) := \mathbf{e}_{n_1+1}$, $\cdots$,
$g_s(u_m) := \mathbf{e}_{n_1+\cdots+n_{m-1}+1}$.
Let $F_s \in \boldsymbol{R}^{n\times n}$ *be the same as in Theorem (6.19).*
Let $h_s \in \boldsymbol{R}^{1\times n}$ *be* $h_s := [h_{1s}, h_{2s}, \cdots, h_{ms}]$,
where $n := n_1 + n_2 + \cdots + n_m$.

For the real time standard system $\sigma_r = ((\boldsymbol{R}^n, F_r), g_r^0, g_r, h_r, h_r^0)$, *its modified impulse responses* $I(u_1)(i) := h_r F_r^i g_r(u_1)$, $I(u_2)(i) := h_r F_r^i g_r(u_2)$ *and* $I(u_3)(i) := h_r F_r^i g_r(u_3)$ *are written by* $I(1)\_(n_1, n_2, n_3)$ *and* $I(2)\_(n_1, n_2, n_3)$ *and* $I(3)\_(n_1, n_2, n_3)$ *respectively.*

[proof] By 1) and 3), the noisy part in the data can be excluded in the sense of the number of dimensions by checking what part is the noisy part and

by using the ratio of Input/output matrix norm, which implies the noise to signal ratio. The matrices A_1 in 2), A_2 in 4), A_3 in 6), $\cdots$ and A_m in 2*m) correspond to the matrix A in Lemma (2.17). Hence, the noisy part of the given finite-sized Input/output matrix is excluded. Therefore, applying Theorem (6.19), we can obtain g_s, F_s and h_s by 2*m+1).

Remark 1: A determination method of the degree of n in the linear system $\sigma = ((\boldsymbol{R}^n, F_s), g, h_s)$ can be found in the Principal Component Method. The method is popular.
Remark 2: Let S and N be the norm of a signal and noise. Then the selected ratio of matrices in the algorithm may be considered as $\frac{N}{S+N}$.
Remark 3: This algebraically noisy realization method is very new.
Remark 4: For the noisy case, the AIC method is famous for determining linear systems including dimensions of the state spaces.

Example 6.28. Let signals be the modified impulse responses of the following 3-dimensional pseudo linear system $\sigma = ((R^3, F), g, h, h^0)$,

$$\text{where } F = \begin{bmatrix} 0.9 & 0.3 & 0.4 \\ 0 & 0.2 & -0.5 \\ 0 & 0 & -0.7 \end{bmatrix}, \; h = [12,\ -8,\ -5],\ g(u_1) = [1,\ 0,\ 0]^T,$$

$g(u_2) = [0,\ 1,\ 0]^T$, $g(u_3) = [0,\ 0,\ 1]^T$, $h^0 = 1$.

Let added noises be given in Fig. 6.5.

Then the algebraically noisy realization problem is solved as follows:

covariance matrix	eigenvalues				
	1	2	3	4	5
$H_{\underline{a}\ (3,50)}(3,0,0)H^T_{\underline{a}\ (3,50)}(3,0,0)$	1968	11.7	3.3		
$H_{\underline{a}\ (4,50)}(4,0,0)H^T_{\underline{a}\ (4,50)}(4,0,0)$	2269	14.2	4.97	1.7	
$H_{\underline{a}\ (3,50)}(1,2,0)H^T_{\underline{a}\ (3,50)}(1,2,0)$	877	130	5.2		
$H_{\underline{a}\ (4,50)}(1,3,0)H^T_{\underline{a}\ (4,50)}(1,3,0)$	963	130	6.3	3.5	
$H_{\underline{a}\ (4,50)}(1,1,2)H^T_{\underline{a}\ (4,50)}(1,1,2)$	841	514	117	2.4	
$H_{\underline{a}\ (5,50)}(1,1,3)H^T_{\underline{a}\ (5,50)}(1,1,3)$	841	619	118	4.2	1.8
covariance matrix	square root of eigenvalues				
	1	2	3	4	5
$H_{\underline{a}\ (4,50)}(4,0,0)H^T_{\underline{a}\ (4,50)}(4,0,0)$	47.6	3.8	2.2	1.3	
$H_{\underline{a}\ (4,50)}(1,3,0)H^T_{\underline{a}\ (4,50)}(1,3,0)$	31	11.4	2.5	1.9	
$H_{\underline{a}\ (5,50)}(1,1,3)H^T_{\underline{a}\ (5,50)}(1,1,3)$	29	24.9	10.9	2	1.3

1) A set $\{3.8,\ 2.2,\ 1.3\}$ is composed of relatively small and equally-sized numbers in the square root of eigenvalues for $H_{\underline{a}\ (4,50)}(4,0.0)H^T_{\underline{a}\ (4,50)}(4,0,0)$.
2) After determining the number n_1 of dimensions which is 1, we will continue the algebraically noisy realization algorithm.

Therefore, the modified impulse response $I(1)$ of a pseudo linear system obtained by the algebraic CLS method is constructed for a 1-dimensional space.
3) A set $\{2.5,\ 1.9\}$ is composed of relatively small and equally-sized numbers in the square root of eigenvalues for $H_{\underline{a}\ (4,50)}(1,3,0)H^T_{\underline{a}\ (4,50)}(1,3,0)$.
4) After determining the number n_2 of dimensions which is 1, we execute the algebraically noisy realization algorithm.
5) A set $\{2,\ 1.3\}$ is composed of relatively small and equally-sized numbers in the square root of eigenvalues for $H_{\underline{a}\ (5,50)}(1,1,3)H^T_{\underline{a}\ (5,50)}(1,1,3)$.
6) After determining the number n_3 of dimensions which is 1, we execute the algebraically noisy realization algorithm.

Therefore, the modified impulse responses $I(1)$, $I(2)$ and $I(3)$ of an approximate pseudo linear system obtained by the algebraic CLS method is realized by a (1,1,1)-dimensional pseudo linear system.

The 3-dimensional pseudo linear system $\sigma_n = ((\boldsymbol{R}^3, F_n), g_n, h_n, h^0)$ obtained by the algebraic CLS method is expressed as follows:

$$F_n = \begin{bmatrix} 0.9 & 0.3 & 0.4 \\ 0 & 0.2 & -0.53 \\ 0 & 0 & -0.72 \end{bmatrix},\ g_n(u_1) = \mathbf{e}_1,\ g_n(u_2) = \mathbf{e}_2,\ g_n(u_3) = \mathbf{e}_3,$$

$h_n = [12.3, -7.9, -4.6]$ and $h^0 = 1$.

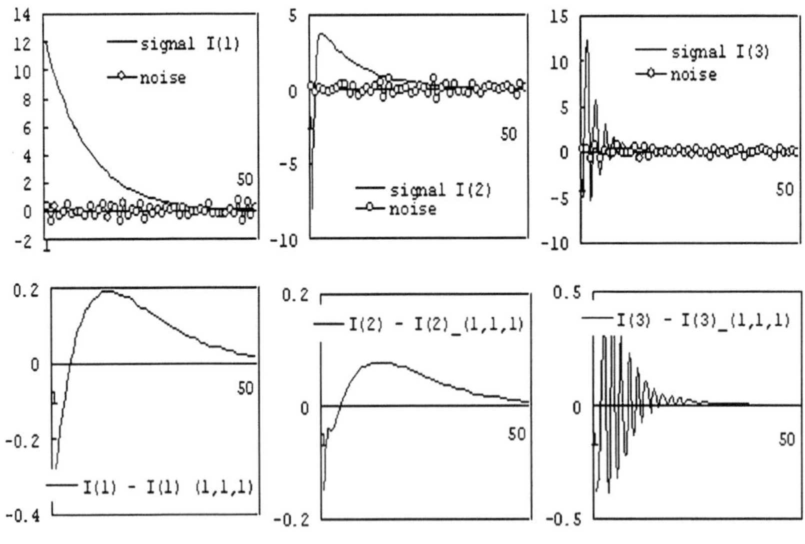

Fig. 6.5 In Example (6.28), the left are the original signal $I(1)$ with noise and the difference between $I(1)$ and the obtained signal $I(1)\_(1,1,1)$. The middle are the original signal $I(2)$ with noise and the difference between $I(2)$ and the obtained signal $I(2)\_(1,1,1)$. The right are the original signal $I(3)$ with noise and the difference between $I(3)$ and the obtained signal $I(3)\_(1,1,1)$.

The obtained modified impulse responses $I(1)$, $I(2)$ and $I(3)$ are illustrated in Fig. 6.5.

We can show that the algorithm for noisy realization given by the analytic CLS method in the reference [Hasegawa, 2008] produces the same system as the above one in the sense of the numerical calculation.

In this example, the original signals $I(1)$, $I(2)$ and $I(3)$ are characterized as the modified impulse responses of a 1-dimensional linear space respectively. The desirable modified impulse responses have been attempted to be obtained by the algebraic CLS method. The model obtained by the algebraic CLS method is a (1,1,1)-dimensional pseudo linear system which has the same number of dimensions as the number of the original system.

Just as we expected, the following table and Fig. 6.5 indicate that the model obtained by the algebraic CLS method is a good (1,1,1)-dimensional noisy realization system for the original (1,1,1)-dimensional system.

dimen-sion	ratio of matrices	mean values of square root for sum of signal	signal by CLS	error	cosine ① and ②	error ratio
		①	②	③	$\cos\theta$	③/①
I(1)_(1,1,1)	0.08	0.551	0.55	0.02	0.999	0.03
I(2)_(1,1,1)	0.08	0.248	0.244	0.007	0.999	0.03
I(3)_(1,1,1)	0.07	0.322	0.333	0.02	0.999	0.06

Example 6.29. Let signals be the modified impulse responses of the following 4-dimensional pseudo linear system $\sigma = ((R^4, F), g, h, h^0)$,

$$\text{where } F = \begin{bmatrix} 0 & 0.5 & 0 & 0 \\ 1 & 0.4 & 0 & 0.3 \\ 0 & 0 & -0.8 & 0.2 \\ 0 & 0 & 0 & 0.8 \end{bmatrix}, h = [12,\ -1,\ -15,\ 6],\ g(u_1) = [1,\ 0,\ 0,\ 0]^T,$$

$g(u_2) = [0,\ 0,\ 1,\ 0]^T,\ g(u_3) = [0,\ 0,\ 0,\ 1]^T,\ h^0 = 1.$

Let added noises be given in Fig. 6.6.

Then the algebraically noisy realization problem is solved by the following algorithm:

1) A set $\{2.4,\ 1.3,\ 0.9\}$ is composed of relatively small and equally-sized numbers in the square root of eigenvalues for $H_{\underline{a}\ (5,50)}(5, 0.0)H^T_{\underline{a}\ (5,50)}(5, 0, 0)$.

2) After determining the number n_1 of dimensions which is 2, we will continue the algebraically noisy realization algorithm.

Therefore, the modified impulse response $I(1)$ of a pseudo linear system obtained by the algebraic CLS method is constructed for a 2-dimensional space.

3) A set $\{1.4,\ 0.8\}$ is composed of relatively small and equally-sized numbers in the square root of eigenvalues for $H_{\underline{a}\ (5,50)}(2, 3, 0)H^T_{\underline{a}\ (5,50)}(2, 3, 0)$.

4) After determining the number n_2 of dimensions which is 1, we execute the algebraically noisy realization algorithm.

5) A set $\{2.4,\ 1.5,\ 1.2\}$ is composed of relatively small and equally-sized numbers in the square root of eigenvalues for $H_{\underline{a}\ (6,50)}(2, 1, 4)H^T_{\underline{a}\ (6,50)}(2, 1, 4)$.

6) After determining the number n_3 of dimensions which is 1, we execute the algebraically noisy realization algorithm. The noisy realization by the algebraic CLS method may not be so good. The reason is likely to be caused by rapid damping in Fig. 6.6.

covariance matrix	eigenvalues						
	1	2	3	4	5	6	7
$H_{\underline{a}\ (4,50)}(4,0,0)H^T_{\underline{a}\ (4,50)}(4,0,0)$	420	120	3.9	1.2			
$H_{\underline{a}\ (5,50)}(5,0,0)H^T_{\underline{a}\ (5,50)}(5,0,0)$	482	124	5.9	1.6	0.8		
$H_{\underline{a}\ (4,50)}(2,2,0)H^T_{\underline{a}\ (4,50)}(2,2,0)$	1149	204	27.9	1.1			
$H_{\underline{a}\ (5,50)}(2,3,0)H^T_{\underline{a}\ (5,50)}(2,3,0)$	1401	205	28.1	2.1	0.7		
$H_{\underline{a}\ (5,50)}(2,1,3)H^T_{\underline{a}\ (5,50)}(2,1,3)$	792	419	211	26	5.2	1.7	
$H_{\underline{a}\ (6,50)}(2,1,4)H^T_{\underline{a}\ (6,50)}(2,1,4)$	801	489	226	26	5.8	2.4	1.4
covariance matrix	square root of eigenvalues						
	1	2	3	4	5	6	7
$H_{\underline{a}\ (5,50)}(5,0,0)H^T_{\underline{a}\ (5,50)}(5,0,0)$	22	11.1	2.4	1.3	0.9		
$H_{\underline{a}\ (5,50)}(2,3,0)H^T_{\underline{a}\ (5,50)}(2,3,0)$	37.4	14.3	5.3	1.4	0.8		
$H_{\underline{a}\ (6,50)}(2,1,4)H^T_{\underline{a}\ (6,50)}(2,1,4)$	28.3	22.1	15	5.1	2.4	1.5	1.2

Therefore, the modified impulse responses $I(1)$, $I(2)$ and $I(3)$ of an approximate pseudo linear system obtained by the algebraic CLS method is realized by a (2,1,1)-dimensional pseudo linear system.

The 4-dimensional pseudo linear system $\sigma_n = ((\boldsymbol{R}^4,\ F_n),\ g_n,\ h_n,\ h^0)$ obtained by the algebraic CLS method is expressed as follows:

$$F_n = \begin{bmatrix} 0 & 0.53 & 0.003 & -0.03 \\ 1 & 0.38 & -0.003 & 0.33 \\ 0 & 0 & -0.8 & 0.14 \\ 0 & 0 & 0 & 0.82 \end{bmatrix},$$

$g_n(u_1) = \mathbf{e}_1$, $g_n(u_2) = \mathbf{e}_3$, $g_n(u_3) = [0.57,\ 2.4,\ -0.1]^T$,
$h_n = [12.2,\ -1.5,\ -14.9,\ -5.8]$ and $h^0 = 1$.

The obtained modified impulse responses $I(1)$, $I(2)$ and $I(3)$ are illustrated in Fig. 6.6.

We can show that the algorithm for noisy realization given by the analytic CLS method in the reference [Hasegawa, 2008] produces the same system as the above one in the sense of the numerical calculation.

In this example, the original signal $I(1)$ is characterized as the modified impulse response of a 2-dimensional linear space. The original signals $I(2)$ and $I(3)$ are characterized as the modified impulse responses of a 1-dimensional linear space respectively. The desirable modified impulse responses have been attempted to be obtained by the algebraic CLS method. The model obtained by the algebraic CLS method is a (2,1,1)-dimensional pseudo linear system which has the same nuumber of dimensions as the number of the original system.

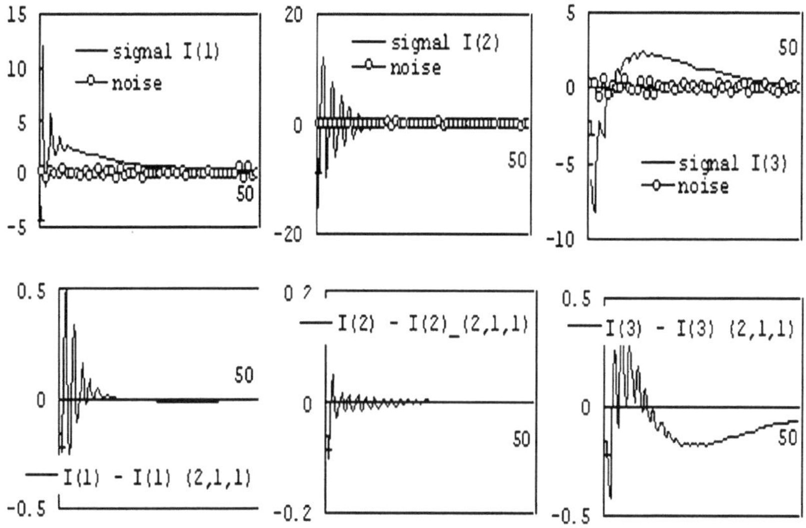

Fig. 6.6 In Example (6.29), the left are the original signal $I(1)$ with noise and the difference between $I(1)$ and the obtained signal $I(1)\_(2,1,1)$. The middle are the original signal $I(2)$ with noise and the difference between $I(2)$ and the obtained signal $I(2)\_(2,1,1)$. The right are the original signal $I(3)$ with noise and the difference between $I(3)$ and the obtained signal $I(3)\_(2,1,1)$.

Just as we thought, the following table and Fig. 6.6 indicate that the model obtained by the algebraic CLS method is not such a good (2,1,1)-dimensional noisy realization system for the original (2,1,1)-dimensional system.

dimension	ratio of matrices	mean values of square root for sum of signal ①	signal by CLS ②	error ③	cosine ① and ② $\cos\theta$	error ratio ③/①
I(1)_(2,1,1)	0.11	0.3133	0.319	0.015	0.999	0.052
I(2)_(2,1,1)	0.04	0.5	0.4977	0.002	0.9999	0.005
I(3)_(2,1,1)	0.08	0.285	0.287	0.02	0.997	0.07

Example 6.30. Let signals be the modified impulse responses of the following 5-dimensional socalled linear system $\sigma = ((R^5, F),\ g,\ h,\ h^0)$,

where $F = \begin{bmatrix} 0 & 0.4 & 0 & 0 & 0.7 \\ 1 & 0.6 & 0 & 0 & 0.3 \\ 0 & 0 & 0 & -0.7 & 0.2 \\ 0 & 0 & 1 & 0 & 0.8 \\ 0 & 0 & 0 & 0 & -0.8 \end{bmatrix}$, $h = [12,\ -1,\ -15,\ 4,\ -2]$,

$g(u_1) = [1,\ 0,\ 0,\ 0,\ 0]^T$, $g(u_2) = [0,\ 0,\ 1,\ 0,\ 0]^T$, $g(u_1) = [0,\ 0,\ 0,\ 0,\ 1]^T$, $h^0 = 1$.

Let added noises be given in Fig. 6.7.

Then the algebraically noisy realization problem is solved by the following algorithm:

covariance matrix	eigenvalues							
	1	2	3	4	5	6	7	8
$H_{\underline{a}\ (4,50)}(4,0,0)H^T_{\underline{a}\ (4,50)}(4,0,0)$	1517	118	2	0.6				
$H_{\underline{a}\ (5,50)}(5,0,0)H^T_{\underline{a}\ (5,50)}(5,0,0)$	1883	120	2.9	0.8	0.3			
$H_{\underline{a}\ (6,50)}(2,4,0)H^T_{\underline{a}\ (6,50)}(2,4,0)$	887	648	364	63	1.7	0.3		
$H_{\underline{a}\ (7,50)}(2,5,0)H^T_{\underline{a}\ (7,50)}(2,5,0)$	956	692	364	64	1.9	1.1	0.2	
$H_{\underline{a}\ (7,50)}(2,2,3)H^T_{\underline{a}\ (7,50)}(2,2,3)$	1206	650	320	234	61	1	0.5	
$H_{\underline{a}\ (8,50)}(2,2,4)H^T_{\underline{a}\ (8,50)}(2,2,4)$	1295	652	398	286	61	1.5	0.7	0.3
covariance matrix	eigenvalues							
	1	2	3	4	5	6	7	8
$H_{\underline{a}\ (5,50)}(5,0,0)H^T_{\underline{a}\ (5,50)}(5,0,0)$	43.4	11	1.7	0.9	0.5			
$H_{\underline{a}\ (7,50)}(2,5,0)H^T_{\underline{a}\ (7,50)}(2,5,0)$	31	26.3	17.9	8	1.4	1	0.4	
$H_{\underline{a}\ (8,50)}(2,2,4)H^T_{\underline{a}\ (8,50)}(2,2,4)$	36	25.5	19.9	16.9	7.8	1.2	0.8	0.5

1) A set $\{1.7,\ 0.9,\ 0.5\}$ is composed of relatively small and equally-sized numbers in the square root of eigenvalues for $H_{\underline{a}\ (5,50)}(5,0.0)H^T_{\underline{a}\ (5,50)}(5,0,0)$.
2) After determining the number n_1 of dimensions which is 2, we will continue the algebraically noisy realization algorithm.

Therefore, the modified impulse response $I(1)$ of a pseudo linear system obtained by the algebraic CLS method is constructed for a 2-dimensional space.
3) A set $\{1.4,\ 1,\ 0.4\}$ is composed of relatively small and equally-sized numbers in the square root of eigenvalues for $H_{\underline{a}\ (7,50)}(2,5,0)H^T_{\underline{a}\ (7,50)}(2,5,0)$.
4) After determining the number n_2 of dimensions which is 2, we execute the algebraically noisy realization algorithm.
5) A set $\{1.2,\ 0.8,\ 0.5\}$ is composed of relatively small and equally-sized numbers in the square root of eigenvalues for $H_{\underline{a}\ (8,50)}(2,2,4)H^T_{\underline{a}\ (8,50)}(2,2,4)$.
6) After determining the number n_3 of dimensions which is 1, we execute the algebraically noisy realization algorithm.

Therefore, the modified impulse responses $I(1)$, $I(2)$ and $I(3)$ of an approximate pseudo linear system obtained by the algebraic CLS method are realized by a (2,2,1)-dimensional pseudo linear system.

The 5-dimensional pseudo linear system $\sigma_n = ((\boldsymbol{R}^5, F_n), g_n, h_n, h^0)$ obtained by the algebraic CLS method may be expressed as follows:

$$F_n = \begin{bmatrix} 0 & 0.43 & 0 & 0.01 & 0.71 \\ 1 & 0.57 & 0 & -0.01 & 0.28 \\ 0 & 0 & 0 & -0.7 & 0.2 \\ 0 & 0 & 1 & -0.002 & 0.78 \\ 0 & 0 & 0 & 0 & -0.78 \end{bmatrix},$$

$g_n(u_1) = \mathbf{e}_1$, $g_n(u_2) = \mathbf{e}_3$,$g_n(u_3) = \mathbf{e}_5$, $h_n = [12.1, -1.33, -14.8, 4.77, -1.8]$ and $h^0 = 1$.

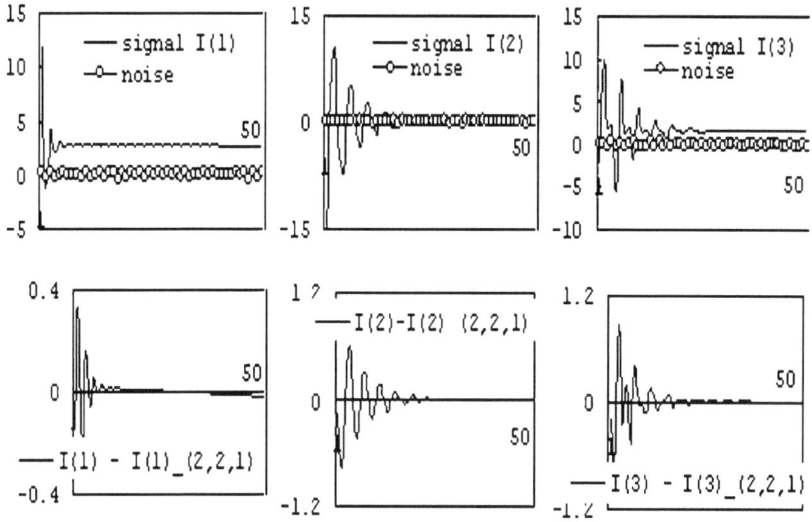

Fig. 6.7 In Example (6.30), the left are the original signal $I(1)$ with noise and the difference between $I(1)$ and the obtained signal $I(1)\_(2,2,1)$. The middle are the original signal $I(2)$ with noise and the difference between $I(2)$ and the obtained signal $I(2)\_(2,2,1)$. The right are the original signal $I(3)$ with noise and the difference between $I(3)$ and the obtained signal $I(3)\_(2,2,1)$.

The obtained modified impulse responses $I(1)$, $I(2)$ and $I(3)$ are illustrated in Fig. 6.7.

We can show that the algorithm for noisy realization given by the analytic CLS method in the reference [Hasegawa, 2008] produces the same system as the above one in the sense of the numerical calculation.

In this example, the original signals $I(1)$ and $I(2)$ are characterized as the modified impulse responses of a 2-dimensional linear space respectively. The original signal $I(3)$ is characterized as the modified impulse responses of a 1-dimensional linear space. The desirable modified impulse responses have been attempted to be obtained by the algebraic CLS method. The model obtained by the algebraic CLS method is a (2,2,1)-dimensional pseudo linear system which has the same number of dimensions as the number of the original system.

Just as we expected, the following table and Fig. 6.7 indicate that the model obtained by the algebraic CLS method is a somewhat good (2,2,1)-dimensional system for the original (2,2,1)-dimensional system.

dimension	ratio of matrices	mean values of square root for sum of: signal	signal by CLS	error	cosine ① and ②	error ratio
		①	②	③	$\cos\theta$	③/①
I(1)_(2,2,1)	0.04	0.45	0.443	0.009	0.999	0.02
I(2)_(2,2,1)	0.05	0.435	0.438	0.02	0.998	0.05
I(3)_(2,2,1)	0.03	0.356	0.368	0.03	0.997	0.07

Example 6.31. Let signals be the modified impulse responses of the following 6-dimensional pseudo linear system $\sigma = ((R^6, F),\ g,\ h,\ h^0)$,

where $F = \begin{bmatrix} 0 & -0.3 & 0 & 0.4 & 0 & 0.1 \\ 1 & 0.7 & 0 & 0.3 & 0 & -0.3 \\ 0 & 0 & 0 & -0.4 & 0 & -0.2 \\ 0 & 0 & 1 & 0.7 & 0 & 0.5 \\ 0 & 0 & 0 & 0 & 0 & -0.8 \\ 0 & 0 & 0 & 0 & 1 & -0.7 \end{bmatrix}$, $h = [15,\ -1,\ -10,\ 6,\ -8,\ 4]$,

$g(u_1) = [1,\ 0,\ 0,\ 0,\ 0,\ 0]^T$, $g(u_2) = [0,\ 0,\ 1,\ 0,\ 0,\ 0]^T$,
$g(u_3) = [0,\ 0,\ 0,\ 0,\ 1,\ 0]^T$, $h^0 = 1$.

Let added noises be given in Fig. 6.8.

Then the algebraically noisy realization problem is solved as follows:

covariance matrix	eigenvalues								
	1	2	3	4	5	6	7	8	9
$H_{\underline{a}\ (4,35)}(4,0,0)H^T_{\underline{a}\ (4,35)}(4,0,0)$	304	61	1.7	0.7					
$H_{\underline{a}\ (5,35)}(5,0,0)H^T_{\underline{a}\ (5,35)}(5,0,0)$	304	61	1.8	1.1	0.25				
$H_{\underline{a}\ (6,35)}(2,4,0)H^T_{\underline{a}\ (6,35)}(2,4,0)$	934	569	42	13	1.4	0.7			
$H_{\underline{a}\ (7,35)}(2,5,0)H^T_{\underline{a}\ (7,35)}(2,5,0)$	936	658	42	16	1.7	0.9	0.4		
$H_{\underline{a}\ (8,35)}(2,2,4)H^T_{\underline{a}\ (8,35)}(2,2,4)$	963	565	164	162	23	8.4	1.4	0.2	
$H_{\underline{a}\ (9,35)}(2,2,5)H^T_{\underline{a}\ (9,35)}(2,2,5)$	979	565	231	163	23	8.6	2.4	0.4	0.09
covariance matrix	square root of eigenvalues								
	1	2	3	4	5	6	7	8	9
$H_{\underline{a}\ (5,35)}(5,0,0)H^T_{\underline{a}\ (5,35)}(5,0,0)$	17.4	7.8	1.3	1	0.5				
$H_{\underline{a}\ (7,35)}(2,5,0)H^T_{\underline{a}\ (6,35)}(2,5,0)$	30.6	25.6	6.5	4	1.3	0.9	0.6		
$H_{\underline{a}\ (9,35)}(2,2,5)H^T_{\underline{a}\ (9,35)}(2,2,5)$	31.3	23.8	15.2	12.7	4.8	2.9	1.5	0.6	0.4

1) A set $\{1.3,\ 1,\ 0.5\}$ is composed of relatively small and equally-sized numbers in the square root of eigenvalues for $H_{\underline{a}\ (5,35)}(5,0.0)H^T_{\underline{a}\ (5,35)}(5,0,0)$.
2) After determining the number n_1 of dimensions which is 2, we will continue the algebraically noisy realization algorithm.

Therefore, the modified impulse response $I(1)$ of a pseudo linear system obtained by the algebraic CLS method is constructed for a 2-dimensional space.
3) A set $\{1.3,\ 0.9,\ 0.6\}$ may be composed of relatively small and equally-sized numbers in the square root of eigenvalues for $H_{\underline{a}\ (7,35)}(2,5,0)H^T_{\underline{a}\ (7,35)}(2,5,0)$.
4) After determining the number n_2 of dimensions which is 2, we execute the algebraically noisy realization algorithm.
5) A set $\{1.5,\ 0.6,\ 0.4\}$ may be composed of relatively small and equally-sized numbers in the square root of eigenvalues for $H_{\underline{a}\ (9,35)}(2,2,5)H^T_{\underline{a}\ (9,35)}(2,2,5)$.
6) After determining the number n_3 of dimensions which is 2, we execute the algebraically noisy realization algorithm.

Therefore, the modified impulse responses $I(1)$, $I(2)$ and $I(3)$ of an approximate pseudo linear system obtained by the algebraic CLS method are realized by a (2,2,2)-dimensional pseudo linear system.

The 6-dimensional pseudo linear system $\sigma_n = ((\boldsymbol{R}^6,\ F_n),\ g_n,\ h_n,\ h^0)$ obtained by the algebraic CLS method is expressed as follows:

$$F_n = \begin{bmatrix} 0 & -0.28 & 0 & 0.42 & 0 & 0.04 \\ 1 & 0.7 & 0 & 0.26 & 0 & -0.29 \\ 0 & 0 & 0 & -0.4 & 0 & -0.24 \\ 0 & 0 & 1 & 0.68 & 0 & 0.56 \\ 0 & 0 & 0 & 0 & 0 & -0.84 \\ 0 & 0 & 0 & 0 & 1 & -0.73 \end{bmatrix},\ g_n(u_1) = \mathbf{e}_1,\ g_n(u_2) = \mathbf{e}_3,\ g_n(u_3) = \mathbf{e}_5,$$

$h_n = [15.1,\ -1.26,\ -9.8,\ 6.1,\ -7.8,\ 4.2]$ and $h^0 = 1$.

We can show that the algorithm for noisy realization given by the analytic CLS method in the reference [Hasegawa, 2008] produces the same system as the above one in the sense of the numerical calculation.

In this example, the original signal $I(1)$, $I(2)$ and $I(3)$ are characterized as the modified impulse responses of a 2-dimensional linear space respectively. The desirable modified impulse responses have been attempted to be obtained by the algebraic CLS method. The model obtained by the algebraic CLS

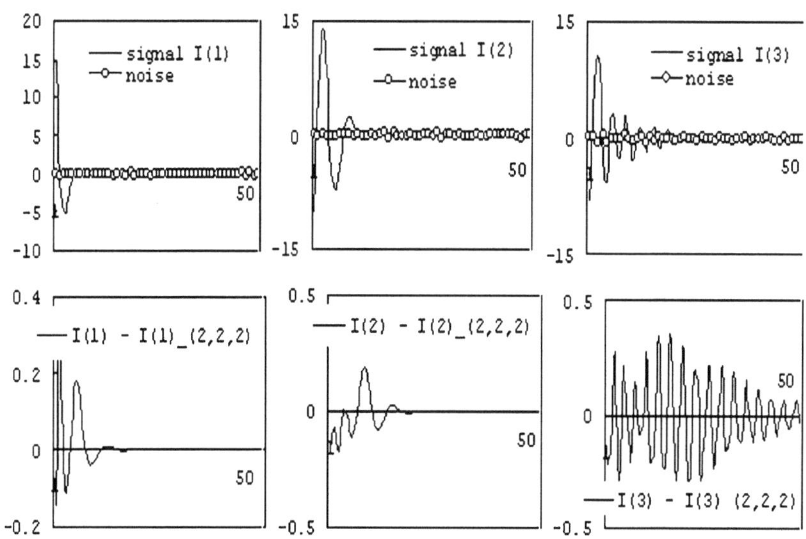

Fig. 6.8 In Example (6.31), the left are the original signal $I(1)$ with noise and the difference between $I(1)$ and the obtained signal $I(1)\_(2,2,2)$. The middle are the original signal $I(2)$ with noise and the difference between $I(2)$ and the obtained signal $I(2)\_(2,2,2)$. The right are the original signal $I(3)$ with noise and the difference between $I(3)$ and the obtained signal $I(3)\_(2,2,2)$.

method is a (2,2,2)-dimensional pseudo linear system which has the same number of dimensions as the number of the original system.

Just as we expected, the following table and Fig. 6.8 indicate that the model obtained by the algebraic CLS method is a good (2,2,2)-dimensional noisy realization system for the original (2,2,2)-dimensional system.

dimension	ratio of matrices	mean values of square root for sum of signal ①	signal by CLS ②	error ③	cosine ① and ② $\cos\theta$	error ratio ③/①
I(1)-(2,2,2)	0.07	0.326	0.328	0.008	0.999	0.03
I(2)-(2,2,2)	0.04	0.4249	0.4244	0.008	0.999	0.02
I(3)-(2,2,2)	0.05	0.3315	0.3331	0.02	0.997	0.08

6.7 Historical Notes and Concluding Remarks

We have proposed algebraically approximate and noisy realization problems of pseudo linear systems, which are close to linear systems. In a previous monograph [Matsuo and Hasegawa, 2003], fundamental facts about pseudo linear systems were first established. These important facts were a representation of their behavior and the partial realization algorithm in Definition (6.1) and Theorem (6.19), where the representation of their behavior means that any pseudo linear system can be completely characterized by the modified impulse responses. The algebraically approximate realization problem was attempted to be solved by presenting an algebraically approximate realization algorithm. The algorithm is made up of a ratio of input/output matrix norm and the algebraic CLS method, i.e., the algebraically constrained least square method. Through the introduction of the ratio of the matrix norm which is the square norm, we can decrease the dimensional number of state spaces while keeping information loss in mind. By using the algebraic CLS method, we can make full efforts to characterize the relation of a linear combination. By applying this algorithm to several examples of pseudo linear systems, we have shown that this algorithm is practical, useful and easy. In the case where the ratio of input/output matrix norm is within some percent, we have shown that this algebraically approximate realization algorithm produces good results with the exception of pseudo linear systems whose modified impulse responses have small changing values and rapid damping as in Examples (6.23) and (6.24). Our several examples show that the changing relations among the ratio of the matrix norm and the error to signal ratio are proportional relations and the ratio is 0.01 for the Input/output matrix norm while the error to signal ranges from 0.02 to 0.03. This approximate realization algorithm appears to be very promising.

We treated algebraically noisy realization problems and attempted to solve them by presenting a noisy realization algorithm. The algorithm is composed by making a set of singular values of a matrix and applying the algebraic CLS method, i.e., the constrained least square method. By producing a set of singular values of a matrix, we can determine the dimensional number of state spaces by drawing a distinction between a noiseless part and a noisy part in the given signal. By using the algebraic CLS method, we can make full efforts to characterize a relation of a linear combination in the noiseless part.

By applying this algorithm to several examples of pseudo linear systems, we have shown that this algorithm is practical and useful. In the case that we can make a set composed of relatively small and equally-sized numbers in the square root of eigenvalues for an Input/output matrix, we have shown that this noisy realization algorithm produces good results. Our several examples in noisy realizations show that the changing relations among the ratio of the matrix norm and the error to signal ratio are proportional relations and the ratio is 0.01 for the Input/output matrix norm while the error to signal ratio ranges from 0.003 to 0.02. This noisy realization algorithm also appears to be very promising.

As we mentioned before, concrete discussions of algebraically approximate and noisy realization for non linear systems are very new.

For both algebraically approximate and noisy realization problems, we could propose a new law which says that pseudo linear systems obtained by the algebraic CLS method are the same as ones obtained by the analytic CLS method proposed in the reference [Hasegawa, 2008]. The law is called a law of a constrained least square.

The analtic CLS method for determing n variables is reduced to the minimization of the following rational polynomial $p(x_1, x_2, \cdots, x_n)$ in n variables:

$$p(x_1, x_2, \cdots, x_n) = \frac{\sum_{c_1, c_2, \cdots, c_n} \alpha(c_1, c_2, \cdots, c_n) \times x_1^{c_1} x_2^{c_2} \cdots x_n^{c_n}}{(1 + x_1^2 + x_2^2 + \cdots + x_n^2)^2},$$

where $\sum_{c_1, c_2, \cdots, c_n}$ means all summation of any combination with the conditions $0 \leq c_i \leq 4$ and $c_1 + c_2 + \cdots + c_n \leq 4$ for any $1 \leq i \leq n$ and $\alpha(c_1, c_2, \cdots, c_n) \in \boldsymbol{R}$.

Therefore, our new Law shows that approximate and noisy problems can be solved using only algebraic calculations, namely, without treating partial differential equations.

Chapter 7
Algebraically Approximate and Noisy Realization of Affine Dynamical Systems

In this chapter, we will discuss algebraically approximate and noisy realization problems of affine dynamical systems, which realize any input response map, equivalently, as an input/output map with causality. Affine dynamical systems were proposed and the realization problems of the systems were solved in the reference [Matsuo & Hasegawa, 2003]. We characterized the finite-dimensionality of affine dynamical systems. We obtained the same results as ones established in linear system theory.

A criterion for canonical finite-dimensional affine dynamical systems was given. There uniquely exists a quasi-reachable standard system in the isomorphic class of finite-dimensional canonical affine dynamical systems. We obtained a criterion for the behavior of finite-dimensional affine dynamical systems. We also gave a procedure on how to obtain the quasi-reachable standard system from an input response map.

Approximate and noisy realization problems of affine dynamical systems were proposed and solved with the introduction of a new method called a CLS (constrained least square) method in the reference [Hasegawa, 2008]. The CLS method implies that the minimization of the error or noise in the sense of square norm can be reduced to determine the minimum value of a rational polynomial in n variables. In order to obtain the minimum value of a rational polynomial in n variables, partial differential equations usually must be solved. The work is too much of a bother. In this monograph, we call the CLS method the analytic CLS method.

Also regarding affine dynamical systems, we will propose an algebraically approximate or noisy realization problem, provide an algebraic algorithm to obtain approximate or noisy realization with the introduction of an algebraically constrained least square method, abbreaviated, algebraic CLS method, and discover a new law through the numerical experiments which says that an affine dynamical system obtained by the algorithm based on the algebraic CLS method is the same as a system obtained by the algorithm based on the analytic CLS method. Hence, we can solve an approximate or

Y. Hasegawa: Algebra. Approx. & Noisy Reali. of Discrete-Time Sys., LNEE 50, pp. 147–183.
springerlink.com

noisy realization problem with a very easy calculation. The law may be called a law of a constrained least square.

Firstly, we state many facts obtained in the references [Matsuo and Hasegawa, 2003] and [Hasegawa, 2008] needed for our discussion.

7.1 Basic Facts about Affine Dynamical Systems

Definition 7.1. Affine Dynamical Systems

1) A system given by the following system equation is written as a collection $\sigma = ((X, F), g, h, h^0)$ and it is said to be an affine dynamical system.

$$\begin{cases} x(t+1) = F(\omega(t+1))x(t) + g(\omega(t+1)) \\ x(0) \quad\; = 0 \\ \gamma(t) \quad\;\; = h^0 + hx(t) \end{cases}$$

for any $t \in N$, $x(t) \in X$, $\gamma(t) \in Y$, where X is a linear space over the field $\boldsymbol{R}$ that may be called a state space, F is a map $F : U \to L(X); u \mapsto F(u)$, a map $g : U \to X$, a linear map $h : X \to Y$ and $h^0 \in Y$.
2) The input response map $a_\sigma : U^* \to Y; \omega \mapsto a_\sigma(\omega)$
$= h^0 + h(\sum_{j=1}^{|\omega|}(F(\omega(|\omega|))F(\omega(|\omega|-1))\cdots F(\omega(|\omega|-j))g(\omega(j))))$ is said to be a behavior of σ.

An affine dynamical system σ which satisfies $a_\sigma = a$ is said to be a realization of an input response map a.
3) An affine dynamical system σ is said to be quasi-reachable if the linear hull of the reachable set $\{\sum_{j=1}^{|\omega|}(F(\omega(|\omega|))F(\omega(|\omega|-1))\cdots$
$F(\omega(|\omega|-j))g(\omega(j)); \omega \in U^*\}$ is equal to X and an affine dynamical system σ is said to be distinguishable if $h(F(\omega(|\omega|))F(\omega(|\omega|-1))\cdots F(\omega(|\omega|-j))x_1$
$= h(F(\omega(|\omega|))F(\omega(|\omega|-1))\cdots F(\omega(|\omega|-j))x_2$ implies $x_1 = x_2$ for any $\omega \in U^*$.
4) An affine dynamical system σ is said to be canonical if σ is quasi-reachable and distinguishable.

Remark 1: It is meant for σ to be a faithful model for the input response map a such that σ realizes a.
Remark 2: Notice that a canonical affine dynamical system:
$\sigma = ((X, F), g, h, h^0)$ is a system which has the most reduced state set X among systems that have the behavior a_σ.

In order to show intuitively that affine dynamical systems are general dynamical systems, we will state a relation between affine dynamical systems and inhomogeneous bilinear systems.

We will consider the following dynamical system:

$$\begin{cases} x(t+1) &= (A+\sum_{i=1}^{m}\mathbf{N}_i\cdot\omega_i(t+1)x(t)+\sum_{i=1}^{m}\bar{g}\cdot\omega_i(t+1) \\ x(0) &= 0 \\ \gamma(t) &= h^0+hx(t) \end{cases}$$

$\omega_i(t)\in \boldsymbol{R}$, $x(\mathrm{t})$, $\bar{g}\in\boldsymbol{R}^n$, $\mathbf{A}$, $\mathbf{N}_i\in\boldsymbol{R}^{n\times n}$ and $\gamma(t)\in\boldsymbol{R}$.

Let $F(\omega(t+1)) = \mathbf{A}+\sum_{i=1}^{m}\omega_i(t+1)\mathbf{N}_i$, $g(\omega(t+1)) = \sum_{i=1}^{m}\bar{g}\omega_i(t+1)$. Then the above dynamical system is an affine dynamical system. Therefore, the inhomogeneous bilinear system is an example of our affine dynamical systems.

Definition 7.2. Let $\sigma_1 = ((X_1,F_1),g_1,h_1,h^0)$ and $\sigma_2 = ((X_2,F_2),g_2,h_2,h^0)$ be affine dynamical systems. Then a linear operator $T: X_1\to X_2$ is called an affine dynamical system morphism $T:\sigma_1\to\sigma_2$ if T is a linear map : $X_1\to X_2$ that satisfies $fF_1(u) = F_2(u)f$, $fg_1 = g_2$ and $h_1 = h_2 f$.

A bijective affine dynamical system morphism $T:\sigma_1\to\sigma_2$ is called an isomorphism.

Corollary 7.3. *Let σ_1 and σ_2 be affine dynamical systems and $T:\sigma_1\to\sigma_2$ be an affine dynamical system morphism. Then $a_{\sigma_1} = a_{\sigma_2}$ holds.*

Example 7.4. Let $U^+ := U\setminus 1$ and $V(U^+) := \{\lambda = \sum_{\omega\in U^+}\lambda(\omega)\mathbf{e}_\omega(\text{finite sum})\ ;\ \lambda(\omega)\in\boldsymbol{R}\}$, where $\mathbf{e}_\omega(\bar{\omega}) = 1$ for $\omega = \bar{\omega}$ and $\mathbf{e}_\omega(\bar{\omega}) = 0$ for $\omega\neq\omega$. Let ψ be a map : $U\to L(V(U^+)); u\mapsto\psi(u)[;\mathbf{e}_\omega\mapsto\mathbf{e}_{\mathbf{u}|\omega}-\mathbf{e}_{\mathbf{u}}]$.

And let a map $e: U\to V(U^+); u\mapsto\mathbf{e}_{\mathbf{u}}$, where $e(1) = 0$. In addition, we consider a linear map $a_l : V(U^+)\to Y;\mathbf{e}_\omega\mapsto a(\omega)-a(1)$ for any input response map $a\in F(U^*,Y)$. Then $((V(U^+),\psi),e,a_l,a(1))$ is a quasi-reachable affine dynamical system that realizes $a\in F(U^*,Y)$.

Example 7.5. Let $a\in F(U^*,Y)$ be any input response map and S_l be defined by $S_l(u)a : U^*\to Y;\omega\mapsto a(\omega|u)$. Then $S_l(u)\in L(F(U^*,Y)$ for any $u\in U$. Let a map $\xi: U\to F(U^*,Y)$ be $u\mapsto\xi(u)[;\omega\mapsto a(\omega|u)-a(\omega)]$. And let 1 be a linear map : $F(U^*,Y)\to Y; a\mapsto a(1)$. Then $((F(U^*,Y),S_l),\xi,1,a(1))$ is a distinguishable affine dynamical system that realizes $a\in F(U^*,Y)$.

Remark: Examples (7.4) and (7.5) imply that there exist many affine dynamical systems that realize a given input response map $a\in F(U^*,Y)$. However, there is no relation between them. Therefore, we introduce canonical affine dynamical systems, and we will make a clear relation between them.

Theorem 7.6. *For any input response $a\in F(U^*,Y)$, there exist the following two canonical affine dynamical systems that realize it.*

1) $((V(U^+)/_{=a},\tilde{\psi}),\tilde{e},\tilde{a}_l,a(1))$,
where $V(U^+)/_{=a}$ is a quotient space derived by equivalence relation:
$\sum_\omega\lambda(\omega)\mathbf{e}_\omega = \sum_{\bar{\omega}}\lambda(\bar{\omega})\mathbf{e}_{\bar{\omega}}\iff$
$\sum_\omega\lambda(\omega)(a(\omega)-a(1)) = \sum_{\bar{\omega}}\lambda(\bar{\omega})(a(\bar{\omega})-a(1))$,
$\tilde{\psi}$ is given by a map $:U\to L(V(U^+)/_{=a}); u\mapsto\tilde{\psi}(u)[;\lambda\mapsto$

$\sum_\omega \lambda(\omega)(\mathbf{e}_{\mathbf{u}|\omega} - \mathbf{e}_\mathbf{u})$, $\tilde{e}$ *is given by* $\tilde{e} : U \to V(U^+)/_{=a}; u \mapsto [e_u]$ *and* $\tilde{a}_l$ *is given by* $\tilde{a}_l : V(U^+)/_{=a} \to Y; [\lambda] \mapsto \tilde{a}_l([\lambda]) = \sum_\omega \lambda(\omega)(a(\omega) - a(1))$.
2) $((\ll S_l(U^*)a - a \gg, S_l), \xi, 1, a(1))$,
where $S_l(U^*)a - a = \{S_l(\omega)a - a; \omega \in U^*\}$ *and* $\ll S_l(U^*)a - a \gg$ *denotes the smallest linear space which contains* $S_l(U^*)a - a$.

We conclude that there exist at least two canonical affine dynamical systems that realize any input response map in Theorem (7.6). Next, we will insist on the uniqueness of the systems that have the same behavior.

Theorem 7.7. *Realization Theorem*

For any input response map $a \in F(U^*, Y)$, *there exist at least two canonical affine dynamical systems that realize it.*

Let $\sigma_1 = ((X_1, F_1), g_1, h_1, h^0)$ *and* $\sigma_2 = ((X_2, F_2), g_2, h_2, h^0)$ *be canonical affine dynamical systems that realize any* $a \in F(U^*, Y)$, *then there exists a unique isomorphism* $T : \sigma_1 \to \sigma_2$.

7.2 Finite Dimensional Affine Dynamical Systems

Based on Realization Theorem (7.7), we clarified the finite-dimensionality of the systems. Therefore, we obtained the same results as obtained in the linear systems by R. E. Kalman.

As previously described, we introduce finite dimensional affine dynamical systems needed for our algebraically approximate and noisy realization problems.

Firstly, we assume that the set U of input's values is finite, and we show that the assumption of finiteness is not so special. Namely, affine dynamical systems with an assumption include biaffine systems as a subclass. Biaffine systems were discussed by Tarn and Nonoyama [1979].

The following results have been obtained for the systems. It is given as a criterion for canonical finite dimensional affine dynamical systems. We give a criterion for the behavior of finite dimensional affine dynamical systems. The companion form for canonical finite-dimensional affine dynamical systems is also given. Moreover, a procedure to obtain the companion form from a given input/output map has been obtained.

Therefore, it is obvious that the theory of these affine dynamical systems is the extension of the linear system theory established by Kalman et al for the non-linear case.

An affine dynamical system is different from a state-affine system in [Sontag, 1979a]. Our system is introduced on the basis of Theorem (2.6) and Definition (2.7) in [Matsuo and Hasegawa 2003], which is the representation theorem for any input/output map with causality. Hence, our systems are more general than state-affine systems.

If the state space X of an affine dynamical system $\sigma = ((X, F), g, h, h^0)$ is finite dimensional (n-dimensional), then σ is said to be a finite dimensional (n-dimensional) affine dynamical system.

There is the following fact about n dimensional linear space in [Halmos, 1958].

Fact: [Every n dimensional linear space over the field $\boldsymbol{R}$ is isomorphic to $\boldsymbol{R}^n$. Moreover, every linear operator from $\boldsymbol{R}^n$ to $\boldsymbol{R}^m$ is isomorphic to a matrix $F \in \boldsymbol{R}^{m\times n}$.]

Therefore, without loss of generality, a n dimensional affine dynamical system can be represented by $\sigma = ((\boldsymbol{R}^n, F), g, h, h^0)$,
where, F is a map : $U \to \boldsymbol{R}^{n\times n}$, g is a map : $U \to \boldsymbol{R}^n$ and $h \in \boldsymbol{R}^{p\times n}$ and $h^0 \in \boldsymbol{R}^p$.

According to the above discussion, we can treat an n-dimensional affine dynamical system $\sigma = ((\boldsymbol{R}^n, F), g, h, h^0)$ which is easily embodied by computer programs or electrical circuits.

From now on, we assume that the set U of input's values is finite. Let $U = \{u_1, u_2, \cdots, u_m\}$. Now, we demonstrate that the assumption is not so special.

Biaffine Systems 7.8

We will consider the following system:

$$\begin{cases} x(t+1) = (A + \sum_{i=1}^{m} N_i \cdot \omega_i(t+1))x(t) + \sum_{i=1}^{m} \mathbf{b}_i \cdot \omega_i(t+1) + \mathbf{a} \\ x(0) \quad\;\; = 0 \\ \gamma(t) \quad\;\;\, = h^0 + hx(t) \end{cases}$$

$\omega_i(t) \in \mathbf{R}$, $x(\mathrm{t})$, $\mathbf{b}_i$ and $\mathbf{a} \in \mathbf{R}^n, N_i \in \mathbf{R}^{n\times n}$ and $\gamma(t) \in Y$.

Transferring time in input, we will conclude that the above system is a biaffine system as treated in [Tarn and Nonoyama, 1979],
where maps $\tilde{F} : \mathbf{R}^m \to \mathbf{R}^{n\times n}$ and $g : \mathbf{R}^m \to \mathbf{R}^{n\times n}$ are affine, namely,
$\tilde{F}(\sum_{i=1}^{m} \omega_i(t+1)\mathbf{e_i}) = A + \sum_{i=1}^{m} N_i\omega_i(t+1)$,
$\tilde{g}((\sum_{i=1}^{m} \omega_i(t+1)\mathbf{e_i}) = \mathbf{a} + \sum_{i=1}^{m} \mathbf{b}_i\omega_i(t+1)$.

Then we can obtain an affine dynamical system $\sigma = ((R^n, F), g, h, h^0)$,
where F and g are given by the following relations:
$F(0) = A,\ F(\mathbf{e_i}) = A + N_i (1 \le i \le m)$,
$g(0) = \mathbf{a},\ g(\mathbf{e_i}) = \mathbf{a} + b_i (1 \le i \le m)$.

And U is given by $U = \{\mathbf{0}, \mathbf{e_1}, \mathbf{e_2}, \cdots, \mathbf{e_m}\}$ and $e_i = [0, 0, \cdots, 0, \overset{i}{1}, 0, \cdots, 0]^T$,
where T denotes the transpose.

Therefore, we can conclude that the assumption for the set U to be finite is not so special.

Proposition 7.9. *Let $\sigma = ((\boldsymbol{R}^n, F), g, h, h^0)$ be an affine dynamical system. σ is canonical if and only if*

1) rank $[(g(u_1), g(u_2), \cdots, g(u_m), F(u_1)g(u_1), \cdots, F(u_1)g(u_m), \cdots, F^{n-1}(u_m)g(u_1), \cdots, F^{n-1}(u_m)g(u_m)] = n.$
2) rank $[h^T, (hF(u_1))^T, (hF(u_2))^T, \cdots, (hF(u_m))^T, \cdots, (hF^2(u_1))^T, \cdots, (hF^2(u_m))^T, (hF^{n-1}(u_1)g(u_m))^T, \cdots, (hF^{n-1}(u_m)g(u_m))^T] = n.$

Definition 7.10. Let the input value's set U be $U := \{u_i; 1 \leq i \leq m\}$ and let a map $\| \ \| : U \to N$ be $u_i \mapsto \|u_i\| = i$. And let a numerical value $|||\omega|||$ of an input $\omega \in U^*$ be $|||\omega||| = \|\omega(|\omega|)\| + \|\omega(|\omega|-1)\| \times m + \cdots + \|\omega(1)\| \times m^{|\omega|-1}$ and $|||\mathit{1}||| = 0$.

Then we can define a totally ordered relation by this numerical value in U^*.

Namely, $\omega_1 \leq \omega_2 \Longleftrightarrow |||\omega_1||| \leq |||\omega_2|||$.

Definition 7.11. Let $\sigma_s = ((\boldsymbol{R}^n, F_s), g_s, h_s, h^0)$ be a canonical affine dynamical system. If input sequences $\{\omega_i \in U^*; 1 \leq i \leq n\}$ satisfy the following conditions, then σ_s is said to be a quasi-reachable standard system.

1) $\mathbf{e}_\mathrm{i} = \sum_{j=1}^{i} F_s(\omega_j(|\omega_j|)F_s(\omega_j(|\omega_j|-1)F_s(\omega_j(|\omega_j|-j)g_s(\omega_j(j))$
2) $\mathit{1} = \omega_1 < \omega_2 < \cdots < \omega_n$ and $|\omega_i| \leq i-1$ for $i(1 \leq i \leq n)$ hold.
3) $\sum_{j=1}^{|\omega|} F_s(\omega(|\omega|)F_s(\omega(|\omega|-1)F_s(\omega(|\omega|-j)g_s(\omega(j)) = \sum_{i=1}^{j} \alpha_i \mathbf{e}_\mathrm{i}, \alpha_i \in \boldsymbol{R}$ holds for any input sequence $\omega \in U^*$ such that $\omega_j < \omega < \omega_{j+1} (1 \leq i \leq n-1)$.

Theorem 7.12. *For any canonical affine dynamical system*

$\sigma = ((\boldsymbol{R}^n, F), g, h, h^0)$, *there exists a unique quasi-reachable standard system* $\sigma_s = ((\boldsymbol{R}^n, F_s), g_s, h_s, h^0)$ *which is isomorphic to it.*

Definition 7.13. For any input response map $a \in F(U^*, Y)$, there uniquely exists a linear operator $A : V(U^+) \to F(U^*, Y)$ such that A satisfies $S_l(u)A = A\psi(u)$ for any $u \in U$. Hence, $A(e_\omega)(\bar{\omega}) = a(\bar{\omega}|\omega) - a(\bar{\omega})$ holds for any $\omega, \ \bar{\omega} \in U^*$.

Therefore, for any $\omega, \ \bar{\omega} \in U^*$, we can consider the following infinite matrix H_a^A.

The H_a^A is called a Hankel matrix of a. The column vector of H_a^A may be written by $S_l(\omega)a - a$.

$$H_a^A = \begin{array}{c} \\ \\ \\ \bar{\omega} \end{array} \left(\begin{array}{ccc} & & \omega \\ & & \vdots \\ & & \vdots \\ \cdots & \cdots & a(\bar{\omega}\,|\omega) - a(\bar{\omega}) \\ & & \end{array} \right)$$

Theorem 7.14. *Theorem for existence criterion*

For an input response map $a \in F(U^*, Y)$, *the following conditions are equivalent:*

1) a is a behavior of an n-dimensional canonical affine dynamical system.
2) $\{S_l(\omega)a - a : \omega \in U^*\}$ *have n linearly independent vectors.*
3) rank of H_a^A *is n,*
where $S_l(\omega)a - a \in F(U^*, Y)$ *is defined by* $S_l(\omega)a - a : U^* \to Y; \bar{\omega} \mapsto a(\bar{\omega}|\omega) - a(\bar{\omega})$.

Theorem 7.15. *Theorem for a realization procedure*

Let an input response map $a \in F(U^*, Y)$ *satisfy the condition of Theorem (7.14). Then the quasi-reachable standard system* $\sigma_s = ((\boldsymbol{R}^n, F_s), g_s, h_s, h^0)$ *which realizes it can be obtained by the following procedure:*

1) Select n linearly independent vectors $\{S_l(\omega_i)a - a : (1 \leq i \leq n)\}$ *from* $\{S_l(\omega)a - a : \omega \in U^*, |\omega| \leq n-1\}$ *in order of the numerical value of* U^*.
2) Let the state space be $\boldsymbol{R}^n$. *For the set* $\{\omega_j : |\omega_j| = 1\}$ *of input sequence, set* $g_s(\omega_j) = \mathbf{e_j}$. *Moreover, let* $g_s(\omega_j) = \sum_{i=1}^{j} \alpha_i \mathbf{e_i}$ *for any* $\omega \in U^*$ *such that* $\omega_j < \omega < \omega_{j+1}$ *and* $|\omega_j| = |\omega_{j+1}| = 1$.
3) Let $h_s = [a(\omega_1) - a(1), a(\omega_2) - a(1), \cdots, a(\omega_n) - a(1)]$.
4) For any $i (1 \leq i \leq n)$, *let* ${}_if_j$ *in* $F_s(u_i) = [{}_if_1, {}_if_2, \cdots, {}_if_n] \in \boldsymbol{R}^{n \times n}$ *be* ${}_if_j = [{}_if_{j1}, {}_if_{j2}, \cdots, {}_if_{jn}]^T$,
where $S_l(u_i)(S_l(\omega_j)a - a) = \sum_{k=1}^{n} {}_if_{jk}\ (S_l(\omega_k)a - a)$ *holds for any* $j (1 \leq j \leq n)$.
5) Set $h^0 = a(1)$.

7.3 Partial Realization Theory of Affine Dynamical Systems

Here we consider a partial realization problem by multi-experiment. Let $\underline{a}$ be an $\underline{N}$ sized input response map($\in F(U^*_{\underline{N}}, Y)$), where $\underline{N} \in N$ and $U^*_{\underline{N}} := \{\omega \in U^*; |\omega| \leq \underline{N}\}$. The $\underline{a}$ is said to be a partial input response map. A finite dimensional affine dynamical system $\sigma = ((X, F), g, , h, h^0)$ is called a partial realization of $\underline{a}$ if $h^0 + h(\sum_{j=1}^{|\omega|}(F(\omega(|\omega|))F(\omega(|\omega|-1)) \cdots F(\omega(|\omega| - j))g(\omega(j))) = \underline{a}(\omega)$ holds for any $\omega \in U^*_{\underline{N}}$.

A partial realization problem of affine dynamical systems can be stated as follows:

< For any given $\underline{a} \in F(U^*_{\underline{N}}, Y)$, find a partial realization σ of $\underline{a}$ such that the dimensions of state space X of σ is minimum, where the σ is said to be a minimal partial realization of $\underline{a}$. Moreover, show when the minimal realizations are isomorphic.>
We will state facts about it.

Proposition 7.16. *For any given* $\underline{a} \in F(U^*_{\underline{N}}, Y)$, *there always exists a minimal partial realization of it.*

Minimal partial realizations are, in general, not unique modulo isomorphisms. Therefore, we introduce a natural partial realization, and we show that natural partial realizations exist if and only if they are isomorphic.

Definition 7.17. For an affine dynamical system $\sigma = ((X,F),g,h,h^0)$ and some $p \in N$, if $X = \ll \{\sum_{j=1}^{|\omega|}(F(\omega(|\omega|))F(\omega(|\omega|-1))\cdots F(\omega(|\omega|-j))g(\omega(j)); \omega \in U_p^*\} \gg$, then σ is said to be p-quasi-reachable, where $\ll S \gg$ denotes the smallest linear space which contains a set S.

Let q be some integer. If $hF(\omega(|\omega|))F(\omega(|\omega|-1))\cdots F(\omega(1))x = 0$ implies x=0 for any $\omega \in U_q^*$, then σ is said to be q-distinguishable.

For a given $\underline{a} \in F(U_L^*, Y)$, if there exist p and $q \in N$ such that $p+q<L$ and σ is p-quasi-reachable and q-distinguishable, then σ is said to be a natural partial realization of $\underline{a}$.

For a partial input response map $\underline{a} \in F(U_L^*, Y)$, the following matrix $H^A_{\underline{a}\,(p,L-p)}$ is said to be a finite-sized Hankel matrix of $\underline{a}$.

The column vector of $H^A_{\underline{a}\,(p,L-p)}$ may be written by $S_l(\omega)\underline{a} - \underline{a}$.

$$H^A_{\underline{a}\,(p,L-p)} = \begin{pmatrix} & & \omega & \\ & & \vdots & \\ & & \vdots & \\ \bar{\omega} \quad \cdots & \cdots & \underline{a}(\bar{\omega}|\omega) - \underline{a}(\bar{\omega}) & \\ & & & \end{pmatrix}$$

where $\omega \in U_p^*$ and $\bar{\omega} \in U_{L-p}^*$.

In discussion of approximate and noisy realization of affine dynamical systems, the notation of $H^A_{\underline{a}\,(p,L-p)}(|||\omega_1|||, |||\omega_2|||, |||\omega_3|||, |||\omega_4|||)$ is used as follows:

$H^A_{\underline{a}\,(p,L-p)}(|||\omega_1|||, |||\omega_2|||, |||\omega_3|||, |||\omega_4|||) := [\underline{S}_l(\omega_1)\underline{a}-\underline{a}, \underline{S}_l(\omega_2)\underline{a}-\underline{a}, \underline{S}_l(\omega_3)\underline{a}-\underline{a}, \underline{S}_l(\omega_4)\underline{a}-\underline{a}]$.

Theorem 7.18. *Let $H^L_{\underline{a}\,(p,L-p)}$ be the finite Hankel matrix of $\underline{a} \in F(U_L^*, Y)$. Then there exists a natural partial realization of $\underline{a}$ if and only if the following conditions hold:*

rank $H^L_{\underline{a}\,(p,L-p)}$= rank $H^L_{\underline{a}\,(p,L-p-1)}$ =rank $H^L_{\underline{a}\,(p+1,L-p-1)}$ *for some* $p \in N$.

Theorem 7.19. *There exists a natural partial realization of a given partial input response map $\underline{a} \in F(U_L^*, Y)$ if and only if the minimal partial realization of $\underline{a}$ are unique modulo isomorphisms.*

Theorem 7.20. *Let a partial input response $\underline{a} \in F(U_L^*, Y)$ satisfy the condition of Theorem (4.26), then the quasi-reachable standard system $\sigma_s = ((X,F_s),g_s,h_s,h^0$ which realizes $\underline{a}$ can be obtained by the following algorithm. Set $n :=$ rank $H^L_{\underline{a}\,(p,L-p)}$, where $H^L_{\underline{a}\,(p,L-p)}$ is the finite Hankel-matrix of $\underline{a} \in F(U_L^*, Y)$.*

1) Select the linearly independent vectors $\{S_l(\omega_i)\underline{a} - \underline{a} \in F(U^*_{L-p}, Y); 1 \leq i \leq n\}$ *from* $H^L_{\underline{a}\ (p,L-p)}$ *in order of their numerical value.*
2) Let the state space be $\boldsymbol{R}^n$*, the map* $g_s : U \to X$ *be* $g_s(u_i) = \mathbf{e}_i$*, where* $\mathbf{e}_i := [0, \cdots, 0, \overset{i}{1}, 0, \cdots, 0]^T$.
3) Let the output map $h_s = [\underline{a}(\omega_1) - \underline{a}(1), \underline{a}(\omega_2) - \underline{a}(1), \underline{a}(\omega_3) - \underline{a}(1), \cdots, \underline{a}(\omega_n) - \underline{a}(1)]$.
4) Let ${}_i\mathbf{f_j} \in \boldsymbol{R}^n$ *in* $F_s(u_i) := [{}_i\mathbf{f_1}\ {}_i\mathbf{f_2} \cdots {}_i\mathbf{f_n}]$ *be* ${}_i\mathbf{f_j} := [{}_i\mathbf{f}_{j,1}, {}_i\mathbf{f}_{j,2} \cdots {}_i\mathbf{f}_{j,n}]^T$ *for* $1 \leq i \leq n$*, where* ${}_i\mathbf{f}_j$ *is given by the following:*
$\underline{S_l}(u_i)(\underline{S_l}(\omega_j)\underline{a} - \underline{a}) = \sum_{k=1}^{n} {}_if_{j,k}(\underline{S_l}(\omega_k)\underline{a} - \underline{a})$*,* ${}_if_{j,k} \in \boldsymbol{R}$ *in the sense of* $F(U^*_{L-p}, Y)$ *and* $\underline{S_l}(\omega) : F(U^*_s, Y) \to F(U^*_{s-|\omega|}, Y)$ *;* $\underline{a} \mapsto \underline{S_l}(\omega)\underline{a}[; \bar{\omega} \mapsto \underline{a}(\bar{\omega}|\omega)]$.

7.4 Algebraically Approximate Realization of Affine Dynamical Systems

In this section, we discuss algebraically approximate realization problems of affine dynamical systems.

We will discuss an algebraically approximate realization problem under the assumption that the set U of input values is a finite set $U = \{u_j : 1 \leq j \leq m\}$ for a finite integer $m \in N$. In the reference [Matsuo and Hasegawa, 2003], we showed that this assumption is not so special. However, for simplicity of our discussion, we assume that the set U of input values is $U = \{u_1,\ u_2,\ u_3\}$.

Roughly speaking, the algebraically approximate realization of affine dynamical systems can be stated as follows:

< For any given partial data of an affine dynamical system, find, using only algebraic calculations, an affine dynamical system which approximates the given data. >

In order to make our discussion simple, we assume that the set Y of output is the set $\boldsymbol{R}$ of real numbers, namely 1-output.

Theorem 7.21. *Algebraic algorithm for approximate realization*

Let an input response map $\underline{a}$ *be a considered object which is an affine dynamical system. Then an approximate realization* $\sigma = ((\boldsymbol{R}^n, F_s), g_s, h_s, h^0)$ *of* $\underline{a}$ *is given by the following algorithm:*

1) Based on the ratio of the square root of eigenvalues for a matrix $H_{\underline{a}\ (n,\bar{p})}(|||\omega_1|||, \cdots, |||\omega_n|||)H_{\underline{a}\ (n,\bar{p})}(|||\omega_1|||, \cdots, |||\omega_n|||)^T$*, determine the value* n *of rank for the matrix, where* $|||\omega_1|||$*,* $|||\omega_2|||$*,* $\cdots$ *and* $|||\omega_n|||$ *are selected in the order of numerical value of input and* $\{\underline{S_l}(\omega_i)\underline{a} - \underline{a}; 1 \leq i \leq n,\ \omega_i \in U^*\}$ *is a set of independent vectors.*
Namely, determine the value n *of rank for the matrix* $H_{\underline{a}\ (n,\bar{p})}(|||\omega_1|||, \cdots, |||\omega_n|||)$ *such that the ratio of the square root of eigenvalues for the covariance matrix becomes very small. The small ratio*

means the nearness of approximation degree.

2) *In order to determine* g_s*, the algebraic CLS method is used as follows:*

① *In particular, set* $g_s(\omega_i) := \mathbf{e}_i$ *for* $\omega_i \in U$*. Namely,* $g_s(\omega_1) := \mathbf{e}_1$*,* $g_s(\omega_2) := \mathbf{c}_2, \cdots, g_s(\omega_k) := \mathbf{e}_k$ *for some* $k \in N$*.*
For $u \in U$ *such that* $u \notin \{\omega_i; 1 \leq i \leq n\}$ *and* $\omega_r < u$*,*
$g_s(u) = \sum_{j=1}^{r} b_{u,j}(\underline{S_l}(\omega_j)\underline{a} - \underline{a})$ *is obtained as follows:*
Based on Proposition (2.14), determine coefficients $\{b_{u,j} : 1 \leq j \leq r\}$*.*
The Q *in Proposition (2.14) can be considered as the matrix composed from the eigenvectors of* $H_{\underline{a}\ (r+1,L)}(|||\omega_1|||, |||\omega_2|||, \cdots, |||\omega_r|||, |||u|||)$
$\times H^T_{\underline{a}\ (r+1,L)}(|||\omega_1|||, |||\omega_2|||, \cdots, |||\omega_r|||, |||u|||)$.
Let a matrix $A_u \in \boldsymbol{R}^{1\times(r+1)}$ *be* $A_u := [b_{u,1}, b_{u,2}, \cdots, b_{u,r}, -1]$*.*

② *Determine the error vectors* $\{\underline{S_l}(\omega_j)\underline{\bar{a}} - \underline{\bar{a}} \in \boldsymbol{R}^{L\times 1} : 0 \leq j \leq r\}$ *by using the equation*
$[\underline{S_l}(\omega_1)\underline{\bar{a}} - \underline{\bar{a}}, \underline{S_l}(\omega_2)\underline{\bar{a}} - \underline{\bar{a}}, \cdots, \underline{S_l}(\omega_r)\underline{\bar{a}} - \underline{\bar{a}}, \underline{S_l}(u)\underline{\bar{a}} - \underline{\bar{a}}]^T :=$
$A_u^T[A_uA_u^T]^{-1}A_uH^T_{\underline{a}\ (r+1,L)}(|||\omega_1|||, |||\omega_2|||, \cdots, |||\omega_r|||, |||u|||)$
and $H^T_{\underline{a}\ (|||u|||+1,L)}(|||\omega_1|||, |||\omega_2|||, \cdots, |||\omega_r|||, |||u|||) :=$
$[\underline{S_l}(\omega_1)\underline{a} - \underline{a}, \cdots, \underline{S_l}(\omega_r)\underline{a} - \underline{a}, \underline{S_l}(u)\underline{a} - \underline{a}]$.

3) *In order to obtain* F_s*, the algebraic CLS method is used as follows:*

① *Let* ${}_i\mathbf{f}_j \in \boldsymbol{R}^n$ *in* $F_s(u_i) := [{}_i\mathbf{f}_1\ {}_i\mathbf{f}_2 \cdots {}_i\mathbf{f}_n]$ *be* ${}_i\mathbf{f}_j := [{}_if_{j,1}\ {}_if_{j,2}, \cdots {}_if_{j,n}]^T$ *for* $1 \leq i \leq n$*, where* ${}_if_{j,k}$ *is given by the following:*
$\underline{S_l}(u_i)(\underline{S_l}(\omega_j)\underline{a} - \underline{a}) = \sum_{k=1}^{n} {}_if_{j,k}(\underline{S_l}(\omega_k)\underline{a} - \underline{a})$, ${}_if_{j,k} \in \boldsymbol{R}$ *in the sense of* $F(U^*_{L-p}, Y)$*.*
We cannot directly obtain the coefficients $\{{}_i\mathbf{f}_j; 1 \leq i \leq 3,\ 1 \leq j \leq n\}$*.*
Firstly, we will determine coefficients $\{{}_i\bar{f}_{j,k}; 1 \leq k \leq n\}$ *from the equation* $\underline{S_l}(u_i)\underline{S_l}(\omega_j)\underline{a} - \underline{a} = \sum_{k=1}^{n} {}_i\bar{f}_{j,k}(\underline{S_l}(\omega_k)\underline{a} - \underline{a})$ *by using the algebraic CLS method.*

②*For* i $(1 \leq i \leq 3)$*,* j $(1 \leq j \leq n)$ *and for the maximum number* r $(1 \leq r \leq n)$ *such that* $\omega_r, \omega_j \in \{\omega_j; 1 \leq j \leq n\}$ *and* $|||\omega_r||| < |||u_i|\omega_j|||$*, and based on Proposition (2.14), determine coefficients* $\{{}_i\bar{f}_{j,k} = 0; r+1 \leq k \leq n\}$ *and* $\{{}_i\bar{f}_{j,k} : 1 \leq k \leq r\}$*.*
The Q *in Proposition (2.14) can be considered as the matrix composed from the eigenvectors of* $H_{\underline{a}\ (r+1,L)}(|||\omega_1|||, \cdots, |||\omega_r|||, |||u_i|\omega_j|||)$
$\times H^T_{\underline{a}\ (r+1,L)}(|||\omega_1|||, \cdots, |||\omega_r|||, |||u_i|\omega_j|||)$.
Let a matrix ${}_iA_j \in \boldsymbol{R}^{1\times(r+1)}$ *be* ${}_iA_j := [{}_i\bar{f}_{j,1}, {}_i\bar{f}_{j,2}, \cdots, {}_i\bar{f}_{j,r}, -1]$*.*
Determine the error vectors $\{\underline{S_l}(\omega_j)\underline{\bar{a}} - \underline{\bar{a}} \in \boldsymbol{R}^{L\times 1} : 0 \leq j \leq r\}$ *and* $\underline{S_l}(u_i|\omega_j)\underline{\bar{a}} - \underline{\bar{a}}$ *by using the equation*
$[\underline{S_l}(\omega_1)\underline{\bar{a}} - \underline{\bar{a}}, \underline{S_l}(\omega_2)\underline{\bar{a}} - \underline{\bar{a}}, \cdots, \underline{S_l}(\omega_r)\underline{\bar{a}} - \underline{\bar{a}}, \underline{S_l}(u_i|\omega_j)\underline{\bar{a}} - \underline{\bar{a}}]^T :=$
${}_iA_j^T[{}_iA_j\ {}_iA_j^T]^{-1}\ {}_iA_j\ H^T_{\underline{a}\ (r+1,L)}(|||\omega_1|||, |||\omega_2|||, \cdots, |||\omega_r|||, |||u_i|\omega_j|||)$ *and*
$H^T_{\underline{a}\ (r+1,L)}(|||\omega_1|||, \cdots, |||\omega_r|||, |||u_i|\omega_j|||) :=$
$[\underline{S_l}(\omega_1)\underline{a} - \underline{a}, \cdots, \underline{S_l}(\omega_r)\underline{a} - \underline{a}, \underline{S_l}(u_i|\omega_j)\underline{a} - \underline{a}]$.

③*Next, using the equations*
$\underline{S_l}(u_i)(\underline{S_l}(\omega_j)\underline{a} - \underline{a})$
$= \underline{S_l}(u_i)\underline{S_l}(\omega_j)\underline{a} - \underline{a} - \underline{S_l}(u_i)\underline{a} + \underline{a}$
$= \sum_{k=1}^{n} {}_i\bar{f}_{j,k}(\underline{S_l}(\omega_k)\underline{a} - \underline{a}) - \sum_{j=1}^{r_{u_i}} b_{u_ij}(\underline{S_l}(u_j)\underline{a} - \underline{a})$,

we obtain $\underline{S_l}(u_i)(\underline{S_l}(\omega_j)\underline{a} - \underline{a})$
$= \sum_{k=1}^{n} {}_i\bar{f}_{j,k}(\underline{S_l}(\omega_k)\underline{a} - \underline{a}) - \sum_{j=1}^{r_{u_i}} b_{u_i,i}(\underline{S_l}(u_j)\underline{a} - \underline{a})$
$= \sum_{j=1}^{r_{u_i}} ({}_i\bar{f}_{j,k} - b_{u_i,j})(\underline{S_l}(\omega_j)\underline{a} - \underline{a}) + \sum_{j=r_{u_i}+1}^{n} {}_i\bar{f}_{j,k}(\underline{S_l}(\omega_j)\underline{a} - \underline{a})$.
On the other hand, the equation
$\underline{S_l}(u_i)(\underline{S_l}(\omega_j)\underline{a} - \underline{a}) = \sum_{k=1}^{n} {}_i f_{j,k}(\underline{S_l}(\omega_k)\underline{a} - \underline{a})$ holds. Therefore, we obtain the following equation:
$\sum_{k=1}^{n} {}_i f_{j,k}(\underline{S_l}(\omega_k)\underline{a} - \underline{a})$
$= \sum_{j=1}^{r_{u_i}} ({}_i\bar{f}_{j,k} - b_{u_i,j})\underline{S_l}(\omega_j)\underline{a} - \underline{a} + \sum_{j=r_{u_i}+1}^{n} {}_i\bar{f}_{j,k}(\underline{S_l}(\omega_j)\underline{a} - \underline{a})$.
Comparing the coefficients, we obtain the following:
$\sum_{j=1}^{r_{u_i}} ({}_i f_{j,k} - {}_i\bar{f}_{j,k} + b_{u_i,j})(\underline{S_l}(\omega_j)\underline{a} - \underline{a}) + \sum_{j=r_{u_i}+1}^{n} ({}_i f_{j,k} - {}_i\bar{f}_{j,k})(\underline{S_l}(\omega_j)\underline{a} - \underline{a})$
$= 0$.
Finally, coefficients ${}_i f_{j,k}$ of ${}_i\mathbf{f}_j$ in F_s are obtained as follows:
${}_i f_{j,k} = {}_i\bar{f}_{j,k} - b_{u_i,j}$ for $1 \le j \le r_{u_i}$,
${}_i f_{j,k} = {}_i\bar{f}_{j,k}$ for $r_{u_i} + 1 \le j \le n$.

4) In order to determine $h_s \in \boldsymbol{R}^{1\times n}$, the algebraic CLS method is used as follows:
① For the first ω_{r_1+1}, $\omega_{r_1} \in \{\omega_i; 1 \le i \le n\}$ such that $\omega_{r_1+1} > \omega_{r_1}$ and $|||\omega_{r_1+1}|||$ - $|||\omega_{r_1}||| > 1$ when starting out from ω_1,
set $\underline{S_l}(\lambda_1)\underline{a} - \underline{a} := \sum_{i=1}^{r_1} b_{\lambda_1,i}\mathbf{e}_i$ for the obtained equation $\underline{S_l}(\lambda_1)\underline{a} - \underline{a} = \sum_{i=1}^{r_1} b_{\lambda_1,i}(\underline{S_l}(\omega_i)\underline{a} - \underline{a})$ for λ_1 such that $|||\lambda_1||| = |||\omega_{r_1}||| + 1$.
Based on Proposition (2.14), determine coefficients $\{b_{\lambda_1,i} : 1 \le i \le r_1\}$.
The Q in Proposition (2.14) can be considered as the matrix composed from the eigenvectors of $H_{\underline{a}\ (|||\lambda_1|||+1,L)}(|||\omega_1|||, |||\omega_2|||, \cdots, |||\omega_{r_1}|||, |||\lambda_1|||)$
$\times H^T_{\underline{a}\ (|||\lambda_1|||+1,L)}(|||\omega_1|||, |||\omega_2|||, \cdots, |||\omega_{r_1}|||, |||\lambda_1|||)$.
Let a matrix $A_{\lambda_1} \in \boldsymbol{R}^{1\times(r_1+1)}$ be $A_{\lambda_1} := [b_{\lambda_1,1}, b_{\lambda_1,2}, \cdots, b_{\lambda_1,r_1}, -1]$.
Determine the error vectors $\{\underline{S_l}(\omega_i)\underline{\bar{a}} - \underline{\bar{a}} \in \boldsymbol{R}^{L\times 1} :\ 0 \le i \le r_1\}$ and $\underline{S_l}(\lambda_1)\underline{\bar{a}} - \underline{\bar{a}}$ by using the equation
$[\underline{S_l}(\omega_1)\underline{\bar{a}} - \underline{\bar{a}}, \underline{S_l}(\omega_2)\underline{\bar{a}} - \underline{\bar{a}}, \cdots, \underline{S_l}(\omega_{r_1})\underline{\bar{a}} - \underline{\bar{a}}, \underline{S_l}(\lambda_1)\underline{\bar{a}} - \underline{\bar{a}}]^T :=$
$A^T_{\lambda_1}[A_{\lambda_1}A^T_{\lambda_1}]^{-1}A_{\lambda_1} H^T_{\underline{a}\ (r_1+1,L)}(|||\omega_1|||, |||\omega_2|||, \cdots, |||\omega_{r_1}|||, |||\lambda_1|||)$ and
$H^T_{\underline{a}\ (|||\lambda_1|||+1,L)}(|||\omega_1|||, |||\omega_2|||, \cdots, |||\omega_{r_1}|||, |||\lambda_1|||) :=$
$[\underline{S_l}(\omega_1)\underline{a} - \underline{a}, \cdots, \underline{S_l}(\omega_{r_1})\underline{a} - \underline{a}, \underline{S_l}(\lambda_1)\underline{a} - \underline{a}]$.
Then let $h_{\lambda_1 s} \in \boldsymbol{R}^{1\times r_1}$ be
$h_{\lambda_1 s} := [\underline{a}(\omega_1) - \underline{a}(1) - (\underline{\bar{a}}(\omega_1) - \underline{\bar{a}}(1)), \underline{a}(\omega_2) - \underline{a}(1) - (\underline{\bar{a}}(\omega_2) - \underline{\bar{a}}(1)), \cdots,$
$\underline{a}(\omega_{r_1}) - \underline{a}(1) - (\underline{\bar{a}}(\omega_{r_1}) - \underline{\bar{a}}(1))]$.
② For the first ω_{r_2+1}, $\omega_{r_2} \in \{\omega_i; 1 \le i \le n\}$ such that $\omega_{r_2+1} > \omega_{r_2}$ and $|||\omega_{r_2+1}|||$ - $|||\omega_{r_2}||| > 1$ when starting out from ω_{r_1+1},
set $\underline{S_l}(\lambda_2)\underline{a} - \underline{a} := \sum_{i=1}^{r_2} b_{\lambda_2,i}\mathbf{e}_i$ for the obtained equation $\underline{S_l}(\lambda_2)\underline{a} - \underline{a} = \sum_{i=1}^{r_2} b_{\lambda_2,i}(\underline{S_l}(\omega_i)\underline{a} - \underline{a})$ for λ_2 such that $|||\lambda_2||| = |||\omega_{r_2}||| + 1$.
Based on Proposition (2.14), determine coefficients $\{b_{\lambda_2,i} : 1 \le i \le r_2\}$.
The Q in Proposition (2.14) can be considered as the matrix composed from the eigenvectors of $H_{\underline{a}\ (r_2+1,L)}(|||\omega_1|||, |||\omega_2|||, \cdots, |||\omega_{r_2}|||, |||\lambda_2|||)$
$\times H^T_{\underline{a}\ (r_2+1,L)}(|||\omega_1|||, |||\omega_2|||, \cdots, |||\omega_{r_2}|||, |||\lambda_2|||)$.
Let a matrix $A_{\lambda_2} \in \boldsymbol{R}^{1\times(r_2+1)}$ be $A_{\lambda_2} := [b_{\lambda_2,1}, b_{\lambda_2,2}, \cdots, b_{\lambda_2,r_2}, -1]$.

Determine the error vectors $\{\underline{S_l}(\omega_i)\bar{\underline{a}} - \bar{\underline{a}} \in \boldsymbol{R}^{L\times 1} : 0 \le i \le r_2\}$ *and* $\underline{S_l}(\lambda_2)\bar{\underline{a}} - \bar{\underline{a}}$ *by using the equation*
$[\underline{S_l}(\omega_1)\bar{\underline{a}} - \bar{\underline{a}}, \underline{S_l}(\omega_2)\bar{\underline{a}} - \bar{\underline{a}}, \cdots, \underline{S_l}(\omega_{r_2})\bar{\underline{a}} - \bar{\underline{a}}, \underline{S_l}(\lambda_2)\bar{\underline{a}} - \bar{\underline{a}}]^T :=$
$A_{\lambda_2}^T[A_{\lambda_2}A_{\lambda_2}^T]^{-1}A_{\lambda_2}H^T_{\underline{a}\ (r_2+1,L)}(|||\omega_1|||, |||\omega_2|||, \cdots, |||\omega_{r_2}|||, |||\lambda_2|||)$
and $H^T_{\underline{a}\ (|||\lambda_2|||+1,L)}(|||\omega_1|||, |||\omega_2|||, \cdots, |||\omega_{r_2}|||, |||\lambda_2|||) :=$
$[\underline{S_l}(\omega_1)\underline{a} - \underline{a}, \cdots, \underline{S_l}(\omega_{r_2})\underline{a} - \underline{a}, \underline{S_l}(\lambda_2)\underline{a} - \underline{a}]$.
Then let $h_{\lambda_2 s} \in \boldsymbol{R}^{1\times r_2}$ *be*
$h_{\lambda_2 s} := [\underline{a}(\omega_{r_1+1}) - \underline{a}(1) - (\bar{\underline{a}}(\omega_{r_1+1}) - \bar{\underline{a}}(1)),$
$\underline{a}(\omega_{r_1+2}) - \underline{a}(1) - (\bar{\underline{a}}(\omega_{r_1+2}) - \bar{\underline{a}}(1)), \cdots, \ \underline{a}(\omega_{r_2}) - \underline{a}(1) - (\bar{\underline{a}}(\omega_{r_2}) - \bar{\underline{a}}(1))]$.

⋮

ⓣ *For* $\omega \in U^*$ *such that* $|||\omega||| = |||\omega_n||| + 1$ *and based on Proposition (2.14), determine coefficients* $\{b_{\omega,i} : 1 \le i \le n\}$.
The Q in Proposition (2.14) can be considered as the matrix composed from the eigenvectors of $H_{\underline{a}\ (n+1,L)}(|||\omega_1|||, |||\omega_2|||, \cdots, |||u|||)$
$\times H^T_{\underline{a}\ (n+1,L)}(|||\omega_1|||, |||\omega_2|||, \cdots, |||u|||)$.
Let a matrix $A_\omega \in \boldsymbol{R}^{1\times(n+1)}$ *be* $A_\omega := [b_{\omega,1}, b_{\omega,2}, \cdots, b_{\omega,n}, -1]$.
Determine the error vectors $\{\underline{S_l}(\omega_i)\bar{\underline{a}} - \bar{\underline{a}} : 0 \le i \le n\}$ *and* $\underline{S_l}(\omega)\bar{\underline{a}} - \bar{\underline{a}}$ *by using the equation*
$[\underline{S_l}(\omega_1)\bar{\underline{a}} - \bar{\underline{a}}, \underline{S_l}(\omega_2)\bar{\underline{a}} - \bar{\underline{a}}, \cdots, \underline{S_l}(\omega_n)\bar{\underline{a}} - \bar{\underline{a}}, \underline{S_l}(\omega)\bar{\underline{a}} - \bar{\underline{a}}]^T :=$
$A_\omega^T[A_\omega A_\omega^T]^{-1}A_\omega H^T_{\underline{a}\ (n+1,L)}(|||\omega_1|||, |||\omega_2|||, \cdots, |||\omega_n|||, |||\omega|||)$
and $H^T_{\underline{a}\ (n+1,L)}(|||\omega_1|||, |||\omega_2|||, \cdots, |||u|||) :=$
$[\underline{S_l}(\omega_1)\underline{a} - \underline{a}, \cdots, \underline{S_l}(\omega_n)\underline{a} - \underline{a}, \underline{S_l}(\omega)\underline{a} - \underline{a}]$,
Then let $h_{\omega s}$ *be*
$h_{\omega s} := [\underline{a}(\omega_{r_t+1}) - \underline{a}(1) - (\bar{\underline{a}}(\omega_{r_t+1}) - \bar{\underline{a}}(1)),$
$\underline{a}(\omega_{r_t+2}) - \underline{a}(1) - (\bar{\underline{a}}(\omega_{r_t+2}) - \bar{\underline{a}}(1)), \cdots, \ \underline{a}(\omega_n) - \underline{a}(1) - (\bar{\underline{a}}(\omega_n) - \bar{\underline{a}}(1))]$.
Finally, let $h_s \in \boldsymbol{R}^{1\times n}$ *be*
$h_s := [h_{\lambda_1 s}, h_{\lambda_2 s}, \cdots, h_{\omega s}]$.

[proof] By 1), 2), 3) and 4), the reduction part in the data can be excluded in the sense of the number of dimensions by using the ratio of the matrix norm, which produces a degree of information loss. The matrices A_u in 2), ${}_iA_j$ in 3) A_{λ_1}, A_{λ_2}, $\cdots$, and A_ω in 4) correspond to the matrix A in Lemma (2.17). Hence, the reduced part of the given finite-sized Input/output matrix was obtained. Therefore, applying Theorem (7.20), we can obtain g_s, F_s and h_s by 2*m+1) .

In the figures of this chapter, we use a notation *Signal_n-d* as an input response map obtained by a n-dimensional affine dynamical system.

In the examples of this chapter, a notation $H^T_{\underline{a}\ (r,40)}(1, \cdots, r)$ is used in place of $H^T_{\underline{a}\ (r,40)}(1, 2, 3, \cdots, r-1, r)$.

Example 7.22. Let the signals be the input response map of the following 3-dimensional affine dynamical system $\sigma = ((\boldsymbol{R}^3, F), g, h, h^0)$, $F(u_1) =$

$$\begin{bmatrix} -1 & 1.5 & 0.1 \\ 0 & 0 & -0.1 \\ 0 & -0.2 & 0.1 \end{bmatrix},\ F(u_2) = \begin{bmatrix} -0.3 & 0 & -0.4 \\ 0 & 1 & 0 \\ -0.7 & 0 & 0.2 \end{bmatrix},\ F(u_3) = \begin{bmatrix} 0 & -0.5 & 0 \\ 0 & 1.1 & 0.3 \\ 0 & 0 & 0.1 \end{bmatrix},$$

$g(u_1) = \mathbf{e}_1,\ g(u_2) = \mathbf{e}_2,\ g(u_3) = \mathbf{e}_3,\ h = [15,\ -3,\ -1],\ h^0 = 1.$

Then the algebraically approximate realization problem is solved as follows:

covariance matrix	eigenvalues						
	1	2	3	4	5	···	12
$H^T_{\underline{a}\ (2,40)}(1,2)H_{\underline{a}\ (2,40)}(1,2)$	9577	874					
$H^T_{\underline{a}\ (3,40)}(1,2,3)H_{\underline{a}\ (3,40)}(1,2,3)$	9908	918	109				
$H^T_{\underline{a}\ (4,40)}(1,\cdots,4)H_{\underline{a}\ (4,40)}(1,\cdots,4)$	9908	918	109	0			
$H^T_{\underline{a}\ (5,40)}(1,\cdots,5)H_{\underline{a}\ (5,40)}(1,\cdots,5)$	17800	978	147	0	0		
$H^T_{\underline{a}\ (12,40)}(1,\cdots,12)H_{\underline{a}\ (12,40)}(1,\cdots,12)$	89200	7292	514	0	0	···	0
covariance matrix	square root of eigenvalues						
$H^T_{\underline{a}\ (2,40)}(1,2)H_{\underline{a}\ (2,40)}(1,2)$	97.9	29.6					
$H^T_{\underline{a}\ (3,40)}(1,2,3)H_{\underline{a}\ (3,40)}(1,2,3)$	99.5	30.3	10.4				
$H^T_{\underline{a}\ (4,40)}(1,\cdots,4)H_{\underline{a}\ (4,40)}(1,\cdots,4)$	99.5	30.3	10.4	0			
$H^T_{\underline{a}\ (5,40)}(1,\cdots,5)H_{\underline{a}\ (5,40)}(1,\cdots,5)$	133	31.3	12.1	0	0		
$H^T_{\underline{a}\ (12,40)}(1,\cdots,12)H_{\underline{a}\ (12,40)}(1,\cdots,12)$	299	85.4	22.7	0	0	···	0

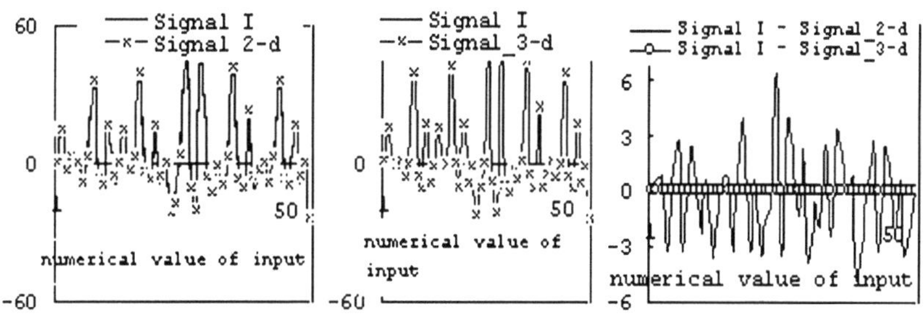

Fig. 7.1 The left is the original input response map and the behavior of a 2-dimensional affine dynamical system obtained by the algebraic CLS method. The middle is the original input response map and the behavior of a 3-dimensional affine dynamical system obtained by the algebraic CLS method. The right is the difference between the original one and the behavior of the 2-dimensional affine dynamical system obtained by the algebraic CLS method or the 3-dimensional affine dynamical system obtained by the algebraic CLS method in Example (7.22).

1) Since the ratio $\frac{10.4}{99.5} = 0.1$ obtained by the square root of $H^T_{\underline{a}\ (3,40)}(1,2,3)H_{\underline{a}\ (3,40)}(1,2,3)$ is a little large, the approximate 2-dimensional affine dynamical system obtained by the algebraic CLS method may not be good.
2) After determining the independent vectors $\underline{S}_l(u_1)\underline{a} - \underline{a}$ and $\underline{S}_l(u_2)\underline{a} - \underline{a}$ whose numerical value of input are 1 and 2, we will continue the approximate realization algorithm by the algebraic CLS method.

Therefore, an approximate 2-dimensional affine dynamical system $\sigma_1 = ((\boldsymbol{R}^2, F_1), g_1, h_1, h^0)$ obtained by the algebraic CLS method is constructed as follows:

$$F_1(u_1) = \begin{bmatrix} -1 & 1.5 \\ 0 & -0.04 \end{bmatrix}, \; F_1(u_2) = \begin{bmatrix} -0.44 & 0 \\ -0.15 & 1 \end{bmatrix}, \; F_1(u_3) = \begin{bmatrix} 0 & -0.54 \\ 0 & 1.1 \end{bmatrix},$$
$$g_1(u_1) = \mathbf{e}_1, \; g_1(u_2) = \mathbf{e}_2, \; g_1(u_3) = [0.21, \; 0.2], \; h_1 = [14.31, \; -3.7], \; h^0 = 1.$$

For reference, a 3-dimensional affine dynamical system $\sigma_2 = ((\boldsymbol{R}^3, F_2), g_2, h_2, h^0)$ obtained by the algebraic CLS method can be expressed as follows:

$$F_2(u_1) = \begin{bmatrix} -1 & 1.5 & 0.1 \\ 0 & 0 & -0.1 \\ 0 & -0.2 & 0.1 \end{bmatrix}, \; F_2(u_2) = \begin{bmatrix} -0.3 & 0 & -0.4 \\ 0 & 1 & 0 \\ -0.7 & 0 & 0.2 \end{bmatrix},$$
$$F_2(u_3) = \begin{bmatrix} 0 & -0.5 & 0 \\ 0 & 1.1 & 0.3 \\ 0 & 0 & 0.1 \end{bmatrix}, \; g_2(u_1) = \mathbf{e}_1, \; g_2(u_2) = \mathbf{e}_2, \; g_2(u_3) = \mathbf{e}_3,$$
$$h_2 = [15, \; -3, \; -1], \; h^0 = 1.$$

The system is completely reconstructed by the algebraic CLS method.

We can show that the algorithm for approximate realization given by the analytic CLS method in the reference [Hasegawa, 2008] produces the same systems as the above ones in the sense of the numerical calculation.

In this example, the original signals are considered as the input response map of a 3-dimensional affine dynamical system and the desirable input response map is obtained by the algebraic CLS method with our bad feeling. The model obtained by the algebraic CLS method is a 2-dimensional affine dynamical system.

For reference, a 3-dimensional affine dynamical system is also given by the algebraic CLS method. The system completely reconstructs the original system.

Just as we thought, the following table and Fig. 7.1 truly indicate that the 2-dimensional affine dynamical system obtained by the algebraic CLS method is a bad approximation. For reference, the input response map of the same dimensional affine dynamical system as the original system is shown. Hence, there does not exist a good approximation for the given system.

dimen-sion	ratio of matrices	mean values of square root for sum of signal ①	signal by CLS ②	error ③	cosine ① and ② $\cos\theta$	error ratio ③/①
$a_{1,2}$	0.1	2.687	2.557	0.351	0.992	0.13
$a_{1,2,3}$	0	2.687	2.687	0	1	0

Example 7.23. Let the signals be the input response map of the following 4-dimensional affine dynamical system $\sigma = ((\boldsymbol{R}^4, F), g, h, h^0)$, $F(u_1) = \begin{bmatrix} 0 & -0.5 & -0.6 & -0.4 \\ 0 & 0.6 & 0 & -0.3 \\ -0.6 & 0 & -0.1 & 0.4 \\ 0 & -0.5 & 0.5 & -0.6 \end{bmatrix}$, $F(u_2) = \begin{bmatrix} 0.6 & 0 & -0.8 & 0.3 \\ 0 & 0.8 & 0 & 0 \\ -0.3 & 0 & 0.6 & 0.8 \\ 0.3 & 0 & 0.3 & -0.7 \end{bmatrix}$, $F(u_3) = \begin{bmatrix} -1 & -0.3 & 0 & -0.7 \\ 0 & 0 & 1 & 0.6 \\ 0 & -0.3 & -0.3 & 0.7 \\ 0.8 & 0.3 & 0.4 & 0.6 \end{bmatrix}$, $g(u_1) = \mathbf{e}_1$, $g(u_2) = \mathbf{e}_2$, $g(u_3) = \mathbf{e}_3$, $h = [16,\ 6,\ 0,\ 1]$, $h^0 = 1$.

Then the algebraically approximate realization problem is solved as follows:

covariance matrix	eigenvalues						
	1	2	3	4	5	$\cdots$	12
$H^T_{\underline{a}\ (2,50)}(1,2)H_{\underline{a}\ (2,50)}(1,2)$	3353	719					
$H^T_{\underline{a}\ (3,50)}(1,2,3)H_{\underline{a}\ (3,50)}(1,2,3)$	4848	2013	685				
$H^T_{\underline{a}\ (4,50)}(1,\cdots,4)H_{\underline{a}\ (4,50)}(1,\cdots,4)$	11082	2065	689	0			
$H^T_{\underline{a}\ (5,50)}(1,\cdots,5)H_{\underline{a}\ (5,50)}(1,\cdots,5)$	14034	2648	1289	106	0		
$H^T_{\underline{a}\ (12,50)}(1,\cdots,12)H_{\underline{a}\ (12,50)}(1,\cdots,12)$	32932	10805	2657	2093	0	$\cdots$	0
covariance matrix	square root of eigenvalues						
$H^T_{\underline{a}\ (2,50)}(1,2)H_{\underline{a}\ (2,50)}(1,2)$	57.9	26.8					
$H^T_{\underline{a}\ (3,50)}(1,2,3)H_{\underline{a}\ (3,50)}(1,2,3)$	69.6	44.9	26.2				
$H^T_{\underline{a}\ (4,50)}(1,\cdots,4)H_{\underline{a}\ (4,50)}(1,\cdots,4)$	105	45.4	26.2	0			
$H^T_{\underline{a}\ (5,50)}(1,\cdots,5)H_{\underline{a}\ (5,50)}(1,\cdots,5)$	118	51.4	35.9	10.3	0		
$H^T_{\underline{a}\ (12,50)}(1,\cdots,12)H_{\underline{a}\ (12,50)}(1,\cdots,12)$	181	104	51.5	45.8	0	$\cdots$	0

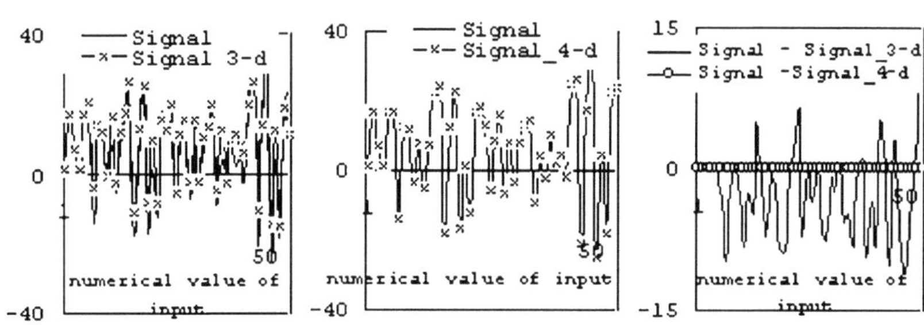

Fig. 7.2 The left is the original input response map and the behavior of a 3-dimensional affine dynamical system obtained by the algebraic CLS method. The middle is the original input response map and the behavior of a 4-dimensional affine dynamical system obtained by the algebraic CLS method. The right is the difference between the original one and the behavior of the 3-dimensional affine dynamical system obtained by the algebraic CLS method or the 4-dimensional affine dynamical system obtained by the algebraic CLS method in Example (7.23).

1) Since the ratio $\frac{10.3}{118} = 0.09$ obtained by the square root of $H^T_{\underline{a}\ (5,50)}(1,\cdots,5) \times H_{\underline{a}\ (5,50)}(1,\cdots,5)$ is not so small and the ratio $\frac{45.8}{181} = 0.25$ obtained by the square root of $H^T_{\underline{a}\ (12,50)}(1,\cdots,12)H_{\underline{a}\ (12,50)}(1,\cdots,12)$ is rather large, the approximate 3-dimensional affine dynamical system obtained by the algebraic CLS method may not be good.
2) After determining the independent vectors $\underline{S_l}(u_1)\underline{a} - \underline{a}$, $\underline{S_l}(u_2)\underline{a} - \underline{a}$ and $\underline{S_l}(u_3)\underline{a} - \underline{a}$ whose numerical value of input are 1, 2 and 3, we will continue the approximate realization algorithm by the algebraic CLS method.

Therefore, an approximate 3-dimensional affine dynamical system $\sigma_1 = ((\boldsymbol{R}^3,\ F_1),\ g_1,\ h_1,\ h^0)$ obtained by the algebraic CLS method is constructed as follows:

$$F_1(u_1) = \begin{bmatrix} 0 & -0.79 & -0.52 \\ 0 & 1.6 & 1.26 \\ -0.6 & -0.22 & -0.05 \end{bmatrix},\ F_1(u_2) = \begin{bmatrix} 0.66 & 0 & -0.77 \\ 0.43 & 0.8 & 0.42 \\ -0.26 & 0 & 0.65 \end{bmatrix},$$
$$F_1(u_3) = \begin{bmatrix} -1.17 & -0.24 & 0.07 \\ 2.24 & 0.32 & 1.7 \\ 0.12 & -0.21 & -0.22 \end{bmatrix},\ g_1(u_1) = \mathbf{e}_1,\ g_1(u_2) = \mathbf{e}_2,$$

$$g_1(u_3) = \mathbf{e}_3,\ h_1 = [16,\ 6,\ 0],\ h^0 = 1.$$

For reference, a 4-dimensional affine dynamical system $\sigma_2 = ((\boldsymbol{R}^4, F_2), g_2, h_2, h^0)$ obtained by the algebraic CLS method can be expressed as follows:

$$F_2(u_1) = \begin{bmatrix} 0 & 0.5 & -1.6 & 1.22 \\ 0 & 2.3 & -1.67 & 3.3 \\ -0.6 & -0.5 & 0.4 & -1.04 \\ 0 & -1.67 & 1.67 & -2.77 \end{bmatrix},\ F_2(u_2) = \begin{bmatrix} 0 & 0 & -1.4 & 0.93 \\ -1 & 0.8 & -1 & 1.2 \\ 0 & 0 & 0.9 & -0.24 \\ 1 & 0 & 1 & -0.4 \end{bmatrix},$$
$$F_2(u_3) = \begin{bmatrix} -2.6 & -0.9 & -0.8 & -2.8 \\ -2.67 & -1 & -0.3 & -2.92 \\ 0.8 & 0 & 0.1 & 0.84 \\ 2.67 & 1 & 1.3 & 2.8 \end{bmatrix},\ g_2(u_1) = \mathbf{e}_1,\ g_2(u_2) = \mathbf{e}_2,$$
$$g_2(u_3) = \mathbf{e}_3,\ h_2 = [16,\ 6,\ 0,\ 15.9],\ h^0 = 1.$$

The system is completely reconstructed by the algebraic CLS method.

We can show that the algorithm for approximate realization given by the analytic CLS method in the reference [Hasegawa, 2008] produces the same systems as the above ones in the sense of the numerical calculation.

In this example, the original signals are considered as the input response map of a 4-dimensional affine dynamical system and the desirable input response map is obtained by the algebraic CLS method with our bad feeling. The model obtained by the algebraic CLS method is a 3-dimensional affine dynamical system.

For reference, a 4-dimensional affine dynamical system is also given by the algebraic CLS method. The system completely reconstructs the original system.

Just as we thought, the following table and Fig. 7.2 truly indicate that the 3-dimensional affine dynamical system obtained by the algebraic CLS method is a bad approximation. For reference, the input response map of the same dimensional affine dynamical system as the original system is shown.

dimen- sion	ratio of matrices	mean values of square root for sum of signal ①	signal by CLS ②	error ③	cosine ① and ② $\cos\theta$	error ratio ③/①
$a_{1,2,3}$	0.08	2.104	2.019	0.756	0.933	0.36
$a_{1,2,3,5}$	0	2.104	2.104	0	1	0

Example 7.24. Let the signals be the input response map of the following 4-dimensional affine dynamical system: $\sigma = ((\boldsymbol{R}^4, F), g, h, h^0)$, where

$$F(u_1) = \begin{bmatrix} -1.5 & -0.3 & -1 & -0.2 \\ 0 & 0 & 1 & 0.6 \\ 0 & 2.3 & -0.3 & 0 \\ 0.8 & 0.3 & 0.2 & 0.4 \end{bmatrix}, \; F(u_2) = \begin{bmatrix} -1.6 & 0 & -0.3 & 0.3 \\ 0 & 1 & 0 & 0 \\ -0.8 & 0 & 0.6 & 0.1 \\ 0.3 & 0 & 0.3 & 0.2 \end{bmatrix},$$

$$F(u_3) = \begin{bmatrix} 0 & -0.5 & 0 & -0.1 \\ 0 & 0.5 & 0 & 0 \\ -1 & 0 & -0.1 & 0 \\ 0 & 0.2 & 0.5 & 0.2 \end{bmatrix}, \; g(u_1) = \mathbf{e}_1, \; g(u_2) = \mathbf{e}_2, \; g(u_3) = \mathbf{e}_3,$$

$h = [6,\ 4,\ -2,\ -1],\ h^0 = 1.$

Then the algebraically approximate realization problem is solved as follows:

covariance matrix	eigenvalues 1	2	3	4	5	$\cdots$	12
$H^T_{\underline{a}\,(2,40)}(1,2)H_{\underline{a}\,(2,40)}(1,2)$	7456	1580					
$H^T_{\underline{a}\,(3,40)}(1,2,3)H_{\underline{a}\,(3,40)}(1,2,3)$	8352	1929	675				
$H^T_{\underline{a}\,(4,40)}(1,\cdots,4)H_{\underline{a}\,(4,40)}(1,\cdots,4)$	10308	2038	848	11			
$H^T_{\underline{a}\,(5,40)}(1,\cdots,5)H_{\underline{a}\,(5,40)}(1,\cdots,5)$	39374	2489	855	11	0		
$H^T_{\underline{a}\,(12,40)}(1,\cdots,12)H_{\underline{a}\,(12,40)}(1,\cdots,12)$	61489	10097	6578	14	0	$\cdots$	0
covariance matrix	square root of eigenvalues						
$H^T_{\underline{a}\,(2,40)}(1,2)H_{\underline{a}\,(2,40)}(1,2)$	86.3	39.7					
$H^T_{\underline{a}\,(3,40)}(1,2,3)H_{\underline{a}\,(3,40)}(1,2,3)$	91.4	43.9	26				
$H^T_{\underline{a}\,(4,40)}(1,\cdots,4)H_{\underline{a}\,(4,40)}(1,\cdots,4)$	102	45.1	29.1	3.3			
$H^T_{\underline{a}\,(5,40)}(1,\cdots,5)H_{\underline{a}\,(5,40)}(1,\cdots,5)$	198	50	29.2	3.3	0		
$H^T_{\underline{a}\,(12,40)}(1,\cdots,12)H_{\underline{a}\,(12,40)}(1,\cdots,12)$	248	100	81.1	3.7	0	$\cdots$	0

1) Since the ratio $\frac{3.3}{102} = 0.03$ obtained by the square root of $H^T_{\underline{a}\,(4,40)}(1,\cdots,4) \times H_{\underline{a}\,(4,40)}(1,\cdots,4)$ and the ratio $\frac{3.7}{248} = 0.01$ obtained by the square root of $H^T_{\underline{a}\,(12,40)}(1,\cdots,12)H_{\underline{a}\,(12,40)}(1,\cdots,12)$ are somewhat small, the approximate 3-dimensional affine dynamical system obtained by the algebraic CLS method may be somewhat good.

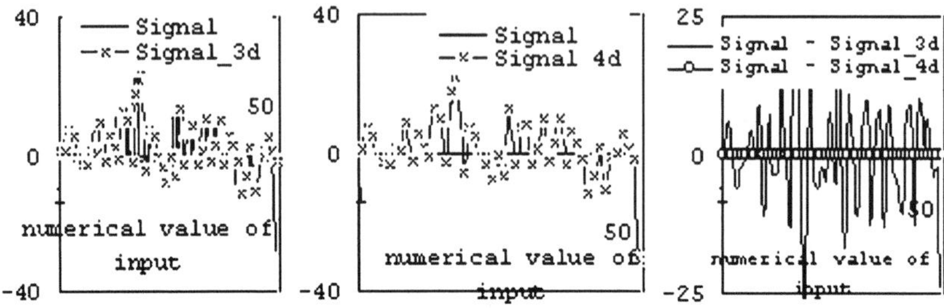

Fig. 7.3 The left is the original input response map and the behavior of a 3-dimensional affine dynamical system obtained by the algebraic CLS method. The middle is the original input response map and the behavior of a 4-dimensional affine dynamical system obtained by the algebraic CLS method. The right is the difference between the original one and the behavior of the 3-dimensional affine dynamical system obtained by the algebraic CLS method or the 4-dimensional affine dynamical system obtained by the algebraic CLS method in Example (7.24).

2) After determining the independent vectors $\underline{S_l}(u_1)\underline{a} - \underline{a}$, $\underline{S_l}(u_2)\underline{a} - \underline{a}$ and $\underline{S_l}(u_3)\underline{a} - \underline{a}$ whose numerical value of input are 1, 2 and 3, we will continue the approximate realization algorithm by the algebraic CLS method.

Therefore, an approximate 3-dimensional affine dynamical system $\sigma_1 = ((\boldsymbol{R}^3,\ F_1),\ g_1,\ h_1,\ h^0)$ obtained by the algebraic CLS method is constructed as follows:

$$F_1(u_1) = \begin{bmatrix} -1.66 & -0.36 & -1.04 \\ 0.14 & 0.05 & 1.03 \\ 0.43 & 2.46 & -0.19 \end{bmatrix},\ F_1(u_2) = \begin{bmatrix} -1.66 & 0 & -0.36 \\ 0.05 & 1 & 0.05 \\ -0.64 & 0 & 0.76 \end{bmatrix},$$

$$F_1(u_3) = \begin{bmatrix} 0 & -0.54 & -0.1 \\ 0 & 0.53 & 0.08 \\ -1 & 0.1 & 0.17 \end{bmatrix},\ g_1(u_1) = \mathbf{e}_1,\ g_1(u_2) = \mathbf{e}_2,\ g_1(u_3) = \mathbf{e}_3,$$

$h_1 = [5.8,\ 4,\ -1.9],\ h^0 = 1.$

For reference, a 4-dimensional affine dynamical system $\sigma_2 = ((\boldsymbol{R}^4,\ F_2),\ g_2,\ h_2, h^0)$ obtained by the algebraic CLS method can be expressed as follows:

$$F_2(u_1) = \begin{bmatrix} -1 & -0.1 & -0.88 & 0.54 \\ 0 & 0 & 1 & 0.48 \\ 0 & 2.3 & -0.3 & 0 \\ 1 & 0.38 & 0.25 & -0.1 \end{bmatrix},\ F_2(u_2) = \begin{bmatrix} -1.4 & 0 & -0.1 & 1.05 \\ 0 & 1 & 0 & 0 \\ -0.8 & 0 & 0.6 & 0.48 \\ 0.38 & 0 & 0.37 & 0.01 \end{bmatrix},$$

$$F_2(u_3) = \begin{bmatrix} 0 & -0.4 & 0.3 & 0.02 \\ 0 & 0.5 & 0 & 0 \\ -1 & 0 & -0.1 & 0.5 \\ 0 & 0.25 & 0.63 & 0.2 \end{bmatrix},\ g_2(u_1) = \mathbf{e}_1,\ g_2(u_2) = \mathbf{e}_2,\ g_2(u_3) = \mathbf{e}_3,$$

$h_2 = [6,\ 4,\ -2,\ -3.8],\ h^0 = 1.$

The system is completely reconstructed by the algebraic CLS method.

We can show that the algorithm for approximate realization given by the analytic CLS method in the reference [Hasegawa, 2008] produces the same systems as the above ones in the sense of the numerical calculation.

In this example, the original signals are considered as the input response map of a 4-dimensional affine dynamical system and the desirable input response map is obtained by the algebraic CLS method within our expectations. The model obtained by the algebraic CLS method is a 3-dimensional affine dynamical system.

For reference, a 4-dimensional affine dynamical system is also given by the algebraic CLS method. The system completely reconstructs the original system.

Just as we thought, the following table and Fig. 7.3 truly indicate that the 3-dimensional affine dynamical system obtained by the algebraic CLS method is a somewhat good approximation. For reference, the input response map of the same dimensional affine dynamical system as the original system is shown. Hence, there exists a somewhat good approximation for the given system.

dimen-sion	ratio of matrices	mean values of square root for sum of signal	signal by CLS	error	cosine ① and ②	error ratio
		①	②	③	$\cos\theta$	③/①
$a_{1,2,3}$	0.03	1.1916	1.172	0.045	0.9994	0.038
$a_{1,2,3,4}$	0	1.1916	1.1916	0	1	0

Example 7.25. Let the signals be the input response map of the following 5-dimensional affine dynamical system: $\sigma = ((\boldsymbol{R}^5, F), g, h, h^0)$, where

$$F(u_1) = \begin{bmatrix} -1 & -0.4 & -0.6 & -1 & -1 \\ 0 & -0.2 & 0.5 & 0.4 & -0.2 \\ 0 & -0.2 & 0 & 0.8 & 0 \\ 1 & 0.2 & 0.1 & 0.7 & 1 \\ 0 & 0.2 & 0.7 & 0.3 & 0 \end{bmatrix}, \quad F(u_2) = \begin{bmatrix} 0 & -0.4 & -0.4 & 0 & 0 \\ -1 & 0.21 & -0.2 & -0.5 & 0 \\ 0 & 0.2 & 0.6 & 0.4 & 0 \\ 0 & -0.2 & 0.2 & 0.5 & 0 \\ 1 & 0.7 & 0.2 & 0.5 & 0.9 \end{bmatrix},$$

$$F(u_3) = \begin{bmatrix} 0 & -0.6 & -0.31 & -0.5 & -0.7 \\ 0 & 0.7 & 0.1 & 0.2 & 0.7 \\ -1 & 0 & -0.1 & 0.4 & -0.3 \\ 0 & 0.5 & 0.5 & 0.8 & 0.6 \\ 0 & 0.2 & -0.1 & -0.2 & 0.2 \end{bmatrix}, \quad g(u_1) = \mathbf{e}_1, \ g(u_2) = \mathbf{e}_2,$$

$g(u_3) = \mathbf{e}_3, \ h = [12, \ -1, \ -2, \ 1, \ 7], \ h^0 = 1.$

Then the algebraically approximate realization problem is solved as follows:

covariance matrix	eigenvalues							
	1	2	3	4	5	6	···	12
$H^T_{\underline{a}\ (3,40)}(1,2,3)H_{\underline{a}\ (3,40)}(1,2,3)$	1412	1148	242					
$H^T_{\underline{a}\ (4,40)}(1,\cdots,4)H_{\underline{a}\ (4,40)}(1,\cdots,4)$	3577	1202	243	151				
$H^T_{\underline{a}\ (5,40)}(1,\cdots,5)H_{\underline{a}\ (5,40)}(1,\cdots,5)$	4943	1219	433	151	1.8			
$H^T_{\underline{a}\ (6,40)}(1,\cdots,6)H_{\underline{a}\ (6,40)}(1,\cdots,6)$	4943	1219	433	151	1.6	0		
$H^T_{\underline{a}\ (12,40)}(1,\cdots,12)H_{\underline{a}\ (12,40)}(1,\cdots,12)$	16485	3728	1243	170	1.9	0	···	0
covariance matrix	square root of eigenvalues							
$H^T_{\underline{a}\ (3,40)}(1,2,3)H_{\underline{a}\ (3,40)}(1,2,3)$	37.6	33.9	15.6					
$H^T_{\underline{a}\ (4,40)}(1,\cdots,4)H_{\underline{a}\ (4,40)}(1,\cdots,4)$	59.8	34.7	15.6	12.3				
$H^T_{\underline{a}\ (5,40)}(1,\cdots,5)H_{\underline{a}\ (5,40)}(1,\cdots,5)$	70.3	34.9	20.8	12.3	1.3			
$H^T_{\underline{a}\ (6,40)}(1,\cdots,6)H_{\underline{a}\ (6,40)}(1,\cdots,6)$	70.3	34.9	20.8	12.3	1.3	0		
$H^T_{\underline{a}\ (12,40)}(1,\cdots,12)H_{\underline{a}\ (12,40)}(1,\cdots,12)$	128	61	35.3	13	1.4	0	···	0

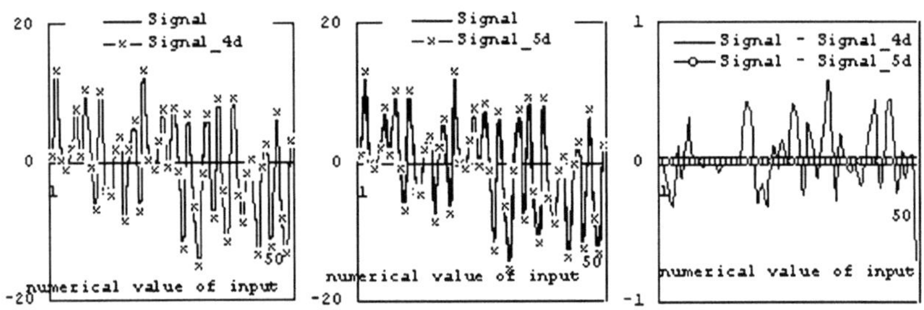

Fig. 7.4 The left is the original input response map and the behavior of a 4-dimensional affine dynamical system obtained by the algebraic CLS method. The middle is the original input response map and the behavior of a 5-dimensional affine dynamical system obtained by the algebraic CLS method. The right is the difference between the original one and the behavior of the 4-dimensional affine dynamical system obtained by the algebraic CLS method or the 5-dimensional affine dynamical system obtained by the algebraic CLS method in Example (7.25).

1) Since the ratio $\frac{1.3}{70.3} = 0.02$ obtained by the square root of $H^T_{\underline{a}\ (5,40)}(1,\cdots,5) \times H_{\underline{a}\ (5,40)}(1,\cdots,5)$ and the ratio $\frac{1.4}{128} = 0.01$ obtained by the square root of $H^T_{\underline{a}\ (12,40)}(1,\cdots,12)H_{\underline{a}\ (12,40)}(1,\cdots,12)$ are small, the approximate 4-dimensional affine dynamical system obtained by the algebraic CLS method may be good.
2) After determining the independent vectors $\underline{S_l}(u_1)\underline{a}-\underline{a}$, $\underline{S_l}(u_2)\underline{a}-\underline{a}$ and $\underline{S_l}(u_3)\underline{a}-\underline{a}$ whose numerical value of input are 1, 2, 3 and 4, we will continue the approximate realization algorithm by the algebraic CLS method.

Therefore, an approximate 4-dimensional affine dynamical system $\sigma_1 = ((\boldsymbol{R}^4,\ F_1),\ g_1,\ h_1,\ h^0)$ obtained by the algebraic CLS method is constructed as follows:

$$F_1(u_1) = \begin{bmatrix} -1 & -0.3 & -0.25 & -0.85 \\ 0 & 0 & 1.2 & 0.7 \\ 0 & -0.3 & -0.26 & 0.7 \\ 1 & 0.3 & 0.35 & 0.8 \end{bmatrix}, F_1(u_2) = \begin{bmatrix} 0.5 & 0.05 & -0.3 & 0.25 \\ 0 & 0.9 & 0 & 0 \\ -0.37 & 0.06 & 0.5 & 0.22 \\ 0.36 & 0.05 & 0.27 & 0.7 \end{bmatrix},$$

$$F_1(u_3) = \begin{bmatrix} 0 & -0.5 & -0.36 & -0.6 \\ 0 & 0.9 & 0 & 0 \\ -1 & 0.07 & 0.06 & 0.47 \\ 0 & 0.6 & 0.46 & 0.73 \end{bmatrix}, g_1(u_1) = \mathbf{e}_1,\ g_1(u_2) = \mathbf{e}_2,$$

$g_1(u_3) = \mathbf{e}_3,\ h_1 = [12.1,\ 0.68,\ -2.1,\ 1.1],\ h^0 = 1.$

For reference, a 5-dimensional affine dynamical system $\sigma_2 = ((\boldsymbol{R}^5,\ F_2),\ g_2,\ h_2, h^0)$ obtained by the algebraic CLS method can be expressed as follows:

$$F_2(u_1) = \begin{bmatrix} -1 & -0.4 & -0.6 & -1 & -1 \\ 0 & -0.2 & 0.5 & 0.4 & -0.2 \\ 0 & -0.2 & 0 & 0.8 & 0 \\ 1 & 0.2 & 0.1 & 0.7 & 1 \\ 0 & 0.2 & 0.7 & 0.3 & 0 \end{bmatrix}, F_2(u_2) = \begin{bmatrix} 0 & -0.4 & -0.4 & 0 & 0 \\ -1 & 0.2 & -0.2 & -0.5 & 0 \\ 0 & 0.2 & 0.6 & 0.4 & 0 \\ 0 & -0.2 & 0.2 & 0.5 & 0 \\ 1 & 0.7 & 0.2 & 0.5 & 0.9 \end{bmatrix},$$

$$F_2(u_3) = \begin{bmatrix} 0 & -0.6 & -0.31 & -0.5 & -0.7 \\ 0 & 0.7 & 0.1 & 0.2 & 0.7 \\ -1 & 0 & -0.1 & 0.4 & -0.3 \\ 0 & 0.5 & 0.5 & 0.8 & 0.6 \\ 0 & 0.2 & -0.1 & -0.2 & 0.2 \end{bmatrix}, g_2(u_1) = \mathbf{e}_1,\ g_2(u_2) = \mathbf{e}_2,$$

$g_2(u_3) = \mathbf{e}_3,\ h_2 = [12,\ -1,\ -2,\ 1,\ 7],\ h^0 = 1.$

The system is completely reconstructed by the algebraic CLS method.

We can show that the algorithm for approximate realization given by the analytic CLS method in the reference [Hasegawa, 2008] produces the same systems as the above ones in the sense of the numerical calculation.

In this example, the original signals are considered as the input response map of a 5-dimensional affine dynamical system and the desirable input response map is obtained by the algebraic CLS method within our expectations. The model obtained by the algebraic CLS method is a 4-dimensional affine dynamical system.

For reference, a 5-dimensional affine dynamical system is also given by the algebraic CLS method. The system completely reconstructs the original system.

Just as we expected, the following table and Fig. 7.4 truly indicate that the 4-dimensional affine dynamical system obtained by the algebraic CLS method is a somewhat good approximation. For reference, the input response map of the same dimensional affine dynamical system as the original system is shown. Hence, there exists a good approximation for the given system.

dimen-sion	ratio of matrices	mean values of square root for sum of			cosine	error ratio
		signal	signal by CLS	error	① and ②	
		①	②	③	$\cos\theta$	③/①
$a_{1,2,3,4}$	0.02	1.05	1.04	0.03	0.9996	0.03
$a_{1,2,3,4,5}$	0	1.05	1.05	0	1	0

Example 7.26. Let the signals be the input response map of the following 5-dimensional affine dynamical system: $\sigma = ((\boldsymbol{R}^5, F), g, h, h^0)$, where

$$F(u_1) = \begin{bmatrix} -1 & -0.2 & -0.7 & -0.2 & 0 \\ 0 & 0.2 & -0.4 & 0.2 & 0.4 \\ 0 & 0.1 & -0.8 & 0 & 0 \\ 1 & 0 & 0 & 0 & 0.1 \\ 0 & -0.2 & -0.4 & -0.2 & -0.2 \end{bmatrix}, \; F(u_2) = \begin{bmatrix} 0 & 0.5 & 0.8 & 0.2 & 0.8 \\ -1 & 0.1 & -0.2 & -0.4 & 0.2 \\ 0 & 0 & 0.6 & 0 & 0 \\ 0.5 & 0 & 0.3 & 0 & 0.1 \\ 1 & 0.8 & 0.2 & 0.4 & 0.7 \end{bmatrix},$$

$$F(u_3) = \begin{bmatrix} 0 & -0.4 & 0 & 0 & -0.3 \\ 0 & 0 & 0.2 & 0 & 0 \\ 0 & 0 & 0.1 & 0 & 0 \\ 0 & 0 & 0 & 0 & 0 \\ 0 & 0.1 & 0 & 0 & -0.5 \end{bmatrix}, \; g(u_1) = \mathbf{e}_1, \; g(u_2) = \mathbf{e}_2, \; g(u_3) = \mathbf{e}_3,$$

$h = [12, \; -3, \; -6, \; 1, \; -6], \; h^0 = 1.$

Then the algebraically approximate realization problem is solved as follows:

covariance matrix	eigenvalues							
	1	2	3	4	5	6	$\cdots$	12
$H^T_{\underline{a}\,(4,40)}(1,\cdots,4)H_{\underline{a}\,(4,40)}(1,\cdots,4)$	849	615	171	7.6				
$H^T_{\underline{a}\,(5,40)}(1,\cdots,5)H_{\underline{a}\,(5,40)}(1,\cdots,5)$	1008	671	273	63	5.3			
$H^T_{\underline{a}\,(6,40)}(1,\cdots,6)H_{\underline{a}\,(6,40)}(1,\cdots,6)$	1737	676	314	63	5.3	0		
$H^T_{\underline{a}\,(12,40)}(1,\cdots,12)H_{\underline{a}\,(12,40)}(1,\cdots,12)$	5599	1848	803	68	6	0	$\cdots$	0
covariance matrix	square root of eigenvalues							
$H^T_{\underline{a}\,(4,40)}(1,\cdots,4)H_{\underline{a}\,(4,40)}(1,\cdots,4)$	29.1	24.8	13.1	2.8				
$H^T_{\underline{a}\,(5,40)}(1,\cdots,5)H_{\underline{a}\,(5,40)}(1,\cdots,5)$	31.7	25.9	16.5	7.9	2.3			
$H^T_{\underline{a}\,(6,40)}(1,\cdots,6)H_{\underline{a}\,(6,40)}(1,\cdots,6)$	41.7	26	17.7	7.9	2.3	0		
$H^T_{\underline{a}\,(12,40)}(1,\cdots,12)H_{\underline{a}\,(12,40)}(1,\cdots,12)$	74.8	43	28.3	8.2	2.4	0	$\cdots$	0

1) Since the ratio $\frac{2.3}{31.7} = 0.07$ obtained by the square root of $H^T_{\underline{a}\,(5,40)}(1,\cdots,5) \times H_{\underline{a}\,(5,40)}(1,\cdots,5)$ and the ratio $\frac{2.4}{74.8} = 0.03$ obtained by the square root of $H^T_{\underline{a}\,(12,40)}(1,\cdots,12)H_{\underline{a}\,(12,40)}(1,\cdots,12)$ are not so small, the approximate 4-dimensional affine dynamical system obtained by the algebraic CLS method may not be good.
2) After determining the independent vectors $\underline{S}_l(u_1)\underline{a} - \underline{a}$, $\underline{S}_l(u_2)\underline{a} - \underline{a}$, $\underline{S}_l(u_3)\underline{a} - \underline{a}$ and $\underline{S}_l(u_1|u_1)\underline{a} - \underline{a}$ whose numerical value of input are 1, 2, 3 and 4, we will continue the approximate realization algorithm by the algebraic CLS method.

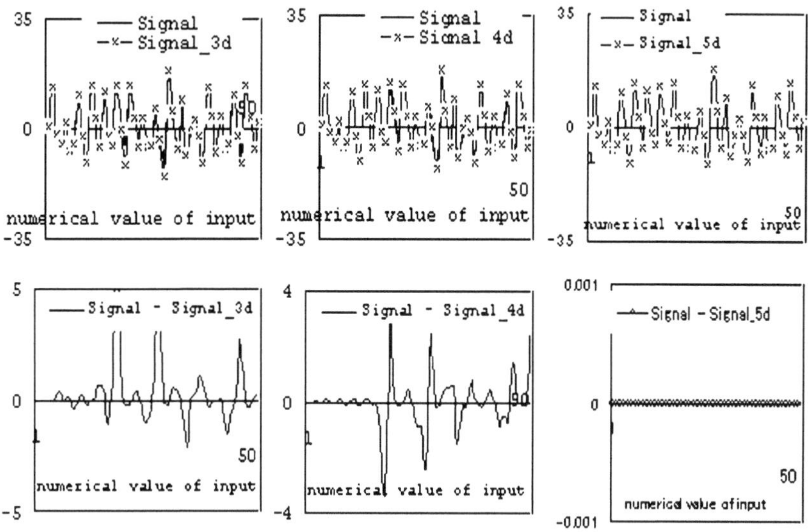

Fig. 7.5 The left are the original input response map and the behavior of a 3-dimensional affine dynamical system obtained by the algebraic CLS method. The middle are the original input response map and the behavior of a 4-dimensional affine dynamical system obtained by the algebraic CLS method. The right are the difference between the original one and the behavior of a 5-dimensional affine dynamical system obtained by the algebraic CLS method in Example (7.26).

Therefore, an approximate 4-dimensional affine dynamical system $\sigma_1 = ((\boldsymbol{R}^4,\ F_1),\ g_1,\ h_1,\ h^0)$ obtained by the algebraic CLS method is constructed as follows:

$$F_1(u_1) = \begin{bmatrix} -1 & -0.07 & -0.4 & -0.07 \\ 0 & 0.07 & -0.65 & 0.07 \\ 0 & 0.1 & -0.77 & 0 \\ 1 & -0.5 & -1.5 & -0.5 \end{bmatrix},\ F_1(u_2) = \begin{bmatrix} -1.2 & -0.4 & 0.69 & -0.16 \\ -0.51 & 0.53 & -0.06 & -0.15 \\ -0.27 & -0.21 & 0.6 & -0.06 \\ 6.9 & 4.9 & 0.71 & 1.77 \end{bmatrix},$$

$$F_1(u_3) = \begin{bmatrix} 0 & -0.45 & 0 & 0 \\ 0 & 0.07 & 0.2 & 0 \\ 0 & 0 & 0.1 & 0 \\ 0 & 0.18 & 0 & 0 \end{bmatrix},\ g_1(u_1) = \mathbf{e}_1,\ g_1(u_2) = \mathbf{e}_2,\ g_1(u_3) = \mathbf{e}_3,$$

$h = [6,\ -3,\ -7,\ 0.4],\ h^0 = 1.$

For reference, a 5-dimensional affine dynamical system $\sigma_2 = ((\boldsymbol{R}^5,\ F_2),\ g_2,\ h_2, h^0)$ obtained by the algebraic CLS method can be expressed as follows:

$$F_2(u_1) = \begin{bmatrix} -1 & -0.2 & -0.7 & -0.2 & 0.2 \\ 0 & 0.2 & -0.4 & 0.2 & 0.4 \\ 0 & 0 & -0.7 & 0 & 0 \\ 1 & 0 & 0 & 0 & 0.1 \\ 0 & -0.2 & -0.4 & -0.2 & -0.2 \end{bmatrix},\ F_2(u_2) = \begin{bmatrix} 0 & 0.5 & 0.8 & 0.2 & 0.7 \\ -1 & 0.1 & -0.2 & -0.4 & 0.2 \\ 0 & 0 & 0.6 & 0 & 0 \\ 0 & 0 & 0.1 & 0 & 0.1 \\ 1 & 0.8 & 0.2 & 0.4 & 0.7 \end{bmatrix},$$

$$F_2(u_3) = \begin{bmatrix} 0 & -0.4 & 0 & 0 & -0.3 \\ 0 & 0 & 0.2 & 0 & 0 \\ 0 & 0 & 0.1 & 0 & 0 \\ 0 & 0 & 0 & 0 & 0 \\ 0 & 0.1 & 0 & 0 & 0.1 \end{bmatrix}, \; g_2(u_1) = \mathbf{e}_1, \; g_2(u_2) = \mathbf{e}_2,$$

$$g_2(u_3) = \mathbf{e}_3, \; h_2 = [12, \; -3, \; -6, \; 1, \; -7.5], \; h^0 = 1.$$

The system is completely reconstructed by the algebraic CLS method.

We can show that the algorithm for approximate realization given by the analytic CLS method in the reference [Hasegawa, 2008] produces the same systems as the above ones in the sense of the numerical calculation.

In this example, the original signals are considered as the input response map of a 5-dimensional affine dynamical system and the desirable input response map is obtained by the algebraic CLS method within our expectations. The model obtained by the algebraic CLS method is a 4-dimensional affine dynamical system.

For reference, a 5-dimensional affine dynamical system is also given by the algebraic CLS method. The system completely reconstructs the original system.

Just as we thought, the following table and Fig. 7.5 truly indicate that the 4-dimensional affine dynamical system obtained by the algebraic CLS method is not a good approximation. For reference, the input response map of the same dimensional affine dynamical system as the original system is shown. Hence, there does not exist a good approximation for the given system.

dimen-sion	ratio of matrices	mean values of square root for sum of signal ①	signal by CLS ②	error ③	cosine ① and ② $\cos\theta$	error ratio ③/①
$a_{1,2,3}$	0.1	1.1004	1.1594	0.1987	0.986	0.18
$a_{1,2,3,4}$	0.03	1.1004	1.13187	0.138	0.992	0.125
$a_{1,2,3,4,5}$	0	1.1004	1.1004	0	1	0

7.5 Algebraically Noisy Realization of Affine Dynamical Systems

In this section, we discuss algebraically noisy realization problems of affine dynamical systems.

For noise $\{\bar{\gamma}(t) : t \in N\}$ added to an unknown affine dynamical system a, we will obtain the observed data $\{\hat{\gamma}(|\omega|) + \bar{\gamma}(|\omega|) : \omega \in U^*\}$.

For any given $\{\hat{\gamma}(|\omega|) + \bar{\gamma}(|\omega|) : \omega \in U^*\}$, σ which satisfies $a_\sigma(\omega) \approx \hat{\gamma}(|\omega|) : \omega \in U^*$ is called a noisy realization of a.

Roughly speaking, we can propose the following algebraically noisy realization problem:

For any given $\{\hat{\gamma}(|\omega|)+\bar{\gamma}(|\omega|) : \omega \in U^*\}$, find, using only algebraic calculations, an affine dynamical system σ which satisfies $a_\sigma(\omega) \approx \hat{\gamma}(|\omega|)$ for any $\omega \in U^*$.

In order to make our discussion simple, we assume that the set Y of output is the set $\boldsymbol{R}$ of real numbers, namely 1-output.

A situation for algebraically noisy realization problem 7.28
Let the observed object be an affine dynamical system and noise be added to the output. Then we will obtain the data $\{\gamma(t)=\hat{\gamma}(t)+\bar{\gamma}(t) : 0 \le t \le \underline{N}\}$ for some integer $\underline{N} \in N$, where $\hat{\gamma}(t)$ is the exact signal which comes from the observed affine dynamical system and $\bar{\gamma}(t)$ is the noise added at the time of observation.

Problem statement of algebraically noisy realization for affine dynamical systems 7.29
Let $H_{\underline{a}\ (p,\bar{p})}$ be the measured finite-sized Input/output matrix. Then find, using only algebraic calculations, the cleaned-up Input/output matrix $\hat{H}_{\underline{a}\ (p,\bar{p})}$ such that $H_{\underline{a}\ (p,\bar{p})} = \hat{H}_{\underline{a}\ (p,\bar{p})} + \bar{H}_{\underline{a}\ (p,\bar{p})}$ holds.

Namely, find, using only algebraic calculations, a minimal dimensional affine dynamical system $\sigma = ((\boldsymbol{R}^n, F_r), g_r, h_r, h^0))$ which realizes $\hat{H}_{\underline{a}\ (p,\bar{p})}$.

Theorem 7.30. *Algebraically algorithm of noisy realization for Affine Danamical systems*

Let an input response map $\underline{a}$ be a considered object which is an affine dynamical system. Then an approximate realization $\sigma = ((\boldsymbol{R}^n, F_s), g_s, h_s, h^0)$ of $\underline{a}$ is given by the following algorithm:

1) Based on the square root of eigenvalues for a matrix $H_{\underline{a}\ (n,\bar{p})}(|||\omega_1|||,\cdots,|||\omega_n|||)H_{\underline{a}\ (n,\bar{p})}(|||\omega_1|||,\cdots,|||\omega_n|||)^T$, determine the value n of rank for the matrix, where $|||\omega_1|||$, $|||\omega_2|||$, $\cdots$ and $|||\omega_n|||$ are suitably selected in order of numerical value of input and $\{\underline{S_l}(\omega_i)\underline{a}-\underline{a}; 1 \le i \le n,\ \omega_i \in U^*\}$ is a set of independent vectors. Namely, determine the value n of rank for the matrix $H_{\underline{a}\ (n,\bar{p})}(|||\omega_1|||,\cdots,|||\omega_n|||)$ such that a set of the square root of eigenvalues for the covariance matrix composed of relatively small and equally-sized numbers is excluded, where the signal part effected by the set may be the noisy part of the observed data.

2) In order to determine g_s, the algebraic CLS method is used as follows:
① In particular, set $g_s(\omega_i) := \mathbf{e}_i$ for $\omega_i \in U$. Namely, $g_s(\omega_1) := \mathbf{e}_1$, $g_s(\omega_2) := \mathbf{e}_2$, $\cdots, g_s(\omega_k) := \mathbf{e}_k$ for some $k \in N$.
For $u \in U$ such that $u \notin \{\omega_i; 1 \le i \le n\}$ and $\omega_r < u$, $g_s(u) = \sum_{j=1}^{r} b_{u,j}(\underline{S_l}(\omega_j)\underline{a}-\underline{a})$ is obtained as follows:
Based on Proposition (2.14), determine coefficients $\{b_{u,j} : 1 \le j \le r\}$. The Q in Proposition (2.14) can be considered as the matrix composed from the eigenvectors of $H_{\underline{a}\ (r+1,L)}(|||\omega_1|||,|||\omega_2|||,\cdots,|||\omega_r|||,|||u|||)$ $\times H^T_{\underline{a}\ (r+1,L)}(|||\omega_1|||,|||\omega_2|||,\cdots,|||\omega_r|||,|||u|||)$.

Let a matrix $A_u \in \boldsymbol{R}^{1\times(r+1)}$ *be* $A_u := [b_{u,1}, b_{u,2}, \cdots, b_{u,r}, -1]$.
② *Determine the error vectors* $\{\underline{S_l}(\omega_j)\bar{\underline{a}} - \bar{\underline{a}} \in \boldsymbol{R}^{L\times 1} :\ 0 \le j \le r\}$ *by using the equation*
$[\underline{S_l}(\omega_1)\bar{\underline{a}} - \bar{\underline{a}}, \underline{S_l}(\omega_2)\bar{\underline{a}} - \bar{\underline{a}}, \cdots, \underline{S_l}(\omega_r)\bar{\underline{a}} - \bar{\underline{a}}, \underline{S_l}(u)\bar{\underline{a}} - \bar{\underline{a}}]^T :=$
$A_u^T[A_u A_u^T]^{-1} A_u H^T_{\underline{a}\ (r+1,L)}(|||\omega_1|||, |||\omega_2|||, \cdots, |||\omega_r|||, |||u|||)$
and $H^T_{\underline{a}\ (|||u|||+1,L)}(|||\omega_1|||, |||\omega_2|||, \cdots, |||\omega_r|||, |||u|||) :=$
$[\underline{S_l}(\omega_1)\underline{a} - \underline{a}, \cdots, \underline{S_l}(\omega_r)\underline{a} - \underline{a}, \underline{S_l}(u)\underline{a} - \underline{a}]$.

3) *In order to obtain* F_s, *the algebraic CLS method is used as follows:*
① *Let* ${}_i\mathbf{f}_j \in \boldsymbol{R}^n$ *in* $F_s(u_i) := [{}_i\mathbf{f}_1\ {}_i\mathbf{f}_2 \cdots {}_i\mathbf{f}_n\ ]$ *be* ${}_i\mathbf{f}_j := [{}_if_{j,1}\ {}_if_{j,2}, \cdots {}_if_{j,n}]^T$ *for* $1 \le i \le n$, *where* ${}_if_{j,k}$ *is given by the following:*
$\underline{S_l}(u_i)(\underline{S_l}(\omega_j)\underline{a} - \underline{a}) = \sum_{k=1}^n {}_if_{j,k}(\underline{S_l}(\omega_k)\underline{a} - \underline{a})$, ${}_if_{j,k} \in \boldsymbol{R}$ *in the sense of* $F(U^*_{L-p}, Y)$.
We cannot directly obtain the coefficients $\{{}_i\mathbf{f}_j; 1 \le i \le 3,\ 1 \le j \le n\}$. *Firstly, we will determine coefficients* $\{{}_i\bar{f}_{j,k}; 1 \le k \le n\}$ *from the equation* $\underline{S_l}(u_i)\underline{S_l}(\omega_j)\underline{a} - \underline{a} = \sum_{k=1}^n {}_i\bar{f}_{j,k}(\underline{S_l}(\omega_k)\underline{a} - \underline{a})$ *by using the algebraic CLS method.*
②*For* i $(1 \le i \le 3)$, j $(1 \le j \le n)$ *and for the maximum number* r $(1 \le r \le n)$ *such that* $\omega_r, \omega_j \in \{\omega_j; 1 \le j \le n\}$ *and* $|||\omega_r||| < |||u_i|\omega_j|||$, *and based on Proposition (2.14), determine coefficients* $\{{}_i\bar{f}_{j,k} = 0;$ $r+1 \le k \le n\}$ *and* $\{{}_i\bar{f}_{j,k} : 1 \le k \le r\}$.
The Q *in Proposition (2.14) can be considered as the matrix composed from the eigenvectors of* $H_{\underline{a}\ (r+1,L)}(|||\omega_1|||, \cdots, |||\omega_r|||, |||u_i|\omega_j|||)$
$\times H^T_{\underline{a}\ (r+1,L)}(|||\omega_1|||, \cdots, |||\omega_r|||, |||u_i|\omega_j|||)$.
Let a matrix ${}_iA_j \in \boldsymbol{R}^{1\times(r+1)}$ *be* ${}_iA_j := [{}_i\bar{f}_{j,1}, {}_i\bar{f}_{j,2}, \cdots, {}_i\bar{f}_{j,r}, -1]$.
Determine the error vectors $\{\underline{S_l}(\omega_j)\bar{\underline{a}} - \bar{\underline{a}} \in \boldsymbol{R}^{L\times 1} :\ 0 \le j \le r\}$ *and* $\underline{S_l}(u_i|\omega_j)\bar{\underline{a}} - \bar{\underline{a}}$ *by using the equation*
$[\underline{S_l}(\omega_1)\bar{\underline{a}} - \bar{\underline{a}}, \underline{S_l}(\omega_2)\bar{\underline{a}} - \bar{\underline{a}}, \cdots, \underline{S_l}(\omega_r)\bar{\underline{a}} - \bar{\underline{a}}, \underline{S_l}(u_i|\omega_j)\bar{\underline{a}} - \bar{\underline{a}}]^T :=$
${}_iA_j^T[{}_iA_j\ {}_iA_j^T]^{-1}\ {}_iA_j\ H^T_{\underline{a}\ (r+1,L)}(|||\omega_1|||, |||\omega_2|||, \cdots, |||\omega_r|||, |||u_i|\omega_j|||)$ *and*
$H^T_{\underline{a}\ (r+1,L)}(|||\omega_1|||, \cdots, |||\omega_r|||, |||u_i|\omega_j|||) :=$
$[\underline{S_l}(\omega_1)\underline{a} - \underline{a}, \cdots, \underline{S_l}(\omega_r)\underline{a} - \underline{a}, \underline{S_l}(u_i|\omega_j)\underline{a} - \underline{a}]$.
③*Next, using the equations*
$\underline{S_l}(u_i)(\underline{S_l}(\omega_j)\underline{a} - \underline{a})$
$= \underline{S_l}(u_i)\underline{S_l}(\omega_j)\underline{a} - \underline{a} - \underline{S_l}(u_i)\underline{a} + \underline{a}$
$= \sum_{k=1}^n {}_i\bar{f}_{j,k}(\underline{S_l}(\omega_k)\underline{a} - \underline{a}) - \sum_{j=1}^{r_{u_i}} b_{u_i j}(\underline{S_l}(u_j)\underline{a} - \underline{a})$,
we obtain $\underline{S_l}(u_i)(\underline{S_l}(\omega_j)\underline{a} - \underline{a})$
$= \sum_{k=1}^n {}_i\bar{f}_{j,k}(\underline{S_l}(\omega_k)\underline{a} - \underline{a}) - \sum_{j=1}^{r_{u_i}} b_{u_i,i}(\underline{S_l}(u_j)\underline{a} - \underline{a})$
$= \sum_{j=1}^{r_{u_i}}({}_i\bar{f}_{j,k} - b_{u_i,j})(\underline{S_l}(\omega_j)\underline{a} - \underline{a}) + \sum_{j=r_{u_i}+1}^n {}_i\bar{f}_{j,k}(\underline{S_l}(\omega_j)\underline{a} - \underline{a})$.
On the other hand, the equation
$\underline{S_l}(u_i)(\underline{S_l}(\omega_j)\underline{a} - \underline{a}) = \sum_{k=1}^n {}_if_{j,k}(\underline{S_l}(\omega_k)\underline{a} - \underline{a})$ *holds. Therefore, we obtain the following equation:*
$\sum_{k=1}^n {}_if_{j,k}(\underline{S_l}(\omega_k)\underline{a} - \underline{a})$
$= \sum_{j=1}^{r_{u_i}}({}_i\bar{f}_{j,k} - b_{u_i,j})\underline{S_l}(\omega_j)\underline{a} - \underline{a} + \sum_{j=r_{u_i}+1}^n {}_i\bar{f}_{j,k}(\underline{S_l}(\omega_j)\underline{a} - \underline{a})$.
Comparing the coefficients, we obtain the following:

$\sum_{j=1}^{r_{u_i}}({}_if_{j,k}-{}_i\bar{f}_{j,k}+b_{u_i,j})(\underline{S_l}(\omega_j)\underline{a}-\underline{a})+\sum_{j=r_{u_i}+1}^{n}({}_if_{j,k}-{}_i\bar{f}_{j,k})(\underline{S_l}(\omega_j)\underline{a}-\underline{a})$
$=0$.

Finally, coefficients ${}_if_{j,k}$ *of* ${}_i\mathbf{f}_j$ *in* F_s *are obtained as follows:*
${}_if_{j,k}={}_i\bar{f}_{j,k}-b_{u_i,j}$ *for* $1\leq j\leq r_{u_i}$,
${}_if_{j,k}={}_i\bar{f}_{j,k}$ *for* $r_{u_i}+1\leq j\leq n$.

4) In order to determine $h_s\in\boldsymbol{R}^{1\times n}$, *the algebraic CLS method is used as follows:*

① *For the first* ω_{r_1+1}, $\omega_{r_1}\in\{\omega_i;1\leq i\leq n\}$ *such that* $\omega_{r_1+1}>\omega_{r_1}$ *and* $|||\omega_{r_1+1}|||$ - $|||\omega_{r_1}|||>1$ *when starting out from* ω_1,
set $\underline{S_l}(\lambda_1)\underline{a}-\underline{a}:=\sum_{i=1}^{r_1}b_{\lambda_1,i}\mathbf{e}_i$ *for the obtained equation* $\underline{S_l}(\lambda_1)\underline{a}-\underline{a}=\sum_{i=1}^{r_1}b_{\lambda_1,i}(\underline{S_l}(\omega_i)\underline{a}-\underline{a})$ *for* λ_1 *such that* $|||\lambda_1|||=|||\omega_{r_1}|||+1$.
Based on Proposition (2.14), determine coefficients $\{b_{\lambda_1,i}:1\leq i\leq r_1\}$.
The Q *in Proposition (2.14) can be considered as the matrix composed from the eigenvectors of* $H_{\underline{a}\ (|||\lambda_1|||+1,L)}(|||\omega_1|||,|||\omega_2|||,\cdots,|||\omega_{r_1}|||,|||\lambda_1|||)$
$\times H^T_{\underline{a}\ (|||\lambda_1|||+1,L)}(|||\omega_1|||,|||\omega_2|||,\cdots,|||\omega_{r_1}|||,|||\lambda_1|||)$.
Let a matrix $A_{\lambda_1}\in\boldsymbol{R}^{1\times(r_1+1)}$ *be* $A_{\lambda_1}:=[b_{\lambda_1,1},b_{\lambda_1,2},\cdots,b_{\lambda_1,r_1},-1]$.
Determine the error vectors $\{\underline{S_l}(\omega_i)\underline{\bar{a}}-\underline{\bar{a}}\in\boldsymbol{R}^{L\times 1}:\ 0\leq i\leq r_1\}$ *and* $\underline{S_l}(\lambda_1)\underline{\bar{a}}-\underline{\bar{a}}$ *by using the equation*
$[\underline{S_l}(\omega_1)\underline{\bar{a}}-\underline{\bar{a}},\underline{S_l}(\omega_2)\underline{\bar{a}}-\underline{\bar{a}},\cdots,\underline{S_l}(\omega_{r_1})\underline{\bar{a}}-\underline{\bar{a}},\underline{S_l}(\lambda_1)\underline{\bar{a}}-\underline{\bar{a}}]^T:=$
$A^T_{\lambda_1}[A_{\lambda_1}A^T_{\lambda_1}]^{-1}A_{\lambda_1}H^T_{\underline{a}\ (r_1+1,L)}(|||\omega_1|||,|||\omega_2|||,\cdots,|||\omega_{r_1}|||,|||\lambda_1|||)$ *and*
$H^T_{\underline{a}\ (|||\lambda_1|||+1,L)}(|||\omega_1|||,|||\omega_2|||,\cdots,|||\omega_{r_1}|||,|||\lambda_1|||):=$
$[\underline{S_l}(\omega_1)\underline{a}-\underline{a},\cdots,\underline{S_l}(\omega_{r_1})\underline{a}-\underline{a},\underline{S_l}(\lambda_1)\underline{a}-\underline{a}]$.
Then let $h_{\lambda_1 s}\in\boldsymbol{R}^{1\times r_1}$ *be*
$h_{\lambda_1 s}:=[\underline{a}(\omega_1)-\underline{a}(1)-(\underline{\bar{a}}(\omega_1)-\underline{\bar{a}}(1)),\underline{a}(\omega_2)-\underline{a}(1)-(\underline{\bar{a}}(\omega_2)-\underline{\bar{a}}(1)),\cdots,$
$\underline{a}(\omega_{r_1})-\underline{a}(1)-(\underline{\bar{a}}(\omega_{r_1})-\underline{\bar{a}}(1))]$.

② *For the first* ω_{r_2+1}, $\omega_{r_2}\in\{\omega_i;1\leq i\leq n\}$ *such that* $\omega_{r_2+1}>\omega_{r_2}$ *and* $|||\omega_{r_2+1}|||$ - $|||\omega_{r_2}|||>1$ *when starting out from* ω_{r_1+1},
set $\underline{S_l}(\lambda_2)\underline{a}-\underline{a}:=\sum_{i=1}^{r_2}b_{\lambda_2,i}\mathbf{e}_i$ *for the obtained equation* $\underline{S_l}(\lambda_2)\underline{a}-\underline{a}=\sum_{i=1}^{r_2}b_{\lambda_2,i}(\underline{S_l}(\omega_i)\underline{a}-\underline{a})$ *for* λ_2 *such that* $|||\lambda_2|||=|||\omega_{r_2}|||+1$.
Based on Proposition (2.14), determine coefficients $\{b_{\lambda_2,i}:1\leq i\leq r_2\}$.
The Q *in Proposition (2.14) can be considered as the matrix composed from the eigenvectors of* $H_{\underline{a}\ (r_2+1,L)}(|||\omega_1|||,|||\omega_2|||,\cdots,|||\omega_{r_2}|||,|||\lambda_2|||)$
$\times H^T_{\underline{a}\ (r_2+1,L)}(|||\omega_1|||,|||\omega_2|||,\cdots,|||\omega_{r_2}|||,|||\lambda_2|||)$.
Let a matrix $A_{\lambda_2}\in\boldsymbol{R}^{1\times(r_2+1)}$ *be* $A_{\lambda_2}:=[b_{\lambda_2,1},b_{\lambda_2,2},\cdots,b_{\lambda_2,r_2},-1]$.
Determine the error vectors $\{\underline{S_l}(\omega_i)\underline{\bar{a}}-\underline{\bar{a}}\in\boldsymbol{R}^{L\times 1}:\ 0\leq i\leq r_2\}$ *and* $\underline{S_l}(\lambda_2)\underline{\bar{a}}-\underline{\bar{a}}$ *by using the equation*
$[\underline{S_l}(\omega_1)\underline{\bar{a}}-\underline{\bar{a}},\underline{S_l}(\omega_2)\underline{\bar{a}}-\underline{\bar{a}},\cdots,\underline{S_l}(\omega_{r_2})\underline{\bar{a}}-\underline{\bar{a}},\underline{S_l}(\lambda_2)\underline{\bar{a}}-\underline{\bar{a}}]^T:=$
$A^T_{\lambda_2}[A_{\lambda_2}A^T_{\lambda_2}]^{-1}A_{\lambda_2}H^T_{\underline{a}\ (r_2+1,L)}(|||\omega_1|||,|||\omega_2|||,\cdots,|||\omega_{r_2}|||,|||\lambda_2|||)$
and $H^T_{\underline{a}\ (|||\lambda_2|||+1,L)}(|||\omega_1|||,|||\omega_2|||,\cdots,|||\omega_{r_2}|||,|||\lambda_2|||):=$
$[\underline{S_l}(\omega_1)\underline{a}-\underline{a},\cdots,\underline{S_l}(\omega_{r_2})\underline{a}-\underline{a},\underline{S_l}(\lambda_2)\underline{a}-\underline{a}]$.
Then let $h_{\lambda_2 s}\in\boldsymbol{R}^{1\times r_2}$ *be*
$h_{\lambda_2 s}:=[\underline{a}(\omega_{r_1+1})-\underline{a}(1)-(\underline{\bar{a}}(\omega_{r_1+1})-\underline{\bar{a}}(1)),$
$\underline{a}(\omega_{r_1+2})-\underline{a}(1)-(\underline{\bar{a}}(\omega_{r_1+2})-\underline{\bar{a}}(1)),\cdots,\ \underline{a}(\omega_{r_2})-\underline{a}(1)-(\underline{\bar{a}}(\omega_{r_2})-\underline{\bar{a}}(1))]$.

⋮

Ⓣ *For* $\omega \in U^*$ *such that* $|||\omega||| = |||\omega_n||| + 1$ *and based on Proposition (2.14), determine coefficients* $\{b_{\omega,i} : 1 \leq i \leq n\}$.
The Q *in Proposition (2.14) can be considered as the matrix composed from the eigenvectors of* $H_{\underline{a}\ (n+1,L)}(|||\omega_1|||, |||\omega_2|||, \cdots, |||u|||)$
$\times H^T_{\underline{a}\ (n+1,L)}(|||\omega_1|||, |||\omega_2|||, \cdots, |||u|||)$.
Let a matrix $A_\omega \in \boldsymbol{R}^{1\times(n+1)}$ *be* $A_\omega := [b_{\omega,1}, b_{\omega,2}, \cdots, b_{\omega,n}, -1]$.
Determine the error vectors $\{\underline{S_l}(\omega_i)\underline{\bar{a}} - \underline{\bar{a}} :\ 0 \leq i \leq n\}$ *and* $\underline{S_l}(\omega)\underline{\bar{a}} - \underline{\bar{a}}$ *by using the equation*
$[\underline{S_l}(\omega_1)\underline{\bar{a}} - \underline{\bar{a}}, \underline{S_l}(\omega_2)\underline{\bar{a}} - \underline{\bar{a}}, \cdots, \underline{S_l}(\omega_n)\underline{\bar{a}} - \underline{\bar{a}}, \underline{S_l}(\omega)\underline{\bar{a}} - \underline{\bar{a}}]^T :=$
$A^T_\omega[A_\omega A^T_\omega]^{-1}A_\omega H^T_{\underline{a}\ (n+1,L)}(|||\omega_1|||, |||\omega_2|||, \cdots, |||\omega_n|||, |||\omega|||)$
and $H^T_{\underline{a}\ (n+1,L)}(|||\omega_1|||, |||\omega_2|||, \cdots, |||u|||) :=$
$[\underline{S_l}(\omega_1)\underline{a} - \underline{a}, \cdots, \underline{S_l}(\omega_n)\underline{a} - \underline{a}, \underline{S_l}(\omega)\underline{a} - \underline{a}]$,
Then let $h_{\omega s}$ *be*
$h_{\omega s} := [\underline{a}(\omega_{r_t+1}) - \underline{a}(1) - (\underline{\bar{a}}(\omega_{r_t+1}) - \underline{\bar{a}}(1)),$
$\underline{a}(\omega_{r_t+2}) - \underline{a}(1) - (\underline{\bar{a}}(\omega_{r_t+2}) - \underline{\bar{a}}(1)), \cdots,\ \underline{a}(\omega_n) - \underline{a}(1) - (\underline{\bar{a}}(\omega_n) - \underline{\bar{a}}(1))]$.
Finally, let $h_s \in \boldsymbol{R}^{1\times n}$ *be*
$h_s := [h_{\lambda_1 s}, h_{\lambda_2 s}, \cdots, h_{\omega s}]$.

[proof] This theorem can be proved to be the same as theorem (7.21)

Remark 1: The number of dimensions is determined by checking what parts are noisy parts and by using the ratio of the Hankel matrix norm, which implies the noise to signal ratio.

Remark 2: According to Theorem (7.20), an affine dynamical system $\sigma = ((\boldsymbol{R}^n, F_s), g_s, h_s, h^0)$ is obtained as follows:

In 2), g_s is obtained directly or by using the algebraic CLS method for A_u corresponding to the matrix A in Lemma (2.17). In 3), F_s is obtained by using the algebraic CLS method for ${}_iA_j$ corresponding to the matrix A in Lemma (2.17). In 4), h_s is obtained by using the algebraic CLS method for A_{λ_1}, A_{λ_2}, A_{λ_ω} corresponding to the matrix A in Lemma (2.17).

Remark 3: Let S and N be the norm of a signal and noise. Then the selected ratio of matrices in the algorithm may be considered as $\frac{N}{S+N}$.

In the figures of this chapter, we use a notation *Signal_n d* as the input response map obtained by a n-dimensional affine dynamical system.

In the examples of this chapter, a notation $H^T_{\underline{a}\ (r,40)}(1, \cdots, r)$ is used in place of $H^T_{\underline{a}\ (r,40)}(1, 2, \cdots, r-1, r)$.

Example 7.31. Let the signals be the input response map of the following 3-dimensional affine dynamical system: $\sigma = ((\boldsymbol{R}^3, F), g, h, h^0)$, where

$$F(u_1) = \begin{bmatrix} -1 & -0.2 & -0.8 \\ 0 & 0 & -0.9 \\ 0 & -0.2 & -0.5 \end{bmatrix}, F(u_2) = \begin{bmatrix} -0.8 & 0 & 0.6 \\ 0 & 1 & 0 \\ -0.5 & 0 & 0.8 \end{bmatrix}, F(u_3) = \begin{bmatrix} 0 & -0.4 & 0 \\ 0 & 0.1 & 0.5 \\ 0 & 0 & 0.5 \end{bmatrix},$$

$g(u_1) = \mathbf{e}_1,\ g(u_2) = \mathbf{e}_2,\ g(u_3) = \mathbf{e}_3,\ h = [12,\ -5,\ -8],\ h^0 = 1.$

Then the algebraically noisy realization problem is solved as follows:

covariance matrix	eigenvalues					
	1	2	3	4	5	6
$H^T_{\underline{a}\ (2,40)}(1,2)H_{\underline{a}\ (2,40)}(1,2)$	1287	356				
$H^T_{\underline{a}\ (3,40)}(1,2,3)H_{\underline{a}\ (3,40)}(1,2,3)$	1725	1270	258			
$H^T_{\underline{a}\ (4,40)}(1,\cdots,4)H_{\underline{a}\ (4,40)}(1,\cdots,4)$	1725	1270	258	5.5		
$H^T_{\underline{a}\ (5,40)}(1,\cdots,5)H_{\underline{a}\ (5,40)}(1,\cdots,5)$	2195	1721	445	7.9	3.1	
$H^T_{\underline{a}\ (6,40)}(1,\cdots,6)H_{\underline{a}\ (6,40)}(1,\cdots,6)$	3380	2124	446	8.3	3.6	2.4
covariance matrix	square root of eigenvalues					
$H^T_{\underline{a}\ (2,40)}(1,2)H_{\underline{a}\ (2,40)}(1,2)$	35.9	18.9				
$H^T_{\underline{a}\ (3,40)}(1,2,3)H_{\underline{a}\ (3,40)}(1,2,3)$	41.5	35.6	16.1			
$H^T_{\underline{a}\ (4,40)}(1,\cdots,4)H_{\underline{a}\ (4,40)}(1,\cdots,4)$	41.5	35.6	16.1	2.3		
$H^T_{\underline{a}\ (5,40)}(1,\cdots,5)H_{\underline{a}\ (5,40)}(1,\cdots,5)$	46.9	41.5	21.1	2.8	1.8	
$H^T_{\underline{a}\ (6,40)}(1,\cdots,6)H_{\underline{a}\ (6,40)}(1,\cdots,6)$	58.1	46.1	21.1	2.9	1.9	1.5

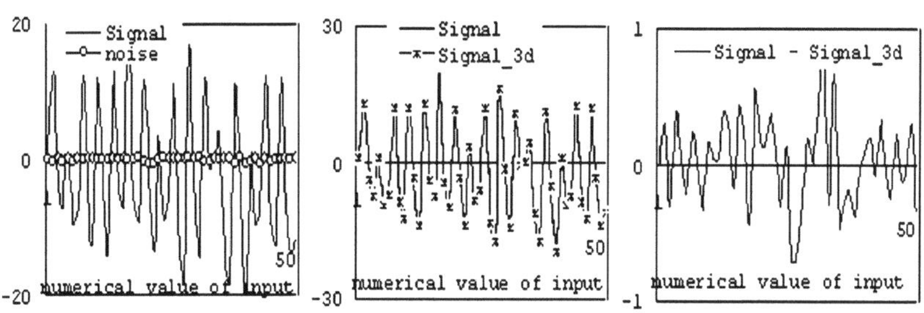

Fig. 7.6 The left is the original input response map and noise added to the original 3-dimensional affine dynamical system. The middle is the original input response map and the behavior of a 3-dimensional affine dynamical system obtained by the algebraic CLS method. The right is the difference between the original input response map and the behavior of the 3-dimensional affine dynamical system obtained by the algebraic CLS method in Example (7.31).

1) A set $\{2.9,\ 1.9,\ 1.5\}$ is composed of relatively small and equally-sized numbers in the square root of eigenvalues for $H^T_{\underline{a}\ (6,40)}(1,\cdots,6)H_{\underline{a}\ (6,40)}(1,\cdots,6)$.
2) After determining the independent vectors $\underline{S}_l(u_1)\underline{a} - \underline{a}$, $\underline{S}_l(u_2)\underline{a} - \underline{a}$ and $\underline{S}_l(u_3)\underline{a} - \underline{a}$ whose numerical value of input are 1, 2 and 3, we will continue the noisy realization algorithm by the algebraic CLS method.

Therefore, a noisy 3-dimensional affine dynamical system $\sigma_1 = ((\boldsymbol{R}^3,\ F_1),\ g_1,\ h_1,\ h^0)$ obtained by the algebraic CLS method is constructed as follows:

$$F_1(u_1) = \begin{bmatrix} -1 & -0.2 & -0.8 \\ 0.03 & 0 & -0.9 \\ -0.02 & -0.22 & -0.5 \end{bmatrix},\ F_1(u_2) = \begin{bmatrix} -0.8 & -0.04 & 0.6 \\ 0.05 & 1 & -0.04 \\ -0.5 & -0.01 & 0.8 \end{bmatrix},$$

$$F_1(u_3) = \begin{bmatrix} 0.01 & -0.4 & -0.01 \\ 0.01 & 0.1 & 0.47 \\ -0.01 & 0 & 0.5 \end{bmatrix},\ g_1(u_1) = \mathbf{e}_1, g_1(u_2) = \mathbf{e}_2, g_1(u_3) = \mathbf{e}_3, h_1 = [11.7, -4.7, -8.4], h^0 = 1.$$

We can show that the algorithm for noisy realization given by the analytic CLS method in the reference [Hasegawa, 2008] produces the same system as the above one in the sense of the numerical calculation.

In this example, the original signals are considered as the input response map of a 3-dimensional affine dynamical system and the desirable input response map is obtained by the algebraic CLS method. The model obtained by the algebraic CLS method is a 3-dimensional affine dynamical system.

Just as we expected, the following table and Fig. 7.6 truly indicate that the 3-dimensional affine dynamical system obtained by the algebraic CLS method is a good noisy realization.

dimen-sion	ratio of matrices	mean values of square root for sum of			cosine	error
		signal	signal by CLS	error	① and ②	ratio
		①	②	③	$\cos\theta$	③/①
$a_{1,2,3}$	0.05	1.4984	1.4966	0.05	0.999	0.03

Example 7.32. Let the signals be the input response map of the following 4-dimensional affine dynamical system: $\sigma = ((\boldsymbol{R}^4, F), g, h, h^0)$, where

$$F(u_1) = \begin{bmatrix} 0 & -0.6 & -0.4 & -0.2 \\ 0 & 0.6 & 0 & -0.9 \\ -0.6 & 0 & -0.1 & 0.8 \\ 0 & -0.5 & 0.5 & -0.2 \end{bmatrix},\ F(u_2) = \begin{bmatrix} 0.6 & 0 & -0.3 & 0.1 \\ 0 & 0.7 & 0 & 0 \\ -0.5 & 0 & 0.6 & 0.4 \\ 0.3 & 0 & 0.3 & 0.7 \end{bmatrix},$$

$$F(u_3) = \begin{bmatrix} -1 & -0.3 & 0 & -0.6 \\ 0 & 0 & 1 & 0.6 \\ 0 & -0.5 & -0.3 & 0.7 \\ 0.8 & 0.3 & 0.5 & 0.7 \end{bmatrix},\ g(u_1) = \mathbf{e}_1,\ g(u_2) = \mathbf{e}_2,\ g(u_3) = \mathbf{e}_3,$$

$$h = [12,\ -10,\ 2,\ 8],\ h^0 = 1.$$

Then the algebraically noisy realization problem is solved as follows:

covariance matrix	eigenvalues					
	1	2	3	4	5	6
$H^T_{\underline{a}\ (3,40)}(1,2,3)H_{\underline{a}\ (3,40)}(1,2,3)$	2027	1226	505			
$H^T_{\underline{a}\ (4,40)}(1,\cdots,4)H_{\underline{a}\ (4,40)}(1,\cdots,4)$	2924	1964	515	0.7		
$H^T_{\underline{a}\ (5,40)}(1,\cdots,5)H_{\underline{a}\ (5,40)}(1,\cdots,5)$	4725	2280	569	31	0.7	
$H^T_{\underline{a}\ (6,40)}(1,\cdots,6)H_{\underline{a}\ (6,40)}(1,\cdots,6)$	8466	2373	1069	172	1.1	0.7
covariance matrix	square root of eigenvalues					
$H^T_{\underline{a}\ (3,40)}(1,2,3)H_{\underline{a}\ (3,40)}(1,2,3)$	45	35	22.5			
$H^T_{\underline{a}\ (4,40)}(1,\cdots,4)H_{\underline{a}\ (4,40)}(1,\cdots,4)$	54.1	44.3	22.7	0.8		
$H^T_{\underline{a}\ (5,40)}(1,\cdots,5)H_{\underline{a}\ (5,40)}(1,\cdots,5)$	68.7	47.7	24.3	5.6	0.8	
$H^T_{\underline{a}\ (6,40)}(1,\cdots,6)H_{\underline{a}\ (6,40)}(1,\cdots,6)$	92	48.7	32.7	13.1	1	0.8

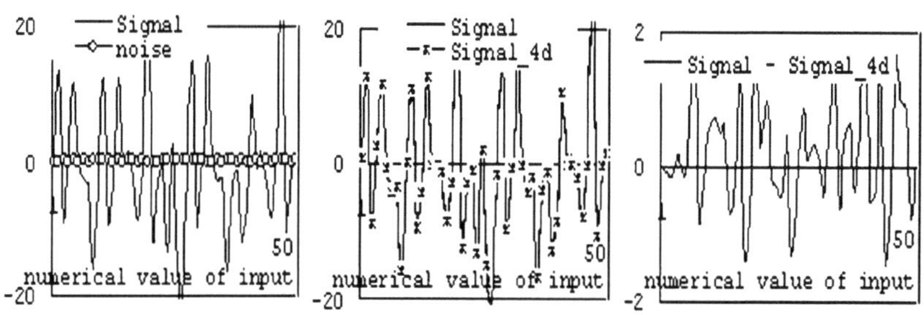

Fig. 7.7 The left is the original input response map and noise added to the original 4-dimensional affine dynamical system. The middle is the original input response map and the behavior of a 4-dimensional affine dynamical system obtained by the algebraic CLS method. The right is the difference between the original input response map and the behavior of the 4-dimensional affine dynamical system obtained by the algebraic CLS method in Example (7.32).

1) A set $\{1,\ 0.8\}$ is composed of relatively small and equally-sized numbers in the square root of eigenvalues for $H^T_{\underline{a}\ (6,40)}(1,\cdots,6)H_{\underline{a}\ (6,40)}(1,\cdots,6)$.
2) After determining the independent vectors $\underline{S_l}(u_1)\underline{a}-\underline{a}$, $\underline{S_l}(u_2)\underline{a}-\underline{a}$, $\underline{S_l}(u_3)\underline{a}-\underline{a}$ and $\underline{S_l}(u_2|u_1)\underline{a}-\underline{a}$ whose numerical value of input are 1, 2, 3 and 5 we will continue the algebraically noisy realization algorithm by the algebraic CLS method.

Therefore, a noisy 4-dimensional affine dynamical system $\sigma_1=((\boldsymbol{R}^4,\ F_1),\ g_1,\ h_1,\ h^0)$ obtained by the algebraic CLS method is constructed as follows:

$$F_1(u_1)=\begin{bmatrix}-0.01 & 0.36 & -1.37 & 1.1\\ -0.01 & 2.2 & -1.6 & 2.9\\ -0.6 & -0.8 & 0.7 & -1.35\\ 0.01 & -1.6 & 1.6 & -2.6\end{bmatrix},\ F_1(u_2)=\begin{bmatrix}0 & -0.03 & -0.92 & 0.06\\ -1 & 0.7 & -1.02 & -0.11\\ 0 & 0.01 & 1.1 & -0.07\\ 1 & 0.04 & 1 & 0.8\end{bmatrix},$$

$$F_1(u_3) = \begin{bmatrix} -2.56 & -0.9 & -0.97 & -2.5 \\ -2.6 & -1 & -0.64 & -2.7 \\ 1.3 & -0.03 & 0.5 & 1.1 \\ 2.58 & 0.96 & 1.6 & 2.4 \end{bmatrix}, \; g_1(u_1) = \mathbf{e}_1, \; g_1(u_2) = \mathbf{e}_2,$$

$g_1(u_3) = \mathbf{e}_3, \; h_1 = [12.1 \; -9.9, \; 1.8, \; -1.5], \; h^0 = 1.$

We can show that the algorithm for noisy realization given by the analytic CLS method in the reference [Hasegawa, 2008] produces the same systems as the above one in the sense of the numerical calculation.

In this example, the original signals are considered as the input response map of a 4-dimensional affine dynamical system and the desirable input response map is obtained by the algebraic CLS method. The model obtained by the algebraic CLS method is a 4-dimensional affine dynamical system.

Just as we expected, the following table and Fig. 7.7 truly indicate that the 4-dimensional affine dynamical system obtained by the algebraic CLS method is a somewhat good noisy realization.

dimen-sion	ratio of matrices	mean values of the square root for sum of			cosine	error
		signal	signal by CLS	error	① and ②	ratio
		①	②	③	$\cos\theta$	③/①
$a_{1,2,3,5}$	0.01	1.3514	1.3511	0.112	0.996	0.08

Example 7.33. Let the signals be the input response map of the following 4-dimensional affine dynamical system: $\sigma = ((\boldsymbol{R}^4, F), g, h, h^0)$, where $F(u_1) =$

$$\begin{bmatrix} 0 & 0.5 & 1 & 0.6 \\ 0 & 2.6 & -0.3 & 0 \\ -0.6 & 0 & -0.1 & 0.4 \\ 0.8 & 0.5 & 0.3 & 0.4 \end{bmatrix}, \; F(u_2) = \begin{bmatrix} -1.6 & 0 & -0.3 & 0.3 \\ 0 & 1 & 0 & 0 \\ -0.8 & 0 & 0.6 & 0.3 \\ 0.3 & 0 & 0.3 & 0.7 \end{bmatrix},$$

$$F(u_3) = \begin{bmatrix} -1 & -0.3 & 0 & -0.7 \\ 0 & 0 & 1 & 0.8 \\ 0 & -0.3 & -0.3 & 0.7 \\ 0.8 & 0.3 & 0.4 & 0.7 \end{bmatrix}, \; g(u_1) = \mathbf{e}_1, \; g(u_2) = \mathbf{e}_2, \; g(u_3) = \mathbf{e}_3,$$

$h = [12, \; -8, \; 1, \; 2], \; h^0 = 1.$

Then the algebraically noisy realization problem is solved as follows:

covariance matrix	eigenvalues						
	1	2	3	4	5	6	7
$H^T_{\underline{a}\,(4,40)}(1,\cdots,4)H_{\underline{a}\,(4,40)}(1,\cdots,4)$	30836	22669	7231	1351			
$H^T_{\underline{a}\,(5,40)}(1,\cdots,5)H_{\underline{a}\,(5,40)}(1,\cdots,5)$	81838	29843	7807	2250	10		
$H^T_{\underline{a}\,(6,40)}(1,\cdots,6)H_{\underline{a}\,(6,40)}(1,\cdots,6)$	85200	35710	23678	3949	11.1	4.4	
$H^T_{\underline{a}\,(7,40)}(1,\cdots,7)H_{\underline{a}\,(7,40)}(1,\cdots,7)$	243148	82603	28542	4266	17.1	8.2	3.3
covariance matrix	square root of eigenvalues						
$H^T_{\underline{a}\,(4,40)}(1,\cdots,4)H_{\underline{a}\,(4,40)}(1,\cdots,4)$	175.6	150.6	85	36.8			
$H^T_{\underline{a}\,(5,40)}(1,\cdots,5)H_{\underline{a}\,(5,40)}(1,\cdots,5)$	286	172.8	88.4	47.4	3.2		
$H^T_{\underline{a}\,(6,40)}(1,\cdots,6)H_{\underline{a}\,(6,40)}(1,\cdots,6)$	291.9	189	153.9	62.8	3.3	2.1	
$H^T_{\underline{a}\,(7,40)}(1,\cdots,7)H_{\underline{a}\,(7,40)}(1,\cdots,7)$	493	287	169	65.3	4.1	2.9	1.8

1) A set $\{4.1, \; 2.9, \; 1.8\}$ is composed of relatively small and equally-sized numbers in the square root of eigenvalues for $H^T_{\underline{a}\,(7,40)}(1,\cdots,7)H_{\underline{a}\,(7,40)}(1,\cdots,7)$.

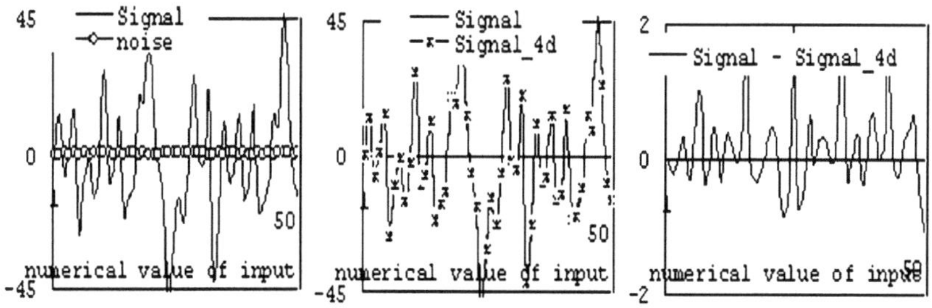

Fig. 7.8 The left is the original input response map and noise added to the original 4-dimensional affine dynamical system. The middle is the original input response map and the behavior of a 4-dimensional affine dynamical system obtained by the algebraic CLS method. The right is the difference between the original input response map and the behavior of the 4-dimensional affine dynamical system obtained by the algebraic CLS method in Example (7.33).

2) After determining the independent vectors $\underline{S_l}(u_1)\underline{a} - \underline{a}$, $\underline{S_l}(u_2)\underline{a} - \underline{a}$, $\underline{S_l}(u_3)\underline{a} - \underline{a}$ and $\underline{S_l}(u_1|u_1)\underline{a} - \underline{a}$ whose numerical value of input are 1, 2, 3 and 5 we will continue the noisy realization algorithm by the algebraic CLS method.

Therefore, a noisy 4-dimensional affine dynamical system $\sigma_1 = ((\boldsymbol{R}^4, F_1), g_1, h_1, h^0)$ obtained by the algebraic CLS method is constructed as follows:

$$F_1(u_1) = \begin{bmatrix} -1 & -0.2 & 0.6 & -1.3 \\ 0 & 2.6 & -0.3 & 0.2 \\ 0 & 0.4 & 0.1 & 0.48 \\ 1 & 0.7 & 0.4 & 1.2 \end{bmatrix}, F_1(u_2) = \begin{bmatrix} -2 & -0.02 & -0.7 & -2 \\ 0.01 & 1 & -0.01 & 0.01 \\ -0.6 & 0.01 & 0.8 & -0.4 \\ 0.36 & 0.01 & 0.4 & 0.8 \end{bmatrix},$$
$$F_1(u_3) = \begin{bmatrix} -2 & -0.7 & -0.5 & -2.9 \\ 0.01 & 0.01 & 1 & 0.06 \\ 0.6 & -0.08 & 0 & 1.6 \\ 1 & 0.4 & 0.5 & 1.4 \end{bmatrix}, g_1(u_1) = \mathbf{e}_1,\ g_1(u_2) = \mathbf{e}_2,$$
$$g_1(u_3) = \mathbf{e}_3,\ h_1 = [12.2,\ -7.9,\ 0.7,\ 13.2],\ h^0 = 1.$$

We can show that the algorithm for noisy realization given by the analytic CLS method in the reference [Hasegawa, 2008] produces the same system as the above one in the sense of the numerical calculation.

In this example, the original signals are considered as the input response map of a 4-dimensional affine dynamical system and the desirable input response map is obtained by the algebraic CLS method. The model obtained by the algebraic CLS method is a 4-dimensional affine dynamical system.

Just as we expected, the following table and Fig. 7.8 truly indicate that the 4-dimensional affine dynamical system obtained by the algebraic CLS method is a good noisy realization.

dimen-sion	ratio of matrices	mean values of the square root for sum of signal	signal by CLS	error	cosine ① and ②	error ratio
		①	②	③	$\cos\theta$	③/①
$a_{1,2,3,4}$	0.01	2.72	2.73	0.09	0.999	0.03

Example 7.34. Let the signals be the input response map of the following 5-dimensional affine dynamical system: $\sigma = ((\boldsymbol{R}^5, F), g, h, h^0)$, where

$$F(u_1)=\begin{bmatrix} -1 & -0.4 & -0.6 & -1 & -1 \\ 0 & -0.2 & 0.5 & 0.4 & -0.2 \\ 0 & -0.2 & 0 & 0.8 & 0 \\ 1 & 0.2 & 0.1 & 0.7 & 1 \\ 0 & 0.2 & 0.7 & 0.3 & 0 \end{bmatrix}, F(u_2)=\begin{bmatrix} 0 & -0.4 & -0.4 & 0 & 0 \\ -1 & 0.21 & -0.2 & -0.5 & 0 \\ 0 & 0.2 & 0.6 & 0.4 & 0 \\ 0 & -0.2 & 0.2 & 0.5 & 0 \\ 1 & 0.7 & 0.2 & 0.5 & 0.9 \end{bmatrix},$$

$$F(u_3) = \begin{bmatrix} 0 & -0.6 & -0.31 & -0.5 & -0.7 \\ 0 & 0.7 & 0.1 & 0.2 & 0.7 \\ -1 & 0 & -0.1 & 0.4 & -0.3 \\ 0 & 0.3 & 0.5 & 0.8 & 0.6 \\ 0 & 0.4 & -0.1 & -0.2 & 0.3 \end{bmatrix}, g(u_1) = \mathbf{e}_1,\ g(u_2) = \mathbf{e}_2,$$

$g(u_3) = \mathbf{e}_3,\ h = [12,\ -1,\ -4,\ 1,\ 8],\ h^0 = 1.$

Then the algebraically noisy realization problem is solved as follows:

covariance matrix	eigenvalues 1	2	3	4	5	6	7	8
$H^T_{\underline{a}\ (4,40)}(1,\cdots,4)H_{\underline{a}\ (4,40)}(1,\cdots,4)$	4341	1378	254	190				
$H^T_{\underline{a}\ (5,40)}(1,\cdots,5)H_{\underline{a}\ (5,40)}(1,\cdots,5)$	6094	1389	451	191	14.6			
$H^T_{\underline{a}\ (6,40)}(1,\cdots,6)H_{\underline{a}\ (6,40)}(1,\cdots,6)$	6094	1390	451	193	16.1	4.6		
$H^T_{\underline{a}\ (7,40)}(1,\cdots,7)H_{\underline{a}\ (7,40)}(1,\cdots,7)$	6818	1844	459	193	16.4	4.6	2.3	
$H^T_{\underline{a}\ (8,40)}(1,\cdots,8)H_{\underline{a}\ (8,40)}(1,\cdots,8)$	8363	3283	590	216	16.5	5.8	2.7	1.5
covariance matrix	square root of eigenvalues							
$H^T_{\underline{a}\ (4,40)}(1,\cdots,4)H_{\underline{a}\ (4,40)}(1,\cdots,4)$	65.9	37.1	15.9	13.8				
$H^T_{\underline{a}\ (5,40)}(1,\cdots,5)H_{\underline{a}\ (5,40)}(1,\cdots,5)$	78.1	37.3	21.2	13.8	3.8			
$H^T_{\underline{a}\ (6,40)}(1,\cdots,6)H_{\underline{a}\ (6,40)}(1,\cdots,6)$	78.1	37.3	21.2	13.9	4	2.1		
$H^T_{\underline{a}\ (7,40)}(1,\cdots,7)H_{\underline{a}\ (7,40)}(1,\cdots,7)$	82.6	42.9	21.4	13.9	4	2.1	1.5	
$H^T_{\underline{a}\ (8,40)}(1,\cdots,8)H_{\underline{a}\ (8,40)}(1,\cdots,8)$	91.4	57.3	24.3	14.7	4.1	2.4	1.6	1.2

1) A set $\{2.4,\ 1.6,\ 1.2\}$ is composed of relatively small and equally-sized numbers in the square root of eigenvalues for $H^T_{\underline{a}\ (8,40)}(1,\cdots,8)H_{\underline{a}\ (8,40)}(1,\cdots,8)$.
2) After determining the independent vectors $\underline{S_l}(u_1)\underline{a} - \underline{a}$, $\underline{S_l}(u_2)\underline{a} - \underline{a}$, $\underline{S_l}(u_3)\underline{a} - \underline{a}$, $\underline{S_l}(u_1|u_1)\underline{a} - \underline{a}$ and $\underline{S_l}(u_2|u_1)\underline{a} - \underline{a}$ whose numerical value of input are 1, 2, 3, 4 and 5 we will continue the noisy realization algorithm by the algebraic CLS method.

Therefore, a noisy 5-dimensional affine dynamical system $\sigma_1 = ((\boldsymbol{R}^5,\ F_1),\ g_1,\ h_1,\ h^0)$ obtained by the algebraic CLS method is constructed as follows:

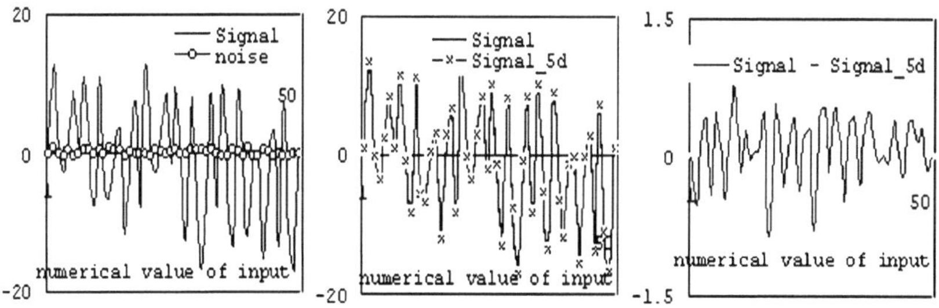

Fig. 7.9 The left is the original input response map and noise added to the original 5-dimensional affine dynamical system. The middle is the original input response map and the behavior of a 5-dimensional affine dynamical system obtained by the algebraic CLS method. The right is the difference between the original input response map and the behavior of the 5-dimensional affine dynamical system obtained by the algebraic CLS method in Example (7.34).

$$F_1(u_1) = \begin{bmatrix} -1 & -0.35 & -0.62 & -1.1 & -1.03 \\ 0 & -0.09 & 0.44 & 0.18 & -0.32 \\ 0 & -0.2 & 0.02 & 0.87 & 0.05 \\ 1 & 0.23 & 0.1 & 0.63 & 0.93 \\ 0 & 0.1 & 0.75 & 0.5 & 0.14 \end{bmatrix},$$

$$F_1(u_2) = \begin{bmatrix} 0 & -0.43 & -0.46 & -0.03 & -0.02 \\ -1 & 0.07 & -0.36 & -0.57 & -0.07 \\ 0 & 0.3 & 0.66 & 0.4 & 0.05 \\ 0 & -0.3 & 0.13 & 0.49 & -0.04 \\ 1 & 0.87 & 0.4 & 0.56 & 0.96 \end{bmatrix},$$

$$F_1(u_3) = \begin{bmatrix} 0.13 & -0.6 & -0.3 & -0.54 & -0.73 \\ 0.28 & 0.62 & 0.07 & 0.09 & 0.61 \\ -1.1 & 0.08 & -0.08 & 0.47 & -0.26 \\ 0.05 & 0.25 & 0.46 & 0.74 & 0.55 \\ -0.22 & 0.5 & -0.05 & -0.1 & 0.4 \end{bmatrix}, \; g_1(u_1) = \mathbf{e}_1, g_1(u_2) = \mathbf{e}_2,$$

$$g_1(u_3) = \mathbf{e}_3, \; h_1 = [12.5, \; -1.1, \; -4.4, \; 1.4, \; 7.5], \; h^0 = 1.$$

We can show that the algorithm for noisy realization given by the analytic CLS method in the reference [Hasegawa, 2008] produces the same system as the above one in the sense of the numerical calculation.

In this example, the original signals are considered as the input response map of a 5-dimensional affine dynamical system and the desirable input response map is obtained by the algebraic CLS method. The model obtained by the algebraic CLS method is a 5-dimensional affine dynamical system.

Just as we expected, the following table and Fig. 7.9 truly indicate that the 5-dimensional affine dynamical system obtained by the algebraic CLS method is a good noisy realization.

dimension	ratio of matrices	mean values of square root for sum of signal	signal by CLS	error	cosine ① and ②	error ratio
		①	②	③	$\cos\theta$	③/①
$a_{1,2,3,4}$	0.05	1.1635	1.1977	0.08	0.998	0.07
$a_{1,2,3,4,5}$	0.03	1.1635	1.187	0.05	0.999	0.04

7.6 Historical Notes and Concluding Remarks

Regarding important facts affecting the affine dynamical systems, we have provided a representation of their behaviors and a partial realization algorithm. The representation of their behaviors means that any affine dynamical system can be completely characterized by the input response map itself.

As for general non-linear systems, approximation and noisy realization problems were proposed by introducing the analytic constrained least square method, abbreviated, analytic CLS method. The problems were attempted to be solved by presenting an algorithm which is made up of the ratio of Hankel matrix norm and the analytic CLS method. The method is reduced to obtain the minimum value of a rational polynomial in multi variables. Therefore, many equations obtained by partial differential equations must be solved with a lot of hard work. In this chapter, the method has been replaced by an easier method which is called the algebraic CLS method.

By applying the algebraic CLS method to several examples of affine dynamical systems in our problems, we could make a law which says that the affine dynamical systems obtained by the algebraic CLS method are the same as the systems obtained by the analytic CLS method in the sense of numerical experiments. The law is called a law of a constrained least square.

By applying this algorithm to several examples of affine dynamical systems, we have shown that this algorithm is practical, useful and easy.

Our several examples in the algebraically approximata realization problem show that the changing relations among the ratio of the matrix norm and the error to signal ratio are proportional relations and the ratio is 0.01 for the Input/output matrix norm while the error to signal ranges from 0.01 to 0.04. This approximate realization algorithm appears to be very promising.

Our several examples in the noisy realization problem show that the changing relations among the ratio of matrices and the error to signal ratio are proportional relations, and the ratio of Hankel matrices is 0.01 while the error to signal ratio ranges from 0.006 to 0.08.

This noisy realization algorithm also appears to be very promising.

As we have mentioned before, the concrete discussions of algebraically approximate and noisy realization for non-linear systems are very new.

The analtic CLS method for determing n variables is reduced to the minimization of the following rational polynomial $p(x_1, x_2, \cdots, x_n)$ in n variables:

$$p(x_1, x_2, \cdots, x_n) = \frac{\sum_{c_1, c_2, \cdots, c_n} \alpha(c_1, c_2, \cdots, c_n) \times x_1^{c_1} x_2^{c_2} \cdots x_n^{c_n}}{(1 + x_1^2 + x_2^2 + \cdots + x_n^2)^2},$$

where $\sum_{c_1, c_2, \cdots, c_n}$ means all summation of any combination with the conditions $0 \le c_i \le 4$ and $c_1 + c_2 + \cdots + c_n \le 4$ for any $1 \le i \le n$ and $\alpha(c_1, c_2, \cdots, c_n) \in \boldsymbol{R}$. Therefore, our new Law shows that approximate and noisy problems can be solved using only algebraic calculations, namely, without treating partial differential equations.

Chapter 8
Algebraically Approximate and Noisy Realization of Linear Representation Systems

Let the set of output's values Y be a linear space over the field $\boldsymbol{R}$. In the reference [Matsuo and Hasegawa, 2003], linear representation systems were presented with the following main theorem. The main theorem says that for any causal input/output map, i.e., any input response map, there exist at least two canonical, namely quasi-reachable and distinguishable linear representation systems which realize, equivalently, faithfully describe it, and any two canonical linear representation systems with the same behavior are isomorphic.

In order for the discussion to be self contained, the results obtained in the reference are stated.

Firstly, the realization theory is listed.

Secondly, the results of finite dimensional linear representation systems are stated. They consist of a criterion for canonical finite dimensional linear representation systems, a representation theorem of isomorphic classes for canonical linear representation systems, a criterions for the behavior of finite dimensional linear representation systems, and a procedure to obtain a canonical linear representation system.

Thirdly, their partial realization is remarked on according to the above results. The existence of minimum partial realization is listed. It rarely happens for minimum partial realizations to be unique up to an isomorphism. To solve the uniqueness problem, we define a notion of natural partial realizations and state the following main results for this partial realization:

1) A necessary and sufficient condition for the existence of natural partial realizations is given by the rank condition of a finite-sized Hankel matrix.
2) The existence condition of natural partial realizations is equivalent to the uniqueness condition of minimum partial realizations.
3) An algorithm to obtain a natural partial realization from a given partial input response map is given.

We can easily understand that the above results of our systems are the same as the ones obtained in linear system theory.

Y. Hasegawa: Algebra. Approx. & Noisy Reali. of Discrete-Time Sys., LNEE 50, pp. 185–212.
springerlink.com

8.1 Basic Facts about Linear Representation Systems

Definition 8.1. Linear Representation System
1) A system given by the following system equation is written as a collection $\sigma = ((X, F), x^0, h)$ and it is said to be a linear representation system.

$$\begin{cases} x(t+1) = F(\omega(t+1))x(t) \\ x(0) \quad\quad = x^0 \\ \gamma(t) \quad\quad = hx(t) \end{cases}$$

for any $t \in N$, $x(t) \in X$, $\gamma(t) \in Y$,
where X is a linear space over the field $\boldsymbol{R}$, F is a linear operator on X, an initial state $x^0 \in X$ and $h : X \to Y$ is a linear operator.
2) The input response map $a_\sigma : U^* \to Y; \omega \mapsto h\phi_F(\omega)x^0$ is said to be the behavior of σ. For an input response map $a \in F(U^*, Y)$, σ which satisfies $a_\sigma = a$ is called a realization of a, where $\phi_F(\omega) := F(\omega(|\omega|))F(\omega(|\omega| - 1)) \cdots F(\omega(1))$.
3) A linear representation system σ is said to be quasi-reachable if the linear hull of the reachable set $\{\phi_F(\omega)x^0; \omega \in U^*\}$ is equal to X and a linear representation system σ is called distinguishable if $h\phi_F(\omega)x_1 = h\phi_F(\omega)x_2$ for any $\omega \in U^*$ implies $x_1 = x_2$.
4) A linear representation system σ is called canonical if σ is quasi-reachable and distinguishable.

Remark 1: The $x(t)$ in the system equation of σ is the state that produces output values of a_σ at the time t, namely the state $x(t)$ and linear operator $h : X \to Y$ generate the output value $a_\sigma(t)$ at the time t.
Remark 2: It is meant for σ to be a faithful model for the input response map a that σ realizes a.
Remark 3: Notice that a canonical linear representation system $\sigma = ((X, F), x^0, h)$ is a system that has the most reduced state space X among systems that have the behavior a_σ.

Example 8.2. $A(U^*) := \{\lambda = \sum_\omega \lambda(\omega)e_\omega$ (finite sum) $\}$, where $\omega = \bar{\omega}$ implies $e_\omega(\bar{\omega}) = 1$, and $\omega \neq \bar{\omega}$ implies $e_\omega(\bar{\omega}) = 0$. Let S_r be a map $: U \to L(A(U^*)); u \mapsto S_r(u)[\lambda \mapsto \sum_\omega \lambda(\omega)e_{u|\omega}$, an initial state be e_1 and a linear output map be $a : A(U^*) \to Y; \lambda \mapsto a(\lambda) = \sum_\omega \lambda(\omega)a(\omega)$. Then a collection $((A(U^*), S_r), e_1, a)$ is a quasi-reachable linear representation system that realizes a.

Let $F(U^*, Y)$ be a set of any input response maps, let $S_l : U \to L(A(U^*)); u \mapsto S_l(u)[a \mapsto [\omega \mapsto a(\omega|u)]]$. Let a linear output map be $1 : F(U^*, Y) \to Y; a \mapsto a(1)$. Then a collection $(F(U^*, Y), S_l), a, 1)$ is a distinguishable linear representation system that realizes a.

Remark: Note that the linear output map $a : A(U^*) \to Y$ is introduced by the fact $F(U^*, Y) = L(A(U^*), Y)$.

Theorem 8.3. *The following two linear representation systems are canonical realizations of any input response map* $a \in F(U^*, Y)$.
1) $((A(U^*)/_a, \hat{S}_r), [e_1], \hat{a})$,
where $A(U^*)/_a$ *is a quotient space obtained by equivalence relation* $\sum_\omega \lambda(\omega)e_\omega = \sum_{\bar{\omega}} \lambda(\bar{\omega})e_{\bar{\omega}} \iff \sum_\omega \lambda(\omega)a(\omega) = \sum_{\bar{\omega}} \lambda(\bar{\omega})a(\bar{\omega})$. $\hat{S}_r$ *is given by a map* $: U \to L(A(U^*)/_a); u \mapsto \hat{S}_r(u)[\lambda \mapsto \sum_\omega \lambda(\omega)[e_{u|\omega}]$, *and* $\hat{a}$ *is given by* $\hat{a} : A(U^*)/_a \to Y; [\lambda] \mapsto \hat{a}([\lambda]) = \sum_\omega \lambda(\omega)a(\omega)]$.
2) $((\ll S_l(U^*)a \gg), S_l), a, 1)$,
where $\ll S_l(U^*)a \gg$ *is the smallest linear space which contains* $S_l(U^*)a := \{S_l(\omega)a; \omega \in U^*\}$.

Definition 8.4. Let $\sigma_1 = ((X_1, F_1), x_1^0, h_1)$ and $\sigma_2 = ((X_2, F_2), x_2^0, h_2)$ be linear representation systems, then a linear operator $T : X_1 \to X_2$ is said to be a linear representation system morphism $T : \sigma_1 \to \sigma_2$ if T satisfies $TF_1(u) = F_2(u)T$ for any $u \in U$, $Tx_1^0 = x_2^0$ and $h_1 = h_2T$. If $T : X_1 \to X_2$ is bijective, then $T : \sigma_1 \to \sigma_2$ is said to be an isomorphism.

Theorem 8.5. *Realization Theorem of linear representation systems*
Existence part: For any input response map $a \in F(U^*, Y)$, *there exist at least two canonical linear representation systems which realize* a.
Uniqueness part: Let σ_1 *and* σ_2 *be any two canonical linear representation systems that realize* $a \in F(U^*, Y)$, *then there exists an isomorphism* $T : \sigma_1 \to \sigma_2$.

8.2 Finite Dimensional Linear Representation Systems

Based on the realization theory (8.5), we again state structures of finite-dimensional linear representation systems in this section that have been previously described. To obtain concrete and meaningful results, we assume that the set U of input values is finite; i.e., $U := \{u_i; 1 \le i \le m \text{ for some } m \in N\}$. This assumption implies that the difference morphism F of a linear representation system $\sigma = ((X, F, x^0, h)$ is completely determined by the finite matrices $\{F(u_i); 1 \le i \le m\}$. But it will be presented that the assumption is not so special. The main results can be stated in the following four steps:

Firstly, we present conditions when finite dimensional linear representation system is canonical.

Secondly, we obtain a representation theorem for finite dimensional canonical linear representation systems, namely, we show a standard system as a representative in their equivalence classes, which is a quasi-reachable standard system.

Thirdly, we give two criteria for the behavior of finite dimensional linear representation systems. One is the rank condition of infinite Hankel matrix, and the other is the application of Kleene's Theorem obtained in automata theory.

Lastly, we give a procedure to obtain the quasi-reachable standard system which realizes a given input response map.

Corollary 8.6. *Let T be a linear representation system morphism $T : \sigma_1 \to \sigma_2$, then $a_{\sigma_1} = a_{\sigma_2}$ holds.*

There is a fact about finite dimensional linear spaces that a n-dimensional linear space over the field $\boldsymbol{R}$ is isomorphic to $\boldsymbol{R}^n$ and $L(\boldsymbol{R}^n, \boldsymbol{R}^m)$ is isomorphic to $\boldsymbol{R}^{m\times n}$ (See Halmos [1958]). Therefore, without loss of generality, we can consider a n-dimensional linear representation system as $\sigma = ((\boldsymbol{R}^n, F), x^0, h)$, where F is a map : $U \to \boldsymbol{R}^{n\times n}$, $x^0 \in \boldsymbol{R}^n$ and $h \in \boldsymbol{R}^{p\times n}$.

Now we will show that the assumption of finiteness for input value's set U is not so special.

$U = \{u_1, u_2\}$ **8.7**

In this case, a linear representation system $\sigma = ((\boldsymbol{R}^n, F), x^0, h)$ can be completely determined by $\{F(u_i); u_i \in U \text{ for } i = 1, 2\}$.

If on-off inputs are applied to a black-box, any non-linear system can be treated in this case.

Moreover, if an optimal solution is a bang-bang control, when a controlled object is in the optimal controlled condition, then it can be treated in this case.

Cases where U is a convex set in R^m 8.8

Let the set U be a convex set in R^m and a set V of the extreme points be a finite set $\{u_j; 1 \le j \le m\}$. Let F in $\sigma = ((\boldsymbol{R}^n, F), x^0, h)$ be a linear operator : $U \to \boldsymbol{R}^{n\times n}$, i.e. $F(\sum_{i=1}^m \alpha_i \mathbf{e_i}) = \sum_{i=1}^m \alpha_i F(u_i)$, $\sum_{i=1}^m \alpha_i = 1$. Then the linear representation system $\sigma = ((\boldsymbol{R}^n, F), x^0, h)$ can be rewritten as a linear representation system $\tilde{\sigma} = ((\boldsymbol{R}^n, \tilde{F}), x^0, h)$,

where $\tilde{F} : V \to \boldsymbol{R}^{n\times n}$ is given by $\tilde{F}(u_i) = F(u_i)$ for any $u_i \in V$.

Note that the quasi-reachability of σ is equivalent to the quasi-reachability of $\tilde{\sigma}$.

$U = R^m$ **8.9**

Let $\boldsymbol{R} = R$ and $V = \{0, \mathbf{e_1}, \mathbf{e_2}, \cdots, \mathbf{e_m}\}$ for basis $\mathbf{e_i}$ in $R^m (1 \le i \le m)$. Let F in $\sigma = ((R^n, F), x^0, h)$ be an affine operator : $U \to R^{n\times n}$, i.e. $F(\sum_{i=1}^m \alpha_i \mathbf{e_i}) = A + (\sum_{i=1}^m \alpha_i \tilde{N}_i)$, A, $\tilde{N}_i \in R^{n\times n}$, $i \in N$. Then the linear representation system $\sigma = ((R^n, F), x^0, h)$ can be rewritten as a linear representation system $\tilde{\sigma} = ((R^n, \tilde{F}), x^0, h)$, where $\tilde{F} : V \to \boldsymbol{R}^{n\times n}$ is given by $F(0) = A$, $F(\mathbf{e_i}) = A + \tilde{N}_i$ for any $i (1 \le j \le m)$. Note that this $\tilde{\sigma}$ is a homogeneous bilinear system investigated by Tarn & Nonoyama [1976]. Note that the quasi-reachability of σ is equivalent to the quasi-reachability of $\tilde{\sigma}$.

Theorem 8.10. *A linear representation system $\sigma = ((K^n, F), x^0, h)$ is canonical if and only if the following conditions 1) and 2) hold.*
1) rank $[x^0, F(u_1)x^0, \cdots, F(u_m)x^0, \cdots, F(u_1)^2x^0, F(u_1)F(u_2)x^0, \cdots, F(u_1)F(u_m)x^0, \cdots, F(u_m)^2x^0, \cdots, F(u_1)^{n-1}x^0, F(u_2)F(u_1)^{n-2}x^0, \cdots, F(u_m)^{n-1}x^0] = n$.
2) rank $[h^T, (hF(u_1))^T, \cdots, (hF(u_m))^T, (hF(u_1)^2)^T, \cdots, (hF(u_1)F(u_m))^T, \cdots, (hF(u_1)^{n-1})^T, (hF(u_1)^{n-2}F(u_m))^T, \cdots, (hF(u_m)^{n-1})^T] = n$.

Definition 8.11. Let the input value's set U be $U := \{u_i; 1 \leq i \leq m\}$ and let a map $\| \ \| : U \to N$ be $u_i \mapsto \|u_i\| = i$. And let a numerical value $|||\omega|||$ of an input $\omega \in U^*$ be $|||\omega||| = \|\omega(|\omega|)\| + \|\omega(|\omega|-1)\| \times m + \cdots + \|\omega(1)\| \times m^{|\omega|-1}$ and $|||\mathit{1}||| = 0$.

Then, we can define a totally ordered relation by this numerical value in U^*. Namely, $\omega_1 \leq \omega_2 \iff |||\omega_1||| \leq |||\omega_2|||$.

Definition 8.12. A canonical linear representation system $\sigma = ((\boldsymbol{R}^n, F_s), \mathbf{e_1}, h_s)$ is said to be a quasi-reachable standard system if input sequences $\{\omega_i; 1 \leq i \leq n\}$ given by $\mathbf{e_i} = \phi_{F_s}(\omega_i)\mathbf{e_1}$ satisfy the following conditions:
1) $\mathit{1} = \omega_1 < \omega_2 < \cdots < \omega_n$ and $|\omega_i| \leq i - 1$ for $i(1 \leq i \leq n)$ hold.
2) $\phi_{F_s}(\omega)\mathbf{e_1} = \sum_{i=1}^{j} \mathbf{e_j}$ holds for any input sequence such that $\omega_j < \omega < \omega_{j+1}(1 \leq i \leq n-1)$, $\omega \in U^*$.

Theorem 8.13. *Representation Theorem for equivalence classes*
For any finite dimensional canonical linear representation system, there exists a uniquely determined isomorphic quasi-reachable standard system.

Definition 8.14. For any input response map $a \in F(U^*, Y)$, the corresponding linear input/output map $A : (A(U^*), S_r) \to (F(U^*, Y), S_l)$ satisfies $A(\mathbf{e}_\omega)(\bar{\omega}) = a(\bar{\omega}|\omega)$ for $\omega, \bar{\omega} \in U^*$.

Hence, A can be represented by the next infinite matrix H_a^L. This H_a^L is said to be a Hankel matrix of a.

$$H_a^L = \begin{array}{c} \\ \\ \\ \bar{\omega} \end{array} \left(\begin{array}{ccc} & & \omega \\ & & \vdots \\ & & \vdots \\ \cdots & \cdots & a(\bar{\omega}\,|\omega) \\ & & \end{array} \right)$$

Theorem 8.15. *Theorem for existence criterion*
For an input response map $a \in F(U^*, Y)$, the following conditions are equivalent:
1) The input response map $a \in F(U^*, Y)$ has the behavior of a n-dimensional canonical linear representation system.
2) There exist n linearly independent vectors and no more than n linearly independent vectors in a set $\{S_l(\omega)a; |\omega| \leq n-1$ for $\omega \in U^*\}$.
3) The rank of the Hankel matrix H_a^L of a is n.

Theorem 8.16. *Theorem for a realization procedure*
Let an input response $a \in F(U^*, Y)$ satisfy the condition of Theorem (8.18), then the quasi-reachable standard system $\sigma_s = ((\boldsymbol{R}^n, F_s), \mathbf{e_1}, h_s)$ which realizes the input response map a can be obtained by the following procedure:

1) Select the linearly independent vectors $\{S_l(\omega_i)a; 1 \le i \le n\}$ *of the set* $\{S_l(\omega)a; |\omega| \le n-1, \omega \in U^*\}$ *in order of their numerical value.*
2) Let the state space be $\boldsymbol{R}^n$*, the initial state be* $\mathbf{e}_1$.
3) Let the output map $h_s = [a(\omega_1), a(\omega_2), a(\omega_3), \cdots, a(\omega_n)]$.
4) Let ${}_if_j \in \boldsymbol{R}^n$ *in* $F_s(u_i) := [{}_if_1\ {}_if_2 \cdots {}_if_n]$ *be* ${}_if_j := [{}_if_{j,1}\ {}_if_{j,2} \cdots {}_if_{j,n}]^T$*, where* $S_l(u_i)S_l(\omega_j)a = \sum_{k=1}^{j} {}_if_{1,k}\ S_l(\omega_k)a$, ${}_if_{j,k} \in \boldsymbol{R}$ *and* $\mathbf{e}_1 = [1, 0, 0, \cdots, 0, 0]^T \in \boldsymbol{R}^n$.

8.3 Partial Realization Theory of Linear Representation Systems

Here we consider a partial realization problem by multi-experiment. Let $\underline{a}$ be an $\underline{N}$ sized input response map($\in F(U^*_{\underline{N}}, Y)$), where $\underline{N} \in N$ and $U^*_{\underline{N}} := \{\omega \in U^*; |\omega| \le \underline{N}\}$. The $\underline{a}$ is said to be a partial input response map. A finite dimensional linear representation system $\sigma = ((X, F), x^0, h)$ is called a partial realization of $\underline{a}$ if $h\phi_F(\omega)x^0 = \underline{a}(\omega)$ holds for any $\omega \in U^*_{\underline{N}}$.

A partial realization problem of linear representation systems can be stated as follows:
< For any given $\underline{a} \in F(U^*_{\underline{N}}, Y)$, find a partial realization σ of $\underline{a}$ such that the dimensions of state space X of σ is minimal, where the σ is said to be a minimal partial realization of $\underline{a}$. Moreover, show when the minimal realizations are isomorphic.>

Proposition 8.17. *For any given* $\underline{a} \in F(U^*_{\underline{N}}, Y)$*, there always exists a minimal partial realization of it.*

[proof] For any $\omega \notin U^*_{\underline{N}}$, set $\underline{a}(\omega) = 0$. Then $\underline{a} \in F(U^*, Y)$, and Theorem (8.18) implies that there exists a finite dimensional partial realization of $\underline{a}$. Therefore, there exists a minimal partial realization.

Minimal partial realizations are, in general, not unique modulo isomorphisms. Therefore, we introduce a natural partial realization, and we show that natural partial realizations exist if and only if they are isomorphic.

Definition 8.18. For a linear representation system $\sigma = ((X, F), x^0, h)$ and some $p \in N$, if $X = \ll \{\phi_F(\omega)x^0; \omega \in U^*_p\} \gg$, then σ is said to be p-quasi-reachable,
where $\ll S \gg$ denotes the smallest linear space which contains a set S.

Let q be some integer. If $h\phi_F(\omega)x = 0$ implies x=0 for any $\omega \in U^*_q$, then σ is said to be q-distinguishable.

For a given $\underline{a} \in F(U^*_L, Y)$, if there exist p and $q \in N$ such that $p + q < L$ and σ is p-quasi-reachable and q-distinguishable then σ is said to be a natural partial realization of $\underline{a}$.

For a partial input response map $\underline{a} \in F(U^*_L, Y)$, the following matrix $H^L_{\underline{a}\,(p,L-p)}$ is said to be a finite-sized Hankel matrix of $\underline{a}$.

$$H^{L}_{\underline{a}\ (p,L-p)} = \begin{pmatrix} & & \omega \\ & & \vdots \\ & & \vdots \\ \bar{\omega} & \cdots \quad \cdots & \underline{a}(\bar{\omega}\,|\omega) \end{pmatrix}$$

where $\bar{\omega} \in U_p^*$ and $\omega \in U_{L-p}^*$.

Theorem 8.19. *Let $H^L_{\underline{a}\ (p,L-p)}$ be the finite Hankel-matrix of $\underline{a} \in F(U_L^*, Y)$. Then there exists a natural partial realization of $\underline{a}$ if and only if the following conditions hold:*
rank $H^L_{\underline{a}\ (p,L-p)}$= rank $H^L_{\underline{a}\ (p,L-p-1)}$ =rank $H^L_{\underline{a}\ (p+1,L-p-1)}$ *for some $p \in N$.*

Theorem 8.20. *There exists a natural partial realization of a given partial input response map $\underline{a} \in F(U_L^*, Y)$ if and only if the minimal partial realization of $\underline{a}$ are unique modulo isomorphisms.*

Theorem 8.21. *Let a partial input response $\underline{a} \in F(U_L^*, Y)$ satisfy the condition of Theorem (8.26), then the quasi-reachable standard system $\sigma_s = ((X, F_s), \mathbf{e_1}, h_s)$ which realizes $\underline{a}$ can be obtained by the following algorithm.*
Set $n :=$ rank $H^L_{\underline{a}\ (p,L-p)}$, where $H^L_{\underline{a}\ (p,L-p)}$ is the finite Hankel matrix of $\underline{a} \in F(U_L^*, Y)$.
1) Select the linearly independent vectors $\{S_l(\omega_i)\underline{a} \in F(U_{L-p}^*, Y); 1 \leq i \leq n\}$ from $H^L_{\underline{a}\ (p,L-p)}$ in order of their numerical value.
2) Let the state space be $\boldsymbol{R}^n$, the initial state be $\mathbf{e_1} = [100, \cdots, 0]^T$.
3) Let the output map $h_s = [\underline{a}(1)\underline{a}(\omega_2)\underline{a}(\omega_3)\cdots\underline{a}(\omega_n)]$.
4) Let ${}_if_j$ in $F_s(u_i) := [{}_if_1\ {}_if_2\ \cdots\ {}_if_n\ ]$ be ${}_if_j := [{}_if_{j,1}\ {}_if_{j,2}\ \cdots\ {}_if_{j,n}]$ for $1 \leq i \leq n$, where ${}_if_j$ is given by the following.
$\underline{S_l}(u_i)\underline{S_l}(\omega_j)\underline{a} = \sum_{k=1}^{j}\ {}_if_{j,k}\underline{S_l}(\omega_i)\underline{a}$, ${}_if_{j,k} \in \boldsymbol{R}$ in the sense of $F(U_{L-p}^*, Y)$ and $\underline{S_l}(\omega) : F(U_s^*, Y) \to F(U_{s-|\omega|}^*, Y)$; $\underline{a} \mapsto \underline{S_l}(\omega)\underline{a}[; \bar{\omega} \mapsto \underline{a}(\bar{\omega}|\omega)]$.

8.4 Algebraically Approximate Realization of Linear Representation Systems

In this section, we discuss approximate realization problems of linear representation systems.

We will discuss the approximate realization problem under the assumption that the set U of input values is a finite set $U = \{u_j : 1 \leq j \leq m\}$ for an finite integer $m \in N$. The reference [Matsuo and Hasegawa, 2003] showed that this assumption is not so special. However, for simplicity of our discussion, we assume that the set U of input values is $U = \{u_1,\ u_2,\ u_3\}$.

Roughly speaking, the approximate realization of linear representation systems can be stated as follows:
< For any given partial data of a linear representation system, find, using only algebraic calculations, a linear representation system which approximates the given data. >

In order to make our discussion simple, we assume that the set Y of output is the set $\boldsymbol{R}$ of real numbers, namely 1-output.

Theorem 8.22. *Algebraic algorithm for approximate realization*
Let an input response map $\underline{a}$ be a considered object which is a linear representation system and $U := \{u_j; 1 \leq j \leq m, m \in N\}$. Then an approximate realization $\sigma = ((\boldsymbol{R}^n, F_s), x_s^0, h_s)$ of $\underline{a}$ is given by the following algorithm:

1) Based on the ratio of the square root of eigenvalues for a matrix $H_{\underline{a}\ (n,\bar{p})}(|||\omega_1|||, \cdots, |||\omega_n|||)H_{\underline{a}\ (n,\bar{p})}(|||\omega_1|||, \cdots, |||\omega_n|||)^T$, determine the value n of rank for the matrix, where $|||\omega_1|||$, $|||\omega_2|||$, $\cdots$ and $|||\omega_n|||$ are selected in order of numerical value of input and $\{\underline{S_l}(\omega_i)\underline{a}; 1 \leq i \leq n,\ \omega_i \in U^*\}$ is a set of independent vectors.
Namely, determine the value n of rank for the matrix $H_{\underline{a}\ (n,\bar{p})}(|||\omega_1|||, \cdots, |||\omega_n|||)$ such that the ratio of the square root of eigenvalues for the covariance matrix becomes very small.
The small ratio means the nearness of approximation degree.

2) In order to determine g_s, the algebraic CLS method is used as follows:
① In particular, set $g_s(\omega_i) := \mathbf{e}_i$ for $\omega_i \in U$. Namely, $g_s(\omega_1) := \mathbf{e}_1$, $g_s(\omega_2) := \mathbf{e}_2$, $\cdots, g_s(\omega_k) := \mathbf{e}_k$ for some $k \in N$.
For $u \in U$ such that $u \notin \{\omega_i; 1 \leq i \leq n\}$ and $\omega_r < u$, $g_s(u) = \sum_{j=1}^{r} b_{u,j}\underline{S_l}(\omega_j)\underline{a}$ is obtained as follows:
Based on Proposition (2.14), determine coefficients $\{b_{u,j} : 1 \leq j \leq r\}$.
The Q in Proposition (2.14) can be considered as the matrix composed from the eigenvectors of $H_{\underline{a}\ (r+1,L)}(|||\omega_1|||, |||\omega_2|||, \cdots, |||\omega_r|||, |||u|||)$ $\times H^T_{\underline{a}\ (r+1,L)}(|||\omega_1|||, |||\omega_2|||, \cdots, |||\omega_r|||, |||u|||)$.
Let a matrix $A_u \in \boldsymbol{R}^{1\times(r+1)}$ be $A_u := [b_{u,1}, b_{u,2}, \cdots, b_{u,r}, -1]$.
② Determine the error vectors $\{\underline{S_l}(\omega_j)\underline{\bar{a}} \in \boldsymbol{R}^{L\times 1} :\ 0 \leq j \leq r\}$ and $\underline{S_l}(u)\underline{\bar{a}}$ by using the equation
$[\underline{S_l}(\omega_1)\underline{\bar{a}}, \underline{S_l}(\omega_2)\underline{\bar{a}}, \cdots, \underline{S_l}(\omega_r)\underline{\bar{a}}, \underline{S_l}(u)\underline{\bar{a}}]^T :=$ $A_u^T[A_uA_u^T]^{-1}A_uH^T_{\underline{a}\ (r+1,L)}(|||\omega_1|||, |||\omega_2|||, \cdots, |||\omega_r|||, |||u|||)$ and $H^T_{\underline{a}\ (|||u|||+1,L)}(|||\omega_1|||, |||\omega_2|||, \cdots, |||\omega_r|||, |||u|||) :=$ $[\underline{S_l}(\omega_1)\underline{a}, \cdots, \underline{S_l}(\omega_r)\underline{a}, \underline{S_l}(u)\underline{a}]$.

3) In order to obtain F_s, the algebraic CLS method is used as follows:
① Let ${}_i\mathbf{f}_j \in \boldsymbol{R}^n$ in $F_s(u_i) := [{}_i\mathbf{f}_1\ {}_i\mathbf{f}_2\ \cdots\ {}_i\mathbf{f}_n]$ be ${}_i\mathbf{f}_j := [{}_if_{j,1}\ {}_if_{j,2}$ $\cdots\ {}_if_{j,n}]^T$ for $1 \leq i \leq m$, where ${}_if_{j,k}$ is given by the following: $\underline{S_l}(u_i)\underline{S_l}(\omega_j)\underline{a} = \sum_{k=1}^{n} {}_if_{j,k}\underline{S_l}(\omega_k)\underline{a}$, ${}_if_{j,k} \in \boldsymbol{R}$ in the sense of $F(U^*_{L-p}, Y)$.
②For i $(1 \leq i \leq m)$, j $(1 \leq j \leq n)$ and for the maximum number r $(1 \leq r \leq n)$ such that $\omega_r, \omega_j \in \{\omega_j; 1 \leq j \leq n\}$ and $|||\omega_r||| < |||u_i|\omega_j|||$.
Based on Proposition (2.14), determine coefficients $\{{}_if_{j,k} = 0;$

$r+1 \le k \le n\}$ *and* $\{{}_i f_{j,k} : 1 \le k \le r\}$.
The Q *in Proposition (2.14) can be considered as the matrix composed from the eigenvectors of* $H_{\underline{a}\ (n+1,L)}(|||\omega_1|||, |||\omega_2|||, \cdots, |||\omega_r|||, |||u_i|\omega_j|||)$
$\times H^T_{\underline{a}\ (n+1,L)}(|||\omega_1|||, |||\omega_2|||, \cdots, |||\omega_r|||, |||u_i|\omega_j|||)$.
Let a matrix ${}_iA_j \in \boldsymbol{R}^{1\times(r+1)}$ *be* ${}_iA_j := [{}_i f_{j,1}, {}_i f_{j,2}, \cdots, {}_i f_{j,r}, -1]$.
Determine the error vectors $\{\underline{S_l}(\omega_j)\bar{\underline{a}}\} \in \boldsymbol{R}^{L\times 1} :\ 0 \le j \le r\}$ *and* $\underline{S_l}(u_i|\omega_j)\bar{\underline{a}}$ *by using the equation*
$[\underline{S_l}(\omega_1)\bar{\underline{a}}, \underline{S_l}(\omega_2)\bar{\underline{a}}, \cdots, \underline{S_l}(\omega_r)\bar{\underline{a}}, \underline{S_l}(u_i|\omega_j)\bar{\underline{a}}]^T :=$
${}_iA_j^T[{}_iA_j\ {}_iA_j^T]^{-1}\ {}_iA_j\ H^T_{\underline{a}\ (n+1,L)}(|||\omega_1|||, |||\omega_2|||, \cdots, |||\omega_r|||, |||u_i|\omega_j|||)$ *and*
$H^T_{\underline{a}\ (r+1,L)}(|||\omega_1|||, |||\omega_2|||, \cdots, |||\omega_r|||, |||u_i|\omega_j|||) :=$
$[\underline{S_l}(\omega_1)\underline{a}, \underline{S_l}(\omega_2)\underline{a}, \cdots, \underline{S_l}(\omega_r)\underline{a}, \underline{S_l}(u_i|\omega_j)\underline{a}]$.

4) In order to determine h_s, *the algebraic CLS method is used as follows:*
① *For the first* $\omega_{r_1+1},\ \omega_{r_1} \in \{\omega_i; 1 \le i \le n\}$ *such that* $\omega_{r_1+1} > \omega_{r_1}$ *and* $|||\omega_{r_1+1}|||$ - $|||\omega_{r_1}||| > 1$ *when starting out from* ω_1,
set $\underline{S_l}(\lambda_1)\underline{a} := \sum_{i=1}^{r_1} b_{\lambda_{1,i}}\mathbf{e}_i$ *for the obtained equation* $\underline{S_l}(\lambda_1)\underline{a} = \sum_{i=1}^{r_1} b_{\lambda_{1,i}}(\underline{S_l}(\omega_i)\underline{a})$ *for* λ_1 *such that* $|||\lambda_1||| = |||\omega_{r_1}||| + 1$.
Based on Proposition (2.14), determine coefficients $\{b_{\lambda_{1,i}} : 1 \le i \le r_1\}$.
The Q *in Proposition (2.14) can be considered as the matrix composed from the eigenvectors of* $H_{\underline{a}\ (r_1+1,L)}(|||\omega_1|||, |||\omega_2|||, \cdots, |||\omega_{r_1}|||, |||\lambda_1|||)$
$\times H^T_{\underline{a}\ (r_1+1,L)}(|||\omega_1|||, |||\omega_2|||, \cdots, |||\omega_{r_1}|||, |||\lambda_1|||)$.
Let a matrix $A_{\lambda_1} \in \boldsymbol{R}^{1\times(r_1+1)}$ *be* $A_{\lambda_1} := [b_{\lambda_1,1}, b_{\lambda_1,2}, \cdots, b_{\lambda_1,r_1}, -1]$.
Determine the error vectors $\{\underline{S_l}(\omega_i)\bar{\underline{a}} \in \boldsymbol{R}^{L\times 1} :\ 0 \le i \le r_1\}$ *and* $\underline{S_l}(\lambda_1)\bar{\underline{a}}$ *by using the equation*
$[\underline{S_l}(\omega_1)\bar{\underline{a}}, \underline{S_l}(\omega_2)\bar{\underline{a}}, \cdots, \underline{S_l}(\omega_{r_1})\bar{\underline{a}}, \underline{S_l}(\lambda_1)\bar{\underline{a}}]^T :=$
$A^T_{\lambda_1}[A_{\lambda_1}A^T_{\lambda_1}]^{-1}A_{\lambda_1}H^T_{\underline{a}\ (r_1+1,L)}(|||\omega_1|||, |||\omega_2|||, \cdots, |||\omega_{r_1}|||, |||\lambda_1|||)$
and $H^T_{\underline{a}\ (|||\lambda_1|||+1,L)}(|||\omega_1|||, |||\omega_2|||, \cdots, |||\omega_{r_1}|||, |||\lambda_1|||) :=$
$[\underline{S_l}(\omega_1)\underline{a}, \cdots, \underline{S_l}(\omega_{r_1})\underline{a}, \underline{S_l}(\lambda_1)\underline{a}]$.
Then let $h_{\lambda_1 s} \in \boldsymbol{R}^{1\times r_1}$ *be*
$h_{\lambda_1 s} := [\underline{a}(\omega_1) - (\bar{\underline{a}}(\omega_1), \underline{a}(\omega_2) - (\bar{\underline{a}}(\omega_2), \cdots, \underline{a}(\omega_{r_1}) - (\bar{\underline{a}}(\omega_{r_1})]$.
②*For the first* $\omega_{r_2+1},\ \omega_{r_2} \in \{\omega_i; 1 \le i \le n\}$ *such that* $\omega_{r_2+1} > \omega_{r_2}$ *and* $|||\omega_{r_2+1}|||$ - $|||\omega_{r_2}||| > 1$ *when starting out from* ω_{r_1+1},
set $\underline{S_l}(\lambda_2)\underline{a} := \sum_{i=1}^{r_2} b_{\lambda_2,i}\mathbf{e}_i$ *for the obtained equation* $\underline{S_l}(\lambda_2)\underline{a} = \sum_{i=1}^{r_2} b_{\lambda_2,i}(\underline{S_l}(\omega_i)\underline{a})$ *for* λ_2 *such that* $|||\lambda_2||| = |||\omega_{r_2}||| + 1$.
Based on Proposition (2.14), determine coefficients $\{b_{\lambda_{2,i}} : 1 \le i \le r_1\}$.
The Q *in Proposition (2.14) can be considered as the matrix composed from the eigenvectors of* $H_{\underline{a}\ (r_2+1,L)}(|||\omega_1|||, |||\omega_2|||, \cdots, |||\omega_{r_2}|||, |||\lambda_2|||)$
$\times H^T_{\underline{a}\ (r_2+1,L)}(|||\omega_1|||, |||\omega_2|||, \cdots, |||\omega_{r_2}|||, |||\lambda_2|||)$.
Let a matrix $A_{\lambda_2} \in \boldsymbol{R}^{1\times(r_2+1)}$ *be* $A_{\lambda_2} := [b_{\lambda_2,1}, b_{\lambda_2,2}, \cdots, b_{\lambda_2,r_2}, -1]$.
Determine the error vectors $\{\underline{S_l}(\omega_i)\bar{\underline{a}} \in \boldsymbol{R}^{L\times 1} :\ 0 \le i \le r_2\}$ *and* $\underline{S_l}(\lambda_2)\bar{\underline{a}}$ *by using the equation*
$[\underline{S_l}(\omega_1)\bar{\underline{a}}, \underline{S_l}(\omega_2)\bar{\underline{a}}, \cdots, \underline{S_l}(\omega_{r_2})\bar{\underline{a}}, \underline{S_l}(\lambda_2)\bar{\underline{a}}]^T :=$
$A^T_{\lambda_2}[A_{\lambda_2}A^T_{\lambda_2}]^{-1}A_{\lambda_2}H^T_{\underline{a}\ (r_2+1,L)}(|||\omega_1|||, |||\omega_2|||, \cdots, |||\omega_{r_2}|||, |||\lambda_2|||)$
and $H^T_{\underline{a}\ (|||\lambda_2|||+1,L)}(|||\omega_1|||, |||\omega_2|||, \cdots, |||\omega_{r_2}|||, |||\lambda_2|||) :=$

$[\underline{S_l}(\omega_1)\underline{a}, \cdots, \underline{S_l}(\omega_{r_2})\underline{a}, \underline{S_l}(\lambda_2)\underline{a}]$.
Then let $h_{\lambda_2 s} \in \boldsymbol{R}^{1\times r_2}$ *be*
$h_{\lambda_2 s} := [\underline{a}(\omega_{r_1+1}) - (\underline{\bar{a}}(\omega_{r_1+1}), \ \underline{a}(\omega_{r_1+2}) - (\underline{\bar{a}}(\omega_{r_1+2}), \cdots, \ \underline{a}(\omega_{r_2}) - (\underline{\bar{a}}(\omega_{r_2})]$.
⋮

ⓣ *For* $\omega \in U^*$ *such that* $|||\omega||| = |||\omega_n||| + 1$,
Based on Proposition (2.14), determine coefficients $\{b_{\omega,i} : 1 \le i \le n\}$.
The Q in Proposition (2.14) can be considered as the matrix composed from the eigenvectors of $H_{\underline{a}\ (r_2+1,L)}(|||\omega_1|||, |||\omega_2|||, \cdots, |||\omega_{r_2}|||, |||\lambda_2|||)$
$\times H^T_{\underline{a}\ (r_2+1,L)}(|||\omega_1|||, |||\omega_2|||, \cdots, |||\omega_{r_2}|||, |||\lambda_2|||)$.
Let a matrix $A_\omega \in \boldsymbol{R}^{1\times(n+1)}$ *be* $A_\omega := [b_{\omega,1}, b_{\omega,2}, \cdots, b_{\omega,n}, -1]$.
Determine the error vectors $\{\underline{S_l}(\omega_i)\underline{\bar{a}} : \ 1 \le i \le n\}$ *and* $\underline{S_l}(\omega)\underline{\bar{a}}$
by using the equation
$[\underline{S_l}(\omega_1)\underline{\bar{a}}, \underline{S_l}(\omega_2)\underline{\bar{a}}, \ \cdots, \ \underline{S_l}(\omega_n)\underline{\bar{a}}, \ \underline{S_l}(\omega)\underline{\bar{a}}]^T :=$
$A^T_\omega[A_\omega A^T_\omega]^{-1} A_\omega H^T_{\underline{a}\ (n+1,L)}(|||\omega_1|||, |||\omega_2|||, \cdots, |||\omega_n|||, |||\omega|||)$
and $H^T_{\underline{a}\ (n+1,L)}(|||\omega_1|||, |||\omega_2|||, \cdots, |||\omega_n|||, |||\omega|||) :=$
$[\underline{S_l}(\omega_1)\underline{a}, \ \cdots, \underline{S_l}(\omega_n)\underline{a}, \ \underline{S_l}(\omega)\underline{a}]$.
Then let $h_{\omega s}$ *be*
$h_{\omega s} := [\underline{a}(\omega_{r_t+1}) - (\underline{\bar{a}}(\omega_{r_t+1}), \ \underline{a}(\omega_{r_t+2}) - \underline{\bar{a}}(\omega_{r_t+2}), \cdots, \ \underline{a}(\omega_n) - \underline{\bar{a}}(\omega_n)]$.
Finally, let $h_s \in \boldsymbol{R}^{1\times n}$ *be*
$h_s := [h_{\lambda_1 s}, h_{\lambda_2 s}, \cdots, h_{\omega s}]$.

[proof] In 1), the number of dimensions is determined by considering the ratio of Hankel matrix norm, which means a degree of information loss. According to Theorem (8.21), a linear representation system $\sigma = ((\boldsymbol{R}^n, F_s), g_s, h_s)$ is obtained as follows. In 2), g_s is obtained directly or by using the CLS method for A_u corresponding to the matrix A in Lemma (2.17). In 3), F_s is obtained by using the CLS method for ${}_iA_j$ corresponding to the matrix A in Lemma (2.17). In 4), h_s is obtained by using the CLS method for A_{λ_1}, A_{λ_2}, A_{λ_ω} corresponding to the matrix A in Lemma (2.17).

In the figures of this chapter, we use a notation *Signal_n d* as an input response map obtained by a n-dimensional linear representation system.

In examples of this chapter, a notation $H^T_{\underline{a}\ (r+1,40)}(0, \cdots, r)$ is used in place of $H^T_{\underline{a}\ (r+1,40)}(0, 1, 2, 3, \cdots, r-1, r)$.

Example 8.23. Let the signals be the behavior of the following 3-dimensional linear representation system: $\sigma = ((\boldsymbol{R}^3, F), x^0, h)$, where

$$F(u_1) = \begin{bmatrix} 0 & -1.4 & -1.3 \\ 1 & -1.5 & -0.3 \\ 0 & 1 & 1 \end{bmatrix}, \ F(u_2) = \begin{bmatrix} 0 & 0.2 & 0.1 \\ 0 & 0.6 & 0.1 \\ 1 & -0.6 & 0.6 \end{bmatrix},$$

$$F(u_3) = \begin{bmatrix} -1.3 & -0.1 & -1.2 \\ 0.6 & -0.5 & 0.1 \\ 0 & 0.8 & 0.4 \end{bmatrix}, \ x^0 = \mathbf{e}_1, \ h = [13, -10.6, 2].$$

Then the algebraically approximate realization problem is solved as follows:

covariance matrix	eigenvalues							
	1	2	3	4	5	6	⋯	13
$H^T_{\underline{a}\ (3,40)}(0,1,2)H_{\underline{a}\ (3,40)}(0,1,2)$	13459	732	41.6					
$H^T_{\underline{a}\ (4,40)}(0,\cdots,3)H_{\underline{a}\ (4,40)}(0,\cdots,3)$	33505	815	45.8	0				
$H^T_{\underline{a}\ (5,40)}(0,\cdots,4)H_{\underline{a}\ (5,40)}(0,\cdots,4)$	33614	1220	181	0	0			
$H^T_{\underline{a}\ (6,40)}(0,\cdots,5)H_{\underline{a}\ (6,40)}(0,\cdots,5)$	35271	1232	206	0	0	0		
$H^T_{\underline{a}\ (13,40)}(0,\cdots,12)H_{\underline{a}\ (13,40)}(0,\cdots,12)$	110984	3437	620	0	0	0	⋯	0
covariance matrix	square root of eigenvalues							
$H^T_{\underline{a}\ (3,40)}(0,1,2)H_{\underline{a}\ (3,40)}(0,1,2)$	116	27.1	6.4					
$H^T_{\underline{a}\ (4,40)}(0,\cdots,3)H_{\underline{a}\ (4,40)}(0,\cdots,3)$	183	28.5	6.8	0				
$H^T_{\underline{a}\ (5,40)}(0,\cdots,4)H_{\underline{a}\ (5,40)}(0,\cdots,4)$	183	34.9	13.4	0	0			
$H^T_{\underline{a}\ (6,40)}(0,\cdots,5)H_{\underline{a}\ (6,40)}(0,\cdots,5)$	188	35.1	14.4	0	0	0		
$H^T_{\underline{a}\ (13,40)}(0,\cdots,12)H_{\underline{a}\ (13,40)}(0,\cdots,12)$	333	58.6	24.9	0	0	0	⋯	0

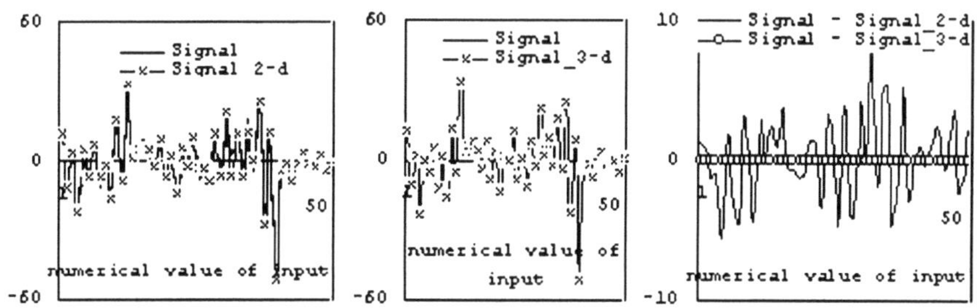

Fig. 8.1 The left is the original input response map and the behavior of a 2-dimensional linear representation system obtained by the algebraic CLS method. The middle is the original input response map and the behavior of a 3-dimensional linear representation system obtained by the algebraic CLS method. The right is the difference between the original one and the behavior of the 2-dimensional linear representation system obtained by the algebraic CLS method or the 3-dimensional linear representation system obtained by the algebraic CLS method in Example (8.23).

1) Since the ratio $\frac{6.4}{116} = 0.06$ obtained by the square root of $H^T_{\underline{a}\ (3,40)}(0,1.2)H_{\underline{a}\ (3,40)}(0,1,2)$ is not small, an approximate 2-dimensional linear representation system obtained by the CLS method may be not good.
2) After determining the independent vectors $\underline{a}$ and $\underline{S_l}(u_1)\underline{a}$ whose numerical value of input are 0 and 1, we will continue the approximate realization algorithm by the algebraic CLS method.

Therefore, an approximate 2-dimensional linear representation system $\sigma_1 = ((\boldsymbol{R}^2,\ F_1),\ x_1^0,\ h_1)$ obtained by the algebraic CLS method is constructed as follows:

$$F_1(u_1) = \begin{bmatrix} 0 & -0.54 \\ 1 & -1 \end{bmatrix}, \ F_1(u_2) = \begin{bmatrix} 0.95 & -0.34 \\ 0.69 & 0.24 \end{bmatrix}, \ F_1(u_3) = \begin{bmatrix} -1.3 & 0.64 \\ 0.6 & 0.004 \end{bmatrix},$$

$x_1^0 = \mathbf{e}_1, h_1 = [11.8, \ -11.5]$.

For reference, a 3-dimensional linear representation system $\sigma_2 = ((\boldsymbol{R}^3, \ F_2), \ x_2^0, \ h_2)$ obtained by the algebraic CLS method can be expressed as follows:

$$F_2(u_1) = \begin{bmatrix} 0 & -1.4 & -1.3 \\ 1 & -1.5 & -0.3 \\ 0 & 1 & 1 \end{bmatrix}, \ F_2(u_2) = \begin{bmatrix} 0 & 0.2 & 0.1 \\ 0 & 0.6 & 0.1 \\ 1 & -0.6 & 0.6 \end{bmatrix},$$

$$F_2(u_3) = \begin{bmatrix} -1.3 & -0.1 & -1.2 \\ 0.6 & -0.5 & 0.1 \\ 0 & 0.8 & 0.4 \end{bmatrix}, \ x_2^0 = \mathbf{e}_1, \ h_2 = [13, \ -10.6, \ 2].$$

The system completely reconstructs the original system.

We can show that the algorithm for approximate realization given by the analytic CLS method in the reference [Hasegawa, 2008] produces the same systems as the above ones in the sense of the numerical calculation.

In this example, the original signals are considered as the behavior of a 3-dimensional linear representation system and the desirable input response map is obtained by the algebraic CLS method. The model obtained by the algebraic CLS method is a 2-dimensional linear representation system.

For reference, a 3-dimensional linear representation system is also obtained by the algebraic CLS method.

Just as we thought, the following table and Fig. 8.1 truly indicate that the 2-dimensional linear representation system obtained by the algebraic CLS method is not a good approximation. For reference, the behavior of the same dimensional linear representation system as the original system is shown. Hence, there does not exist a good approximation for the given system.

dimen-sion	ratio of matrices	mean values of square root for sum of signal ①	signal by CLS ②	error ③	cosine ① and ② $\cos\theta$	error ratio ③/①
$a_{0,1}$	0.06	1.8427	1.885	0.42	0.975	0.23
$a_{0,1,2}$	0	1.8427	1.8427	0	1	0

Example 8.24. Let the signals be the behavior of the following 4-dimensional linear representation system: $\sigma = ((\boldsymbol{R}^4, F), x^0, h)$, where

$$F(u_1) = \begin{bmatrix} 0 & 1.2 & -0.1 & 0.9 \\ 1 & -0.4 & 0 & 0 \\ 0 & -2 & 1.5 & 0 \\ 0 & 0.2 & 0.3 & 0.8 \end{bmatrix}, \ F(u_2) = \begin{bmatrix} 0 & -1 & -0.2 & -0.8 \\ 0 & -0.5 & 0 & 0.2 \\ 1 & 0 & -0.5 & 0.4 \\ 0 & 0 & 0 & -0.6 \end{bmatrix},$$

$$F(u_3) = \begin{bmatrix} 0 & 0.3 & 0 & -0.2 \\ 0 & 0.5 & -0.1 & 1.3 \\ 0 & -0.9 & -0.3 & -0.1 \\ 1 & 0.3 & -0.8 & -0.8 \end{bmatrix},$$

$x^0 = \mathbf{e}_1, \ h = [18, \ 14, \ -9, \ 12]$.

Then the algebraically approximate realization problem is solved as follows:

covariance matrix	eigenvalues							
	1	2	3	4	5	6	···	13
$H^T_{\underline{a}\ (3,40)}(0,1,2)H_{\underline{a}\ (3,40)}(0,1,2)$	59553	4712	2128					
$H^T_{\underline{a}\ (4,40)}(0,\cdots,3)H_{\underline{a}\ (4,40)}(0,\cdots,3)$	82046	6874	3975	504				
$H^T_{\underline{a}\ (5,40)}(0,\cdots,4)H_{\underline{a}\ (5,40)}(0,\cdots,4)$	132501	20219	4307	651	0			
$H^T_{\underline{a}\ (6,40)}(0,\cdots,5)H_{\underline{a}\ (6,40)}(0,\cdots,5)$	181114	20536	4378	968	0	0		
$H^T_{\underline{a}\ (13,40)}(0,\cdots,12)H_{\underline{a}\ (13,40)}(0,\cdots,12)$	350617	37310	17520	1480	0	0	···	0
covariance matrix	square root of eigenvalues							
$H^T_{\underline{a}\ (3,40)}(0,1,2)H_{\underline{a}\ (3,40)}(0,1,2)$	244	68.6	46.1					
$H^T_{\underline{a}\ (4,40)}(0,\cdots,3)H_{\underline{a}\ (4,40)}(0,\cdots,3)$	286	82.9	63	22.4				
$H^T_{\underline{a}\ (5,40)}(0,\cdots,4)H_{\underline{a}\ (5,40)}(0,\cdots,4)$	364	142	65.6	25.5	0			
$H^T_{\underline{a}\ (6,40)}(0,\cdots,5)H_{\underline{a}\ (6,40)}(0,\cdots,5)$	426	143	66.2	31.1	0	0		
$H^T_{\underline{a}\ (13,40)}(0,\cdots,12)H_{\underline{a}\ (13,40)}(0,\cdots,12)$	592	193	132	38.5	0	0	···	0

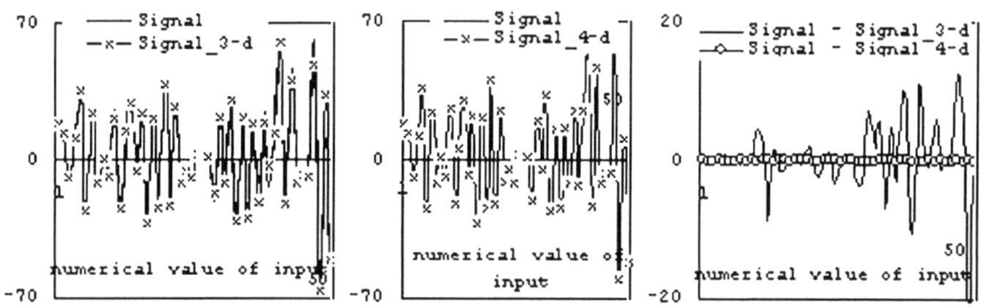

Fig. 8.2 The left is the original input response map and the behavior of a 3-dimensional linear representation system obtained by the algebraic CLS method. The middle is the original input response map and the behavior of a 4-dimensional linear representation system obtained by the algebraic CLS method. The right is the difference between the original one and the behavior of the 3-dimensional linear representation system obtained by the algebraic CLS method or the 4-dimensional linear representation system obtained by the CLS method in Example (8.24).

1) Since the ratio $\frac{22.4}{286} = 0.08$ obtained by the square root of $H^T_{\underline{a}\ (4,40)}(0,\cdots .3)H_{\underline{a}\ (4,40)}(0,\cdots,3)$ is not small, an approximate 3-dimensional linear representation system obtained by the algebraic CLS method may be not good.
2) After determining the independent vectors $\underline{a}$, $\underline{S_l}(u_1)\underline{a}$ and $\underline{S_l}(u_2)\underline{a}$ whose numerical value of input are 0, 1 and 2, we will continue the approximate realization algorithm by the algebraic CLS method.

Therefore, an approximate 3-dimensional linear representation system $\sigma_1 = ((\boldsymbol{R}^3,\ F_1),\ x_1^0,\ h_1)$ obtained by the algebraic CLS method is constructed as follows:

$$F_1(u_1) = \begin{bmatrix} 0 & 1.32 & 0.1 \\ 1 & -0.26 & 0.22 \\ 0 & -1.8 & 1.9 \end{bmatrix}, \; F_1(u_2) = \begin{bmatrix} 0 & -1 & -0.2 \\ 0 & -0.5 & 0 \\ 1 & 0 & -0.5 \end{bmatrix},$$

$$F_1(u_3) = \begin{bmatrix} 0.7 & 0.5 & -0.6 \\ 0.8 & 0.7 & -0.7 \\ 1.5 & -0.57 & -1.5 \end{bmatrix}, \; x_1^0 = \mathbf{e}_1, \; h_1 = [18.3, \; 14.3, \; -8.5].$$

For reference, a 4-dimensional linear representation system $\sigma_2 = ((\boldsymbol{R}^4, \; F_2), \; x_2^0, \; h_2)$ obtained by the algebraic CLS method can be expressed as follows:

$$F_2(u_1) = \begin{bmatrix} 0 & 1.2 & -0.1 & 0.9 \\ 1 & -0.4 & 0 & 0 \\ 0 & -2 & 1.5 & 0 \\ 0 & 0.2 & 0.3 & 0.8 \end{bmatrix}, \; F_2(u_2) = \begin{bmatrix} 0 & -1 & -0.2 & -0.8 \\ 0 & -0.5 & 0 & 0.2 \\ 1 & 0 & -0.5 & 0.4 \\ 0 & 0 & 0 & -0.6 \end{bmatrix},$$

$$F_2(u_3) = \begin{bmatrix} 0 & 0.3 & 0 & -0.2 \\ 0 & 0.5 & -0.1 & 1.3 \\ 0 & -0.9 & -0.3 & -0.1 \\ 1 & 0.3 & -0.8 & -0.8 \end{bmatrix}, \; x_2^0 = \mathbf{e}_1, \; h_2 = [18, \; 14, \; -9, \; 12].$$

We can show that the algorithm for approximate realization given by the analytic CLS method in the reference [Hasegawa, 2008] produces the same systems as the above ones in the sense of the numerical calculation.

In this example, the original signals are considered as the behavior of a 4-dimensional linear representation system and the desirable input response map is obtained by the algebraic CLS method. The model obtained by the algebraic CLS method is a 3-dimensional linear representation system.

For reference, a 4-dimensional linear representation system is also obtained by the algebraic CLS method. The system completely reconstructs the original system.

Just as we thought, the following table and Fig. 8.2 truly indicate that the 3-dimensional linear representation system obtained by the algebraic CLS method is not a good approximation. For reference, the behavior of the same dimensional linear representation system as the original system is shown. Hence, there does not exist a good approximation for the given system.

dimen-sion	ratio of matrices	mean values of square root for sum of signal	signal by CLS	error	cosine ① and ②	error ratio
		①	②	③	$\cos\theta$	③/①
$a_{0,1,2}$	0.08	3.66	3.644	0.744	0.979	0.20
$a_{0,1,2,3}$	0	3.66	3.66	0	1	0

Example 8.25. Let the signals be the behavior of the following 5-dimensional linear representation system: $\sigma = ((\boldsymbol{R}^5, F), x^0, h)$,

$$\text{where } F(u_1) = \begin{bmatrix} 0 & 0 & 0.9 & -0.1 & -0.1 \\ 1 & 0 & 0.7 & -0.1 & 0.8 \\ 0 & 0 & -0.1 & 0.5 & 0.1 \\ 0 & 0 & -0.5 & -1.5 & 0.4 \\ 0 & 1 & 0.8 & -1.5 & 0.9 \end{bmatrix}, \; F(u_2) = \begin{bmatrix} 0 & 0.2 & -1.6 & 0.8 & -0.3 \\ 0 & 0.5 & 0 & -0.2 & 0.2 \\ 1 & -0.5 & 0.2 & -0.4 & 0.2 \\ 0 & -0.5 & 0.8 & 0.4 & 0.3 \\ 0 & 0 & 0.1 & 1.1 & 0.7 \end{bmatrix},$$

$$F(u_3) = \begin{bmatrix} 0 & 0.3 & 0 & -0.5 & 0.3 \\ 0 & 0.8 & 0.5 & -0.2 & 0.6 \\ 0 & 0.1 & 2 & -1 & -0.2 \\ 1 & -1.2 & 0 & 1.8 & 1 \\ 0 & 0 & 0 & 0.9 & 0.8 \end{bmatrix}, \; x^0 = \mathbf{e}_1, \; h = [12, -7, -3, -5, 5].$$

Then the algebraically approximate realization problem is solved as follows:

covariance matrix	eigenvalues								
	1	2	3	4	5	6	7	$\cdots$	13
$H^T_{\underline{a}\ (5,40)}(0,\cdots,4)H_{\underline{a}\ (5,40)}(0,\cdots,4)$	86465	17639	8799	3517	183				
$H^T_{\underline{a}\ (6,40)}(0,\cdots,5)H_{\underline{a}\ (6,40)}(0,\cdots,5)$	99829	24050	9065	3667	193	0			
$H^T_{\underline{a}\ (7,40)}(0,\cdots,6)H_{\underline{a}\ (7,40)}(0,\cdots,6)$	236200	24170	9306	3717	203	0	0		
$H^T_{\underline{a}\ (13,40)}(0,\cdots,12)H_{\underline{a}\ (13,40)}(0,\cdots,12)$	994113	108426	32268	5563	658	0	0	$\cdots$	0
covariance matrix	square root of eigenvalues								
$H^T_{\underline{a}\ (5,40)}(0,\cdots,4)H_{\underline{a}\ (5,40)}(0,\cdots,4)$	294	133	94	59	13.5				
$H^T_{\underline{a}\ (6,40)}(0,\cdots,5)H_{\underline{a}\ (6,40)}(0,\cdots,5)$	316	155	95.2	61	13.9	0			
$H^T_{\underline{a}\ (7,40)}(0,\cdots,6)H_{\underline{a}\ (7,40)}(0,\cdots,6)$	486	155	96	61	14.2	0	0		
$H^T_{\underline{a}\ (13,40)}(0,\cdots,12)H_{\underline{a}\ (13,40)}(0,\cdots,12)$	997	329	180	74.6	25.6	0	0	$\cdots$	0

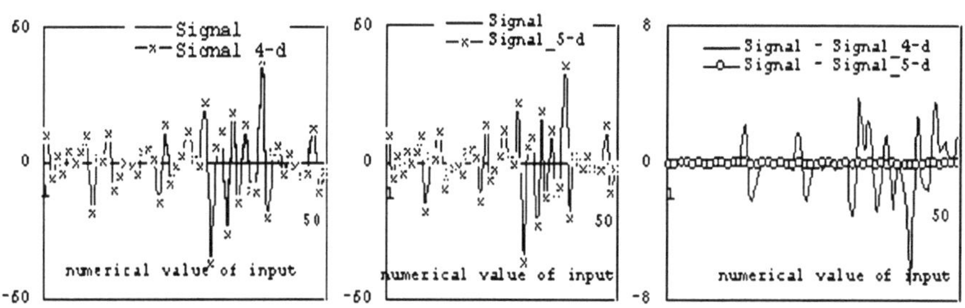

Fig. 8.3 The left is the original input response map and the behavior of a 4-dimensional linear representation system obtained by the algebraic CLS method. The middle is the original input response map and the behavior of a 5-dimensional linear representation system obtained by the algebraic CLS method. The right is the difference between the original one and the behavior of the 4-dimensional linear representation system obtained by the algebraic CLS method or the 5-dimensional linear representation system obtained by the algebraic CLS method in Example (8.25).

1) Since the ratio $\frac{13.5}{294} = 0.05$ obtained by the square root of $H^T_{\underline{a}\ (5,40)}(0,\cdots .4) \times H_{\underline{a}\ (5,40)}(0,\cdots,4)$ is not small, the approximate 4-dimensional linear representation system obtained by the algebraic CLS method may not be good.
2) After determining the independent vectors $\underline{a}$, $\underline{S_l}(u_1)\underline{a}$, $\underline{S_l}(u_2)\underline{a}$ and $\underline{S_l}(u_3)\underline{a}$ whose numerical value of input are 0, 1, 2 and 3, we will continue the approximate realization algorithm by the algebraic CLS method.

Therefore, an approximate 4-dimensional linear representation system $\sigma_1 = ((\boldsymbol{R}^4,\ F_1),\ x_1^0,\ h_1)$ obtained by the algebraic CLS method is constructed as follows:

$$F_1(u_1) = \begin{bmatrix} 0 & -0.4 & 0.7 & 0.5 \\ 1 & -1.3 & 0.25 & 1.8 \\ 0 & -0.2 & -0.23 & 0.76 \\ 0 & -0.22 & -0.66 & -1.2 \end{bmatrix},\ F_1(u_2) = \begin{bmatrix} 0 & 0.2 & -1.6 & 0.4 \\ 0 & 0.5 & -0.1 & -1.6 \\ 1 & -0.5 & 0.2 & -0.6 \\ 0 & -0.5 & 0.8 & 0.2 \end{bmatrix},$$

$$F_1(u_3) = \begin{bmatrix} 0 & 0.3 & 0 & -0.8 \\ 0 & 0.8 & 0.5 & -1.3 \\ 0 & 0.1 & 2 & -1.2 \\ 1 & -1.2 & 0 & 1.6 \end{bmatrix},\ x_1^0 = \mathbf{e}_1,\ h_1 = [12,\ -7,\ 3,\ -5].$$

For reference, a 5-dimensional linear representation system $\sigma_2 = ((\boldsymbol{R}^5,\ F_2),\ x_2^0,\ h_2)$ obtained by the algebraic CLS method can be expressed as follows:

$$F_2(u_1) = \begin{bmatrix} 0 & 0 & 0.9 & -0.1 & -0.1 \\ 1 & 0 & 0.7 & -0.1 & 0.8 \\ 0 & 0 & -0.1 & 0.5 & 0.1 \\ 0 & 0 & -0.5 & -1.5 & 0.4 \\ 0 & 1 & 0.8 & -1.5 & 0.9 \end{bmatrix},\ F_2(u_2) = \begin{bmatrix} 0 & 0.2 & -1.6 & 0.8 & -0.3 \\ 0 & 0.5 & 0 & -0.2 & 0.2 \\ 1 & -0.5 & 0.2 & -0.4 & 0.2 \\ 0 & -0.5 & 0.8 & 0.4 & 0.3 \\ 0 & 0 & 0.1 & 1.1 & 0.7 \end{bmatrix},$$

$$F_2(u_3) = \begin{bmatrix} 0 & 0.3 & 0 & -0.5 & 0.3 \\ 0 & 0.8 & 0.5 & -0.2 & 0.6 \\ 0 & 0.1 & 2 & -1 & -0.2 \\ 1 & -1.2 & 0 & 1.8 & 1 \\ 0 & 0 & 0 & 0.9 & 0.8 \end{bmatrix},\ x_2^0 = \mathbf{e}_1,\ h_2 = [12,\ -7,\ 3,\ -5,\ 5].$$

We can show that the algorithm for approximate realization given by the analytic CLS method in the reference [Hasegawa, 2008] produces the same systems as the above ones in the sense of the numerical calculation.

In this example, the original signals are considered as the behavior of a 5-dimensional linear representation system and the desirable input response map is obtained by the algebraic CLS method. The model obtained by the algebraic CLS method is a 4-dimensional linear representation system.

For reference, a 5-dimensional linear representation system is also obtained by the algebraic CLS method. The system completely reconstructs the original system.

Just as we thought, the following table and Fig. 8.3 truly indicate that the 4-dimensional linear representation system obtained by the algebraic CLS method is not a good approximation. For reference, the behavior of the same dimensional linear representation system as the original system is shown. Hence, there does not exist a good approximation for the given system.

dimen-sion	ratio of matrices	mean values of square root for sum of signal	signal by CLS	error	cosine ① and ②	error ratio
		①	②	③	$\cos\theta$	③/①
$a_{0,1,2,3}$	0.05	1.999	2.082	0.241	0.994	0.12
$a_{0,1,2,3,4}$	0	1.999	1.999	0	1	0

8.5 Algebraically Noisy Realization of Linear Representation Systems

In this section, we discuss an algebraically noisy realization problem of linear representation systems.

For noise $\{\bar{\gamma}(t) : t \in N\}$ added to an unknown linear representation system a, we will obtain observed data $\{\hat{\gamma}(|\omega|) + \bar{\gamma}(|\omega|) : \omega \in U^*\}$.

For the given data $\{\hat{\gamma}(|\omega|) + \bar{\gamma}(|\omega|) : \omega \in U^*\}$, σ which satisfies $a_\sigma(\omega) \approx \hat{\gamma}(|\omega|) : \omega \in U^*$ is called a noisy realization of a.

Roughly speaking, we can propose the following noisy realization problem:

For the given data $\{\hat{\gamma}(|\omega|) + \bar{\gamma}(|\omega|) : \omega \in U^*\}$, find a linear representation system σ which satisfies $a_\sigma(\omega) \approx \hat{\gamma}(|\omega|)$ for any $\omega \in U^*$.

In order to make our discussion simple, we assume that the set Y of output is the set $\boldsymbol{R}$ of real numbers, namely 1-output.

A situation for noisy realization problem 8.26

Let the observed object be a linear representation system and added noise to the output. Then we will obtain the data $\{\gamma(t) = \hat{\gamma}(t) + \bar{\gamma}(t) : 0 \leq t \leq \underline{N}\}$ for some integer $\underline{N} \in N$, where $\hat{\gamma}(t)$ is the exact signal which comes from the observed linear representation system and $\bar{\gamma}(t)$ is the noise added at time of observation.

Problem 8.27. Problem statement of a noisy realization for linear representation systems

Let $H_{\underline{a}\ (p,\bar{p})}$ be the measured finite-sized Input/output matrix. Then find, using only algebraic calculations, the cleaned-up Input/output matrix $\hat{H}_{\underline{a}\ (p,\bar{p})}$ such that $H_{\underline{a}\ (p,\bar{p})} = \hat{H}_{\underline{a}\ (p,\bar{p})} + \bar{H}_{\underline{a}\ (p,\bar{p})}$ holds.

Namely, find a minimal dimensional linear representation system $\sigma = ((\boldsymbol{R}^n, F_r), g_r, h_r, h^0))$ which realizes $\hat{H}_{\underline{a}\ (p,\bar{p})}$.

Theorem 8.28. *Algebraic algorithm of noisy realization for Linear Representation Systems*

Let an input response map $\underline{a}$ be a considered object which is a linear representation system. Then an approximate realization $\sigma = ((\boldsymbol{R}^n, F_s), g_s^0, h_s)$ of $\underline{a}$ is given by the following algorithm:

1) Based on the square root of eigenvalues for a matrix
$H_{\underline{a}\ (n,\bar{p})}(|||\omega_1|||, \cdots, |||\omega_n|||)H_{\underline{a}\ (n,\bar{p})}(|||\omega_1|||, \cdots, |||\omega_n|||)^T$, determine the value n of rank for the matrix, where $|||\omega_1|||$, $|||\omega_2|||$, $\cdots$ and $|||\omega_n|||$ are suitably selected in order of numerical value of input and
$\{\underline{S_l}(\omega_i)\underline{a} - \underline{a}; 1 \leq i \leq n,\ \omega_i \in U^*\}$ is a set of independent vectors.
Namely, determine the value n of rank for the matrix
$H_{\underline{a}\ (n,\bar{p})}(|||\omega_1|||, \cdots, |||\omega_n|||)$ such that a set of the square root of eigenvalues for the covariance matrix composed of relatively small and equally-sized numbers is excluded, where the signal part effected by the set may be the noisy part of the observed data.

2) In order to determine g_s, the algebraic CLS method is used as follows:
① *In particular, set $g_s(\omega_i) := \mathbf{e}_i$ for $\omega_i \in U$. Namely, $g_s(\omega_1) := \mathbf{e}_1$, $g_s(\omega_2) := \mathbf{e}_2, \cdots, g_s(\omega_k) := \mathbf{e}_k$ for some $k \in N$.*
For $u \in U$ such that $u \notin \{\omega_i; 1 \le i \le n\}$ and $\omega_r < u$, $g_s(u) = \sum_{j=1}^{r} b_{u,j}\underline{S_l}(\omega_j)\underline{a}$ is obtained as follows:
Based on Proposition (2.14), determine coefficients $\{b_{u,j} : 1 \le j \le r\}$.
The Q in Proposition (2.14) can be considered as the matrix composed from the eigenvectors of $H_{\underline{a}\ (r+1,L)}(|||\omega_1|||, |||\omega_2|||, \cdots, |||\omega_r|||, |||u|||)$ $\times H^T_{\underline{a}\ (r+1,L)}(|||\omega_1|||, |||\omega_2|||, \cdots, |||\omega_r|||, |||u|||)$.
Let a matrix $A_u \in \boldsymbol{R}^{1\times(r+1)}$ be $A_u := [b_{u,1}, b_{u,2}, \cdots, b_{u,r}, -1]$.
② *Determine the error vectors $\{\underline{S_l}(\omega_j)\underline{\bar{a}} \in \boldsymbol{R}^{L\times 1} : 0 \le j \le r\}$ and $\underline{S_l}(u)\underline{\bar{a}}$ by using the equation*
$[\underline{S_l}(\omega_1)\underline{\bar{a}}, \underline{S_l}(\omega_2)\underline{\bar{a}}, \cdots, \underline{S_l}(\omega_r)\underline{\bar{a}}, \underline{S_l}(u)\underline{\bar{a}}]^T :=$
$A_u^T[A_uA_u^T]^{-1}A_uH^T_{\underline{a}\ (r+1,L)}(|||\omega_1|||, |||\omega_2|||, \cdots, |||\omega_r|||, |||u|||)$
and $H^T_{\underline{a}\ (|||u|||+1,L)}(|||\omega_1|||, |||\omega_2|||, \cdots, |||\omega_r|||, |||u|||):=$
$[\underline{S_l}(\omega_1)\underline{a}, \cdots, \underline{S_l}(\omega_r)\underline{a}, \underline{S_l}(u)\underline{a}]$.

3) In order to obtain F_s, the algebraic CLS method is used as follows:
① *Let ${}_i\mathbf{f}_j \in \boldsymbol{R}^n$ in $F_s(u_i) := [{}_i\mathbf{f}_1\ {}_i\mathbf{f}_2 \cdots {}_i\mathbf{f}_n]$ be ${}_i\mathbf{f}_j := [{}_if_{j,1}\ {}_if_{j,2} \cdots {}_if_{j,n}]^T$ for $1 \le i \le m$, where ${}_if_{j,k}$ is given by the following:*
$\underline{S_l}(u_i)\underline{S_l}(\omega_j)\underline{a} = \sum_{k=1}^{n} {}_if_{j,k}\underline{S_l}(\omega_k)\underline{a}$, ${}_if_{j,k} \in \boldsymbol{R}$ *in the sense of* $F(U^*_{L-p}, Y)$.
②*For i $(1 \le i \le m)$, j $(1 \le j \le n)$ and for the maximum number r $(1 \le r \le n)$ such that $\omega_r, \omega_j \in \{\omega_j; 1 \le j \le n\}$ and $|||\omega_r||| < |||u_i|\omega_j|||$.*
Based on Proposition (2.14), determine coefficients $\{{}_if_{j,k} = 0;$ $r+1 \le k \le n\}$ and $\{{}_if_{j,k} : 1 \le k \le r\}$.
The Q in Proposition (2.14) can be considered as the matrix composed from the eigenvectors of $H_{\underline{a}\ (n+1,L)}(|||\omega_1|||, |||\omega_2|||, \cdots, |||\omega_r|||, |||u_i|\omega_j|||)$ $\times H^T_{\underline{a}\ (n+1,L)}(|||\omega_1|||, |||\omega_2|||, \cdots, |||\omega_r|||, |||u_i|\omega_j|||)$.
Let a matrix ${}_iA_j \in \boldsymbol{R}^{1\times(r+1)}$ be ${}_iA_j := [{}_if_{j,1}, {}_if_{j,2}, \cdots, {}_if_{j,r}, -1]$.
Determine the error vectors $\{\underline{S_l}(\omega_j)\underline{\bar{a}}\} \in \boldsymbol{R}^{L\times 1} : 0 \le j \le r\}$ and $\underline{S_l}(u_i|\omega_j)\underline{\bar{a}}$ by using the equation
$[\underline{S_l}(\omega_1)\underline{\bar{a}}, \underline{S_l}(\omega_2)\underline{\bar{a}}, \cdots, \underline{S_l}(\omega_r)\underline{\bar{a}}, \underline{S_l}(u_i|\omega_j)\underline{\bar{a}}]^T :=$
${}_iA_j^T[{}_iA_j\ {}_iA_j^T]^{-1}\ {}_iA_j\ H^T_{\underline{a}\ (n+1,L)}(|||\omega_1|||, |||\omega_2|||, \cdots, |||\omega_r|||, |||u_i|\omega_j|||)$ *and*
$H^T_{\underline{a}\ (r+1,L)}(|||\omega_1|||, |||\omega_2|||, \cdots, |||\omega_r|||, |||u_i|\omega_j|||):=$
$[\underline{S_l}(\omega_1)\underline{a}, \underline{S_l}(\omega_2)\underline{a}, \cdots, \underline{S_l}(\omega_r)\underline{a}, \underline{S_l}(u_i|\omega_j)\underline{a}]$.

4) In order to determine h_s, the algebraic CLS method is used as follows:
① *For the first ω_{r_1+1}, $\omega_{r_1} \in \{\omega_i; 1 \le i \le n\}$ such that $\omega_{r_1+1} > \omega_{r_1}$ and $|||\omega_{r_1+1}|||$ - $|||\omega_{r_1}||| > 1$ when starting out from ω_1,*
set $\underline{S_l}(\lambda_1)\underline{a} := \sum_{i=1}^{r_1} b_{\lambda_{1,i}}\mathbf{e}_i$ for the obtained equation $\underline{S_l}(\lambda_1)\underline{a} = \sum_{i=1}^{r_1} b_{\lambda_{1,i}}(\underline{S_l}(\omega_i)\underline{a})$ for λ_1 such that $|||\lambda_1||| = |||\omega_{r_1}||| + 1$.
Based on Proposition (2.14), determine coefficients $\{b_{\lambda_{1,i}} : 1 \le i \le r_1\}$.
The Q in Proposition (2.14) can be considered as the matrix composed from the eigenvectors of $H_{\underline{a}\ (r_1+1,L)}(|||\omega_1|||, |||\omega_2|||, \cdots, |||\omega_{r_1}|||, |||\lambda_1|||)$ $\times H^T_{\underline{a}\ (r_1+1,L)}(|||\omega_1|||, |||\omega_2|||, \cdots, |||\omega_{r_1}|||, |||\lambda_1|||)$.

Let a matrix $A_{\lambda_1} \in \boldsymbol{R}^{1\times(r_1+1)}$ *be* $A_{\lambda_1} := [b_{\lambda_1,1}, b_{\lambda_1,2}, \cdots, b_{\lambda_1,r_1}, -1]$.
Determine the error vectors $\{\underline{S_l}(\omega_i)\underline{\bar{a}} \in \boldsymbol{R}^{L\times 1} : 0 \le i \le r_1\}$ *and* $\underline{S_l}(\lambda_1)\underline{\bar{a}}$ *by using the equation*
$[\underline{S_l}(\omega_1)\underline{\bar{a}}, \underline{S_l}(\omega_2)\underline{\bar{a}}, \cdots, \underline{S_l}(\omega_{r_1})\underline{\bar{a}}, \underline{S_l}(\lambda_1)\underline{\bar{a}}]^T :=$
$A^T_{\lambda_1}[A_{\lambda_1}A^T_{\lambda_1}]^{-1}A_{\lambda_1}H^T_{\underline{a}\ (r_1+1,L)}(|||\omega_1|||, |||\omega_2|||, \cdots, |||\omega_{r_1}|||, |||\lambda_1|||)$
and $H^T_{\underline{a}\ (|||\lambda_1|||+1,L)}(|||\omega_1|||, |||\omega_2|||, \cdots, |||\omega_{r_1}|||, |||\lambda_1|||) :=$
$[\underline{S_l}(\omega_1)\underline{a}, \cdots, \underline{S_l}(\omega_{r_1})\underline{a}, \underline{S_l}(\lambda_1)\underline{a}]$.
Then let $h_{\lambda_1 s} \in \boldsymbol{R}^{1\times r_1}$ *be*
$h_{\lambda_1 s} := [\underline{a}(\omega_1) - (\underline{\bar{a}}(\omega_1), \underline{a}(\omega_2) - (\underline{\bar{a}}(\omega_2), \cdots, \underline{a}(\omega_{r_1}) - (\underline{\bar{a}}(\omega_{r_1})]$.
②*For the first* $\omega_{r_2+1},\ \omega_{r_2} \in \{\omega_i; 1 \le i \le n\}$ *such that* $\omega_{r_2+1} > \omega_{r_2}$ *and* $|||\omega_{r_2+1}|||$ - $|||\omega_{r_2}||| > 1$ *when starting out from* ω_{r_1+1},
set $\underline{S_l}(\lambda_2)\underline{a} := \sum_{i=1}^{r_2} b_{\lambda_2,i}\mathbf{e}_i$ *for the obtained equation* $\underline{S_l}(\lambda_2)\underline{a} = \sum_{i=1}^{r_2} b_{\lambda_2,i}(\underline{S_l}(\omega_i)\underline{a})$ *for* λ_2 *such that* $|||\lambda_2||| = |||\omega_{r_2}||| + 1$.
Based on Proposition (2.14), determine coefficients $\{b_{\lambda_{2,i}} : 1 \le i \le r_1\}$.
The Q in Proposition (2.14) can be considered as the matrix composed from the eigenvectors of $H_{\underline{a}\ (r_2+1,L)}(|||\omega_1|||, |||\omega_2|||, \cdots, |||\omega_{r_2}|||, |||\lambda_2|||)$
$\times H^T_{\underline{a}\ (r_2+1,L)}(|||\omega_1|||, |||\omega_2|||, \cdots, |||\omega_{r_2}|||, |||\lambda_2|||)$.
Let a matrix $A_{\lambda_2} \in \boldsymbol{R}^{1\times(r_2+1)}$ *be* $A_{\lambda_2} := [b_{\lambda_2,1}, b_{\lambda_2,2}, \cdots, b_{\lambda_2,r_2}, -1]$.
Determine the error vectors $\{\underline{S_l}(\omega_i)\underline{\bar{a}} \in \boldsymbol{R}^{L\times 1} : 0 \le i \le r_2\}$ *and* $\underline{S_l}(\lambda_2)\underline{\bar{a}}$ *by using the equation*
$[\underline{S_l}(\omega_1)\underline{\bar{a}}, \underline{S_l}(\omega_2)\underline{\bar{a}}, \cdots, \underline{S_l}(\omega_{r_2})\underline{\bar{a}}, \underline{S_l}(\lambda_2)\underline{\bar{a}}]^T :=$
$A^T_{\lambda_2}[A_{\lambda_2}A^T_{\lambda_2}]^{-1}A_{\lambda_2}H^T_{\underline{a}\ (r_2+1,L)}(|||\omega_1|||, |||\omega_2|||, \cdots, |||\omega_{r_2}|||, |||\lambda_2|||)$
and $H^T_{\underline{a}\ (|||\lambda_2|||+1,L)}(|||\omega_1|||, |||\omega_2|||, \cdots, |||\omega_{r_2}|||, |||\lambda_2|||) :=$
$[\underline{S_l}(\omega_1)\underline{a}, \cdots, \underline{S_l}(\omega_{r_2})\underline{a}, \underline{S_l}(\lambda_2)\underline{a}]$.
Then let $h_{\lambda_2 s} \in \boldsymbol{R}^{1\times r_2}$ *be*
$h_{\lambda_2 s} := [\underline{a}(\omega_{r_1+1}) - (\underline{\bar{a}}(\omega_{r_1+1}),\ \underline{a}(\omega_{r_1+2}) - (\underline{\bar{a}}(\omega_{r_1+2}), \cdots,\ \underline{a}(\omega_{r_2}) - (\underline{\bar{a}}(\omega_{r_2})]$.
⋮
ⓣ *For* $\omega \in U^*$ *such that* $|||\omega||| = |||\omega_n||| + 1$,
Based on Proposition (2.14), determine coefficients $\{b_{\omega,i} : 1 \le i \le n\}$.
The Q in Proposition (2.14) can be considered as the matrix composed from the eigenvectors of $H_{\underline{a}\ (r_2+1,L)}(|||\omega_1|||, |||\omega_2|||, \cdots, |||\omega_{r_2}|||, |||\lambda_2|||)$
$\times H^T_{\underline{a}\ (r_2+1,L)}(|||\omega_1|||, |||\omega_2|||, \cdots, |||\omega_{r_2}|||, |||\lambda_2|||)$.
Let a matrix $A_\omega \in \boldsymbol{R}^{1\times(n+1)}$ *be* $A_\omega := [b_{\omega,1}, b_{\omega,2}, \cdots, b_{\omega,n}, -1]$.
Determine the error vectors $\{\underline{S_l}(\omega_i)\underline{\bar{a}} : 1 \le i \le n\}$ *and* $\underline{S_l}(\omega)\underline{\bar{a}}$ *by using the equation*
$[\underline{S_l}(\omega_1)\underline{\bar{a}}, \underline{S_l}(\omega_2)\underline{\bar{a}},\ \cdots,\ \underline{S_l}(\omega_n)\underline{\bar{a}},\ \underline{S_l}(\omega)\underline{\bar{a}}]^T :=$
$A^T_\omega[A_\omega A^T_\omega]^{-1}A_\omega H^T_{\underline{a}\ (n+1,L)}(|||\omega_1|||, |||\omega_2|||, \cdots, |||\omega_n|||, |||\omega|||)$
and $H^T_{\underline{a}\ (n+1,L)}(|||\omega_1|||, |||\omega_2|||, \cdots, |||\omega_n|||, |||\omega|||) :=$
$[\underline{S_l}(\omega_1)\underline{a},\ \cdots, \underline{S_l}(\omega_n)\underline{a},\ \underline{S_l}(\omega)\underline{a}]$.
Then let $h_{\omega s}$ *be*
$h_{\omega s} := [\underline{a}(\omega_{r_t+1}) - (\underline{\bar{a}}(\omega_{r_t+1}),\ \underline{a}(\omega_{r_t+2}) - \underline{\bar{a}}(\omega_{r_t+2}), \cdots,\ \underline{a}(\omega_n) - \underline{\bar{a}}(\omega_n)]$.
Finally, let $h_s \in \boldsymbol{R}^{1\times n}$ *be*
$h_s := [h_{\lambda_1 s}, h_{\lambda_2 s}, \cdots, h_{\omega s}]$.

[proof] In 1), the number of dimensions is determined by checking what part is the noisy part and by using the ratio of Hankel matrix norm, which implies the noise to signal ratio. According to Theorem (8.21), a linear representation system $\sigma = ((\boldsymbol{R}^n, F_s), g_s, h_s)$ is obtained as follows. In 2), g_s is obtained directly or by using the algebraic CLS method for A_u corresponding to the matrix A in Lemma (2.17). In 3), F_s is obtained by using the algebraic CLS method for ${}_iA_j$ corresponding to the matrix A in Lemma (2.17). In 4), h_s is obtained by using the algebraic CLS method for A_{λ_1}, A_{λ_2}, A_{λ_ω} corresponding to the matrix A in Proposition (2.14).

Remark : Let S and N be the norm of a signal and noise. Then the selected ratio of matrices in the algorithm may be considered as $\frac{N}{S+N}$.

In the figures of this chapter, we use a notation *Signal_n d* as an input response map obtained by a n-dimensional linear representation system.

In the examples of this chapter, a notation $H^T_{\underline{a}\ (r+1,40)}(0,\cdots,r)$ is used in place of $H^T_{\underline{a}\ (r+1,40)}(0,1,2,\cdots,r-1,r)$.

Example 8.29. Let the signals be the input response map of the following 3-dimensional linear representation system: $\sigma = ((\boldsymbol{R}^3, F), x^0, h)$, where

$$F(u_1) = \begin{bmatrix} 0 & 0 & 0 \\ 1 & 0 & -0.1 \\ 0 & 1 & 0 \end{bmatrix},\ F(u_2) = \begin{bmatrix} -1 & 1.5 & 0.8 \\ 0 & -1 & 0 \\ 0 & -0.5 & -0.1 \end{bmatrix},\ F(u_3) = \begin{bmatrix} 0 & 0.1 & 0 \\ 0 & 0.2 & 0 \\ 0 & -0.5 & 0.3 \end{bmatrix},$$

$x^0 = \mathbf{e}_1,\ h = [9,\ -8,\ 2]$.

Then the algebraically noisy realization problem is solved as follows:

covariance matrix	eigenvalues					
	1	2	3	4	5	6
$H^T_{\underline{a}\ (2,40)}(0,1)H_{\underline{a}\ (2,40)}(0,1)$	7281	2540				
$H^T_{\underline{a}\ (3,40)}(0,1,2)H_{\underline{a}\ (3,40)}(0,1,2)$	8412	4365	7.7			
$H^T_{\underline{a}\ (4,40)}(0,\cdots,3)H_{\underline{a}\ (4,40)}(0,\cdots,3)$	8412	4365	7.8	6.4		
$H^T_{\underline{a}\ (5,40)}(0,\cdots,4)H_{\underline{a}\ (5,40)}(0,\cdots,4)$	8863	4365	154	7.5	6.4	
$H^T_{\underline{a}\ (6,40)}(0,\cdots,5)H_{\underline{a}\ (6,40)}(0,\cdots,5)$	29904	4397	156	7.5	6.8	4.8
covariance matrix	square root of eigenvalues					
$H^T_{\underline{a}\ (2,40)}(0,1)H_{\underline{a}\ (2,40)}(0,1)$	85	50.4				
$H^T_{\underline{a}\ (3,40)}(0,1,2)H_{\underline{a}\ (3,40)}(0,1,2)$	91.7	66	2.8			
$H^T_{\underline{a}\ (4,40)}(0,\cdots,3)H_{\underline{a}\ (4,40)}(0,\cdots,3)$	91.7	66	2.8	2.5		
$H^T_{\underline{a}\ (5,40)}(0,\cdots,4)H_{\underline{a}\ (5,40)}(0,\cdots,4)$	94	66	12.4	2.7	2.5	
$H^T_{\underline{a}\ (6,40)}(0,\cdots,5)H_{\underline{a}\ (6,40)}(0,\cdots,5)$	173	66.3	12.5	2.7	2.6	2.2

1) A set $\{2.7,\ 2.6,\ 2.2\}$ is composed of relatively small and equally-sized numbers in the square root of eigenvalues for $H^T_{\underline{a}\ (6,40)}(0,\cdots,5)H_{\underline{a}\ (6,40)}(0,\cdots,5)$.
2) After determining the independent vectors $\underline{a}$, $\underline{S_l}(u_1)\underline{a}$ and $\underline{S_l}(u_1|u_1)\underline{a}$ whose numerical value of input are 0, 1 and 4, we will continue the noisy realization algorithm by the CLS method.

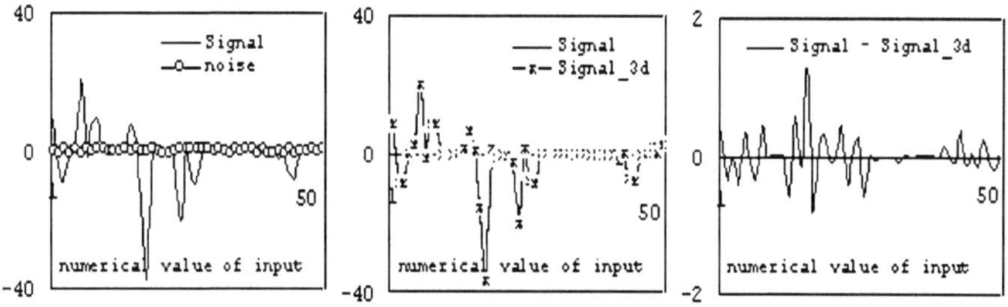

Fig. 8.4 The left is the original input response map and added noise to an original 3-dimensional linear representation system. The middle is the original input response map and the behavior of a 3-dimensional linear representation system obtained by the algebraic CLS method. The right is the difference between the original input response map and the behavior of the 3-dimensional linear representation system obtained by the algebraic CLS method in Example (8.29).

Therefore, a noisy 3-dimensional linear representation system $\sigma_1 = ((\boldsymbol{R}^3,\ F_1),\ x_1^0,\ h_1)$ obtained by the algebraic CLS method is constructed as follows:

$$F_1(u_1) = \begin{bmatrix} 0 & 0 & -0.02 \\ 1 & 0 & -0.12 \\ 0 & 1 & 0.08 \end{bmatrix},\ F_1(u_2) = \begin{bmatrix} -1 & 1.5 & 0.8 \\ -0.002 & -1 & 0.008 \\ 0 & -0.54 & -0.17 \end{bmatrix},$$

$$F_1(u_3) = \begin{bmatrix} 0.003 & 0.12 & 0.002 \\ 0 & 0.18 & 0 \\ 0 & -0.4 & 0.3 \end{bmatrix},\ x_1^0 = \mathbf{e}_1,\ h_1 = [8.7,\ -8.4,\ 2.4].$$

We can show that the algorithm for noisy realization given by the analytic CLS method in the reference [Hasegawa, 2008] produces the same system as the above one in the sense of the numerical calculation.

In this example, the original signals are considered as the behavior of a 3-dimensional linear representation system and the desirable input response map is obtained by the algebraic CLS method. The model obtained by the algebraic CLS method is a 3-dimensional linear representation system.

Just as we expected, the following table and Fig. 8.4 truly indicate that the 3-dimensional linear representation system obtained by the algebraic CLS method is a good noisy realization.

dimen-sion	ratio of matrices	mean values of square root for sum of			cosine	error
		signal	signal by CLS	error	① and ②	ratio
		①	②	③	$\cos\theta$	③/①
$a_{0,1,4}$	0.02	1.1235	1.1123	0.05	0.999	0.04

Example 8.30. Let the signals be the behavior of the following 4-dimensional linear representation system: $\sigma = ((\boldsymbol{R}^4, F), x^0, h)$, where

$$F(u_1) = \begin{bmatrix} 0 & 1.2 & -0.1 & -0.7 \\ 1 & 0.4 & 0 & 0 \\ 0 & -2 & 1.2 & 0.4 \\ 0 & 0.2 & 0.3 & 0.8 \end{bmatrix}, \ F(u_2) = \begin{bmatrix} 0 & -1 & -0.1 & -0.4 \\ 0 & 0.8 & 0 & -0.3 \\ 1 & 0 & -0.5 & 0.4 \\ 0 & 0 & 0 & -0.6 \end{bmatrix},$$

$$F(u_3) = \begin{bmatrix} 0 & 1.4 & 0 & -1.2 \\ 0 & 0.5 & 0.1 & 1.5 \\ 0 & -0.9 & -0.3 & 0.3 \\ 1 & 0.3 & 0.2 & -0.5 \end{bmatrix}, \ x^0 = \mathbf{e}_1, \ h = [12, \ 8, \ 1, \ -2].$$

Then the algebraically noisy realization problem is solved as follows:

covariance matrix	eigenvalues						
	1	2	3	4	5	6	7
$H^T_{\underline{a}\ (3,40)}(0,1,2)H_{\underline{a}\ (3,40)}(0,1,2)$	11998	3717	1385				
$H^T_{\underline{a}\ (4,40)}(0,\cdots,3)H_{\underline{a}\ (4,40)}(0,\cdots,3)$	15398	5891	3717	814			
$H^T_{\underline{a}\ (5,40)}(0,\cdots,4)H_{\underline{a}\ (5,40)}(0,\cdots,4)$	33899	6187	4749	1801	7.5		
$H^T_{\underline{a}\ (6,40)}(0,\cdots,5)H_{\underline{a}\ (6,40)}(0,\cdots,5)$	34024	11362	6159	1908	7.7	5.2	
$H^T_{\underline{a}\ (7,40)}(0,\cdots,6)H_{\underline{a}\ (7,40)}(0,\cdots,6)$	50082	12194	7229	1977	7.9	5.9	5
covariance matrix	square root of eigenvalues						
$H^T_{\underline{a}\ (3,40)}(0,1,2)H_{\underline{a}\ (3,40)}(0,1,2)$	110	61	37				
$H^T_{\underline{a}\ (4,40)}(0,\cdots,3)H_{\underline{a}\ (4,40)}(0,\cdots,3)$	124	77	61	28.5			
$H^T_{\underline{a}\ (5,40)}(0,\cdots,4)H_{\underline{a}\ (5,40)}(0,\cdots,4)$	184	79	69	42	2.7		
$H^T_{\underline{a}\ (6,40)}(0,\cdots,5)H_{\underline{a}\ (6,40)}(0,\cdots,5)$	184	107	78	44	2.8	2.3	
$H^T_{\underline{a}\ (7,40)}(0,\cdots,6)H_{\underline{a}\ (7,40)}(0,\cdots,6)$	224	110	85	44	2.8	2.4	2.2

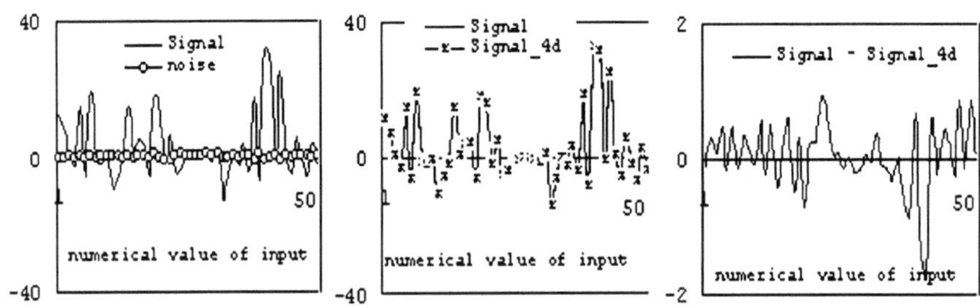

Fig. 8.5 The left is the original input response map and added noise to an original 4-dimensional linear representation system. The middle is the original input response map and the behavior of a 4-dimensional linear representation system obtained by the algebraic CLS method. The right is the difference between the original input response map and the behavior of the 4-dimensional linear representation system obtained by the algebraic CLS method in Example (8.30).

1) A set $\{2.8, \ 2.4, \ 2.2\}$ is composed of relatively small and equally-sized numbers in the square root of eigenvalues for $H^T_{\underline{a}\ (7,40)}(0,\cdots,6)H_{\underline{a}\ (7,40)}(0,\cdots,6)$.
2) After determining the independent vectors $\underline{a}$, $\underline{S}_l(u_1)\underline{a}$, $\underline{S}_l(u_2)\underline{a}$ and $\underline{S}_l(u_3)\underline{a}$ whose numerical value of input are 0, 1, 2 and 3, we will continue the noisy realization algorithm by the algebraic CLS method.

Therefore, a noisy 4-dimensional linear representation system $\sigma_1 = ((\boldsymbol{R}^4,\ F_1),\ x_1^0,\ h_1)$ obtained by the algebraic CLS method is constructed as follows:

$$F_1(u_1) = \begin{bmatrix} 0 & 1.2 & -0.1 & -0.7 \\ 1 & 0.4 & -0.004 & 0.01 \\ 0 & -2 & 1.2 & 0.4 \\ 0 & 0.2 & 0.3 & 0.8 \end{bmatrix},\ F_1(u_2) = \begin{bmatrix} 0 & -1.02 & -0.11 & -0.41 \\ 0 & 0.8 & 0.01 & -0.3 \\ 1 & 0.04 & -0.51 & 0.4 \\ 0 & -0.02 & 0.01 & -0.6 \end{bmatrix},$$

$$F_1(u_3) = \begin{bmatrix} 0 & 1.4 & 0.01 & -1.2 \\ 0 & 0.5 & 0.1 & 1.5 \\ 0 & -0.9 & -0.3 & 0.34 \\ 1 & 0.31 & 0.2 & -0.53 \end{bmatrix},\ x_1^0 = \mathbf{e}_1,\ h_1 = [12.1,\ 7.7,\ 0.93,\ -2.5].$$

We can show that the algorithm for noisy realization given by the analytic CLS method in the reference [Hasegawa, 2008] produces the same system as the above one in the sense of the numerical calculation.

In this example, the original signals are considered as the behavior of a 4-dimensional linear representation system and the desirable input response map is obtained by the algebraic CLS method. The model obtained by the algebraic CLS method is a 4-dimensional linear representation system.

Just as we expected, the following table and Fig. 8.5 truly indicate that the 4-dimensional linear representation system obtained by the algebraic CLS method is a good noisy realization.

dimen-sion	ratio of matrices	mean values of square root for sum of signal	signal by CLS	error	cosine ① and ②	error ratio
		①	②	③	$\cos\theta$	③/①
$a_{0,1,2,3}$	0.01	1.4516	1.49	0.07	0.999	0.05

Example 8.31. Let the signals be the behavior of the following 5-dimensional linear representation system: $\sigma = ((\boldsymbol{R}^5, F), x^0, h)$, where

$$F(u_1) = \begin{bmatrix} 0 & 0 & 0.4 & 1.8 & -0.5 \\ 1 & 0 & 1.7 & 0.1 & -1.1 \\ 0 & 0 & -0.1 & -0.4 & 0.2 \\ 0 & 0 & -0.5 & -0.5 & 0.3 \\ 0 & 1 & -0.5 & -1.5 & 0.8 \end{bmatrix},\ F(u_2) = \begin{bmatrix} 0 & 0.2 & -1.5 & 1.2 & -0.6 \\ 0 & 0.5 & 0 & -0.2 & 0.2 \\ 1 & -0.5 & 0.2 & 0.5 & 0.6 \\ 0 & -0.5 & 0.2 & 0.7 & 0.3 \\ 0 & 0 & 0.1 & 1.1 & 0.7 \end{bmatrix},$$

$$F(u_3) = \begin{bmatrix} 0 & 1.3 & 0 & -0.5 & -0.1 \\ 0 & 1.8 & 0.5 & -0.2 & 0.8 \\ 0 & 0.1 & 2 & -1 & -0.8 \\ 1 & -1.2 & 0 & -0.8 & 1.5 \\ 0 & 0 & 0 & 0 & 0.8 \end{bmatrix},\ x^0 = \mathbf{e}_1,\ h = [12,\ -5,\ 3,\ -4,\ 6].$$

Then the algebraically noisy realization problem is solved as follows:

covariance matrix	eigenvalues							
	1	2	3	4	5	6	7	8
$H^T_{\underline{a}\,(5,40)}(0,\cdots,4)H_{\underline{a}\,(5,40)}(0,\cdots,4)$	31261	15204	14431	2690	1895			
$H^T_{\underline{a}\,(6,40)}(0,\cdots,5)H_{\underline{a}\,(6,40)}(0,\cdots,5)$	33151	20901	14853	2804	2163	4		
$H^T_{\underline{a}\,(7,40)}(0,\cdots,6)H_{\underline{a}\,(7,40)}(0,\cdots,6)$	84394	31910	14858	3356	2662	4.2	3.8	
$H^T_{\underline{a}\,(8,40)}(0,\cdots,7)H_{\underline{a}\,(8,40)}(0,\cdots,7)$	113377	32153	23000	3359	2727	7.4	4.1	3.2
covariance matrix	square root of eigenvalues							
$H^T_{\underline{a}\,(5,40)}(0,\cdots,4)H_{\underline{a}\,(5,40)}(0,\cdots,4)$	177	123	120	52	43.5			
$H^T_{\underline{a}\,(6,40)}(0,\cdots,5)H_{\underline{a}\,(6,40)}(0,\cdots,5)$	182	145	122	58	52	2		
$H^T_{\underline{a}\,(7,40)}(0,\cdots,6)H_{\underline{a}\,(7,40)}(0,\cdots,6)$	291	179	122	58	52	2	1.9	
$H^T_{\underline{a}\,(8,40)}(0,\cdots,7)H_{\underline{a}\,(8,40)}(0,\cdots,7)$	337	179	152	58	52	2.7	2	1.8

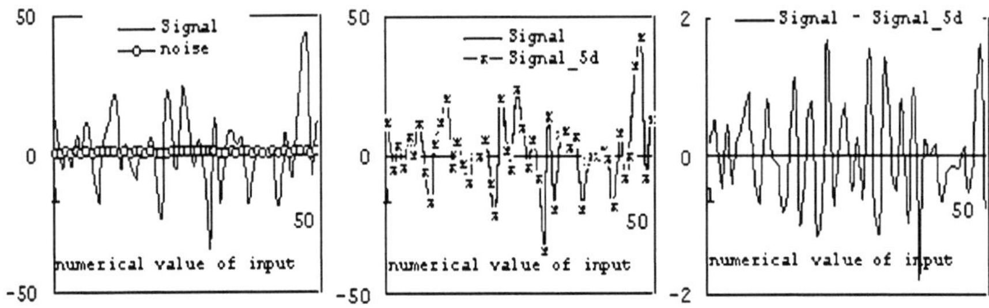

Fig. 8.6 The left is the original input response map and added noise to the original 5-dimensional linear representation system. The middle is the original input response map and the behavior of a 5-dimensional linear representation system obtained by the algebraic CLS method. The right is the difference between the original input response map and the behavior of the 5-dimensional linear representation system obtained by the algebraic CLS method in Example (8.31).

1) A set $\{2.7,\ 2,\ 1.8\}$ is composed of relatively small and equally-sized numbers in the square root of eigenvalues for $H^T_{\underline{a}\,(8,40)}(0,\cdots,7)H_{\underline{a}\,(8,40)}(0,\cdots,7)$.
2) After determining the independent vectors $\underline{a}$, $\underline{S_l}(u_1)\underline{a}$, $\underline{S_l}(u_2)\underline{a}$, $\underline{S_l}(u_3)\underline{a}$ and $\underline{S_l}(u_1|u_1)\underline{a}$ whose numerical value of input are 0, 1, 2, 3 and 4, we will continue the noisy realization algorithm by the algebraic CLS method.

Therefore, a noisy 5-dimensional linear representation system $\sigma_1 = ((\boldsymbol{R}^5,\ F_1),\ x_1^0,\ h_1)$ obtained by the algebraic CLS method is constructed as follows:

$$F_1(u_1) = \begin{bmatrix} 0 & 0 & 0.4 & 1.8 & -0.5 \\ 1 & 0 & 1.7 & 0.1 & -1.1 \\ 0 & 0 & -0.1 & -0.4 & 0.18 \\ 0 & 0 & -0.5 & -0.51 & 0.3 \\ 0 & 1 & -0.5 & -1.5 & 0.8 \end{bmatrix},\ F_1(u_2) = \begin{bmatrix} 0 & 0.2 & -1.5 & 1.2 & -0.6 \\ 0 & 0.5 & -0.01 & -0.2 & 0.2 \\ 1 & -0.5 & 0.2 & 0.5 & 0.6 \\ 0 & -0.5 & 0.2 & 0.7 & 0.3 \\ 0 & 0 & 0.1 & 1.1 & 0.7 \end{bmatrix},$$

$$F_1(u_3) = \begin{bmatrix} 0 & 1.3 & -0.01 & -0.5 & -0.1 \\ 0 & 1.8 & 0.5 & -0.2 & 0.8 \\ 0 & 0.1 & 2 & -1 & -0.8 \\ 1 & -1.2 & 0.001 & -0.8 & 1.5 \\ 0 & 0.02 & 0.01 & -0.01 & 0.8 \end{bmatrix}, \; x_1^0 = \mathbf{e}_1,$$

$$h_1 = [11.8, \; -5.5, \; 3.5, \; -4.5, \; 6.4].$$

We can show that the algorithm for noisy realization given by the analytic CLS method in the reference [Hasegawa, 2008] produces the same systems as the above one in the sense of the numerical calculation.

In this example, the original signals are considered as the behavior of a 5-dimensional linear representation system and the desirable input response map is obtained by the algebraic CLS method. The model obtained by the algebraic CLS method is a 5-dimensional linear representation system.

Just as we expected, the following table and Fig. 8.6 truly indicate that the 5-dimensional linear representation system obtained by the algebraic CLS method is a somewhat good noisy realization.

dimen-sion	ratio of matrices	mean values of square root for sum of signal	signal by CLS	error	cosine ① and ②	error ratio
		①	②	③	$\cos\theta$	③/①
$a_{0,1,2,3,4}$	0.01	1.91	1.914	0.107	0.998	0.06

Example 8.32. Let the signals be the behavior of the following 6-dimensional linear representation system: $\sigma = ((\boldsymbol{R}^6, F), x^0, h)$, where

$$F(u_1) = \begin{bmatrix} 0 & 0 & 0.4 & 1.5 & -0.3 & -0.3 \\ 1 & 0 & -0.1 & 0 & -1.1 & 0.8 \\ 0 & 0 & -0.2 & 0 & 0.2 & 0.1 \\ 0 & 0 & 0 & 0 & 0 & 0 \\ 0 & 1 & -0.5 & 0 & -1 & 0.1 \\ 0 & 0 & 0.2 & 0 & -0.3 & -0.1 \end{bmatrix}, \; F(u_2) = \begin{bmatrix} 0 & 0 & 1.5 & 0 & -1.6 & -0.6 \\ 0 & 0 & -0.9 & 0 & 1.9 & 0.5 \\ 1 & 0 & -0.9 & 0 & 1.5 & -0.7 \\ 0 & 0 & 0 & 0 & 0 & 0 \\ 0 & 0 & -1 & 0 & 2 & 0.6 \\ 0 & 1 & 0.7 & 0 & -1.2 & -0.5 \end{bmatrix},$$

$$F(u_3) = \begin{bmatrix} 0 & 0 & 0.1 & 0 & 0.3 & -0.1 \\ 0 & 0 & 0 & 0 & 0.1 & 0 \\ 0 & 0 & 0 & 0 & 0.1 & 0 \\ 1 & 0 & 0 & 0 & 0 & 0 \\ 0 & 0 & 0 & 0 & 0.1 & 0 \\ 0 & 0 & 0 & 0 & 0 & 0 \end{bmatrix}, \; x^0 = \mathbf{e}_1, \; h = [12, \; -2, \; -6, \; 2, \; 3, \; 5].$$

Then the algebraically noisy realization problem is solved as follows:
1) A set $\{3.2, \; 2.6, \; 2.3\}$ is composed of relatively small and equally-sized numbers in the square root of eigenvalues for $H^T_{\underline{a}\,(9,40)}(0,\cdots,8)H_{\underline{a}\,(9,40)}(0,\cdots,8)$.
2) After determining the independent vectors $\underline{a}$, $\underline{S}_l(u_1)\underline{a}$, $\underline{S}_l(u_2)\underline{a}$, $\underline{S}_l(u_3)\underline{a}$, $\underline{S}_l(u_1|u_1)\underline{a}$ and $\underline{S}_l(u_2|u_1)\underline{a}$ whose numerical value of input are 0, 1, 2, 3, 4 and 5, we will continue the noisy realization algorithm by the algebraic CLS method.

covariance matrix	eigenvalues								
	1	2	3	4	5	6	7	8	9
$H^T_{\underline{a}\,(6,40)}(0,\cdots,5)H_{\underline{a}\,(6,40)}(0,\cdots,5)$	7064	5655	2613	2201	963	326			
$H^T_{\underline{a}\,(7,40)}(0,\cdots,6)H_{\underline{a}\,(7,40)}(0,\cdots,6)$	7064	5655	2613	2202	963	327	6.7		
$H^T_{\underline{a}\,(8,40)}(0,\cdots,7)H_{\underline{a}\,(8,40)}(0,\cdots,7)$	7215	5975	2627	2556	1091	338	10.5	6.5	
$H^T_{\underline{a}\,(9,40)}(0,\cdots,8)H_{\underline{a}\,(9,40)}(0,\cdots,8)$	16649	7174	3423	2624	1261	371	10.5	6.7	5.4
covariance matrix	square root of eigenvalues								
$H^T_{\underline{a}\,(6,40)}(0,\cdots,5)H_{\underline{a}\,(6,40)}(0,\cdots,5)$	84	75.2	51.1	47	31	18			
$H^T_{\underline{a}\,(7,40)}(0,\cdots,6)H_{\underline{a}\,(7,40)}(0,\cdots,6)$	84	75.2	51.1	47	31	18	2.6		
$H^T_{\underline{a}\,(8,40)}(0,\cdots,7)H_{\underline{a}\,(8,40)}(0,\cdots,7)$	85	77.3	51.3	50.6	33	18.4	3.2	2.5	
$H^T_{\underline{a}\,(9,40)}(0,\cdots,8)H_{\underline{a}\,(9,40)}(0,\cdots,8)$	129	84.7	58.5	46.4	35.5	19.3	3.2	2.6	2.3

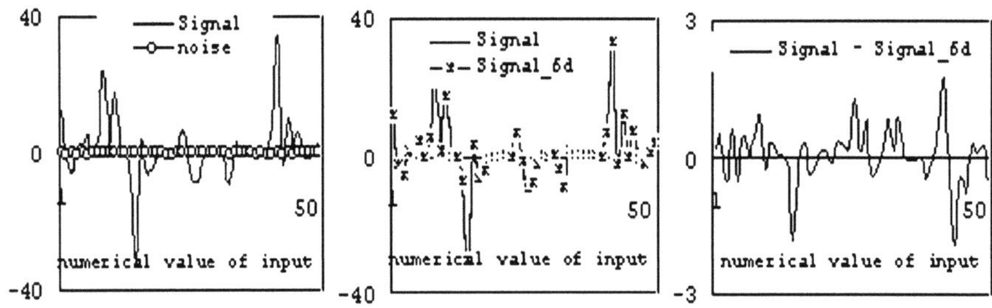

Fig. 8.7 The left is the original input response map and added noise to the original 6-dimensional linear representation system. The middle is the original input response map and the behavior of a 6-dimensional linear representation system obtained by the algebraic CLS method. The right is the difference between the original input response map and the behavior of the 6-dimensional linear representation system obtained by the algebraic CLS method in Example (8.32).

Therefore, a noisy 6-dimensional linear representation system $\sigma_1 = ((\boldsymbol{R}^6, F_1), x_1^0, h_1)$ obtained by the algebraic CLS method is constructed as follows:

$$F_1(u_1) = \begin{bmatrix} 0 & 0 & 0.41 & 1.5 & -0.3 & -0.32 \\ 1 & 0 & -0.07 & -0.02 & -1.1 & 0.8 \\ 0 & 0 & -0.2 & -0.01 & 0.2 & 0.1 \\ 0 & 0 & 0 & 0.01 & -0.01 & 0 \\ 0 & 1 & -0.5 & 0.02 & -1 & 0.1 \\ 0 & 0 & 0.2 & 0.01 & -0.3 & -0.1 \end{bmatrix}, F_1(u_2) = \begin{bmatrix} 0 & 0 & 1.5 & 0.001 & -1.6 & -0.56 \\ 0 & 0 & -0.8 & 0.02 & 1.8 & 0.5 \\ 1 & 0 & -0.8 & 0.02 & 1.4 & -0.7 \\ 0 & 0 & -0.001 & 0.001 & 0.02 & -0.01 \\ 0 & 0 & -0.9 & 0.02 & 1.9 & 0.6 \\ 0 & 1 & 0.7 & 0.005 & -1.18 & -0.5 \end{bmatrix},$$

$$F_1(u_3) = \begin{bmatrix} 0 & 0.01 & 0.13 & 0 & 0.3 & -0.1 \\ 0 & 0.02 & 0.02 & 0.02 & 0.09 & 0 \\ 0 & 0.02 & 0.02 & 0.01 & 0.08 & -0.03 \\ 1 & -0.01 & 0.02 & 0 & 0 & -0.01 \\ 0 & 0.02 & 0.01 & -0.01 & 0.1 & -0.01 \\ 0 & 0 & 0 & 0.01 & 0.01 & -0.01 \end{bmatrix},$$

$x_1^0 = \mathbf{e}_1, h_1 = [11.8,\ -2.5,\ -5.5,\ 1.4,\ 3.5,\ 4.54].$

We can show that the algorithm for noisy realization given by the analytic CLS method in the reference [Hasegawa, 2008] produces the same systems as the above one in the sense of the numerical calculation.

In this example, the original signals are considered as the behavior of a 6-dimensional linear representation system and the desirable input response map is obtained by the algebraic CLS method. The model obtained by the algebraic CLS method is a 6-dimensional linear representation system.

Just as we expected, the following table and Fig. 8.7 truly indicate that the 6-dimensional linear representation system obtained by the algebraic CLS method is a good noisy realization.

dimen-sion	ratio of matrices	mean values of square root for sum of			cosine	error
		signal	signal by CLS	error	① and ②	ratio
		①	②	③	$\cos\theta$	③/①
$a_{0,1,2,3,4,5}$	0.02	1.2818	1.2449	0.09	0.998	0.07

8.6 Historical Notes and Concluding Remarks

Algebraically approximate and noisy realization problems of linear representation systems were proposed from the notion of the Hankel matrix norm and the algebraic CLS method. The matrix norm is used for determining the number of dimensions of state space and the algebraic CLS method is used for determining the parameters of linear representation systems, which are general non-linear systems. Note that there are homogeneous bilinear systems as a subclass of linear representation systems.

Our solutions of approximate and noisy realization problems, like we discussed, are the same as with other linear and non-linear systems. In the past, a unified solution of non-linear systems was proposed in the reference [Hasegawa, 2008] by using the analytic CLS method.

In order to insist that our method for algebraically approximate and noisy realization is effective for our systems, we gave several examples. As a result of the examples, we have shown that the ratio of Hankel matrix norm implies the degree of approximation. For our noisy realization problem, we have shown that we can determine the number of dimensions of linear representation systems when a set of relatively small and equally-sized numbers of the square root of eigenvalues for a finite-sized Hankel matrix can be found.

As stated in the Historical notes and concluding remarks of the other chapters, our methods can be roughly summarized as follows:

Intuitively, our several examples for the algebraically approximate realization problem show that the smaller the ratio of matrices is, the smaller the error to signal ratio is.

The changing relations among the ratio of matrices and the error to signal ratio are proportial relations.

The ratio of a 0.01 input/output matrix ratio ranges from 0.02 to 0.04 for the error to signal ratio.

The several examples suggest that our three features can be expressed as follows:

(1) : The ratio of the matrix norm determines the degree of the crossed angle between directions of the approximated signal and the original signal.

(2) : We could propose a new law which says that linear representation systems obtained by the algebraic CLS method are the same as ones obtained by the analtytic CLS method proposed in the reference [Hasegawa, 2008].
The law is said to be a law of a constrained least square.

(3) : The algebraic CLS method determines the coefficients of linearly dependent vectors such that the error between the approximate signal and the original signal has a minimum value in the sense of the square norm while conserving the crossed angle.

Intuitively, our several examples for the algebraically noisy realization problem show that the smaller the ratio of matrices is, the smaller the error to signal ratio is. The ratio of a 0.01 input/output matrix ratio ranges from 0.02 to 0.06 for the error to signal ratio.

The several examples suggest that our three features can be expressed as follows:

(1) : The ratio of the matrix norm determines the degree of the crossed angle between directions of the obtained signal and the original signal.

(2) : We could propose a new law which says that linear representation systems obtained by the algebraic CLS method are the same as ones obtained by the analtytic CLS method proposed in the reference [Hasegawa, 2008].

(3) : The algebraic CLS method determines the coefficients of linearly dependent vectors such that the error between the obtained signal and original signal has a minimum value in the sense of the square norm while conserving the crossed angle.

The analtic CLS method for determing n variables is reduced to the minimization of the following rational polynomial $p(x_1, x_2, \cdots, x_n)$ in n variables:

$$p(x_1, x_2, \cdots, x_n) = \frac{\sum_{c_1, c_2, \cdots, c_n} \alpha(c_1, c_2, \cdots, c_n) \times x_1^{c_1} x_2^{c_2} \cdots x_n^{c_n}}{(1 + x_1^2 + x_2^2 + \cdots + x_n^2)^2},$$

where $\sum_{c_1, c_2, \cdots, c_n}$ means all summation of any combination with the conditions $0 \le c_i \le 4$ and $c_1 + c_2 + \cdots + c_n \le 4$ for any $1 \le i \le n$ and $\alpha(c_1, c_2, \cdots, c_n) \in \boldsymbol{R}$.

Therefore, our new Law shows that approximate and noisy problems can be solved using only algebraic calculations, namely, without treating partial differential equations.

Part II
Algebraically Approximate and Noisy Realization of Digital Images

Chapter 9
Algebraically Approximate and Noisy Realization of Two-Dimensional Images

Let the set of output values Y be a linear space over the field K.

Commutative linear representation systems can be presented according to the following main theorem [Hasegawa and Suzuki, 2006]:

For any two-dimensional image, there exist at least two canonical, that is, quasi-reachable and distinguishable commutative linear representation systems which realize, that is, which faithfully describe it, and any two canonical commutative linear representation systems with the same behavior are isomorphic.

By using an application of non-linear realization theory [Matsuo and Hasegawa, 2003], we can obtain a realization theory of two-dimensional images.

Finite-dimensional commutative linear representation systems were also investigated in detail. We derived a criterion for the canonicality of finite-dimensional commutative linear representation systems, the representation theorems of isomorphic classes for canonical commutative linear representation systems, and criteria for the behavior of finite-dimensional commutative linear representation systems. In addition, a procedure to obtain a canonical commutative linear representation system was also given.

Partial realization was discussed according to the results stated above. Existence of minimum partial realization was clearly presented. Minimum partial realizations are rarely unique up to an isomorphism. To solve the uniqueness problem, the notion of natural partial realizations was introduced. The main results for partial realization are the following three items:

1) A necessary and sufficient condition for the existence of the natural partial realizations is given by the rank condition of the finite-sized Hankel matrix.
2) The existence condition for natural partial realization is equivalent to the uniqueness condition for minimum partial realizations.
3) An algorithm to obtain a natural partial realization from a given partial two-dimensional image is given.

It is evident that the results for our systems are the same as those obtained in linear system theory.

Y. Hasegawa: Algebra. Approx. & Noisy Reali. of Discrete-Time Sys., LNEE 50, pp. 215–247.
springerlink.com

In order to be self-contained, we will state the main results obtained in [Hasegawa and Suzuki, 2006].

In this chapter, we discuss approximate and noisy realization problems over a finite field of residue class. For the real number field, we can discuss the problems in the same way as in previous chapters by using the facts obtained in the reference [Hasegawa and Suzuki, 2006]. Hence, we will omit the problems over the real number field.

Notations

N : the set of non-negative integers.
$N^2 := N \times N$: the product set in two sets of non negative integers.
N/pN : a finite field of residue class, where p is a prime number.
K : a field.
$K[z_\alpha,\ z_\beta]$: the commutative K-algebra of polynomials in two variables.
$K(z_\alpha,\ z_\beta)$: the field of rational function in two variables.
$F(X,\ Y)$: the set of all functions from X to Y.
$L(X,\ Y)$: the set of all linear maps from X to Y.
$L(X)$: the set of all linear maps from X to X.
K^n : the K-linear space of all n-vectors.
$K^{n\times n}$: the set of all $n \times n$-matrices.
im f : the image of a map f.
ker f : the kernel of a map f.
$\ll S \gg$: the smallest linear space which contains a set S.
$\mid A \mid$: the determinant of a square matrix A.

9.1 Commutative Linear Representation Systems

Definition 9.1. Commutative Linear Representation System
(1) A system given by the following equations is written as a collection $\sigma = ((X, F_\alpha, F_\beta), x^0, h)$ and it is called a commutative linear representation system.

$$\begin{cases} x(i+1,j) = F_\alpha x(i,j) \\ x(i,j+1) = F_\beta x(i,j) \\ x(0,0) \quad\ = x^0 \\ \gamma(i,j) \qquad = hx(i,j) \end{cases}$$

for any $i,j \in N$, $x(i,j) \in X$, $\gamma(i,j) \in Y$, where X is a linear space over the field K. F_α and F_β are linear operators on X which satisfy $F_\alpha F_\beta = F_\beta F_\alpha$. $x^0 \in X$ is an initial state. $h : X \to Y$ is a linear operator.
(2) The two-dimensional image $a_\sigma : N \times N \to Y; (i,j) \mapsto hF_\alpha^i F_\beta^j x^0$ is called the behavior of σ.
(3) For a two-dimensional image $a \in F(N \times N, Y)$, σ which satisfies $a_\sigma = a$ is called a realization of a.
(4) A commutative linear representation system σ is said to be quasi reachable if the linear hull of the reachable set $\{F_\alpha^i F_\beta^j x^0;\ i,j \in N\}$ equals X.

(5) A commutative linear representation system σ is called distinguishable if $hF_\alpha^i F_\beta^j x_1 = hF_\alpha^i F_\beta^j x_2$ for any $i, j \in N$ implies $x_1 = x_2$.
(6) A commutative linear representation system σ is called canonical if σ is quasi-reachable and distinguishable.

Remark 1: The $x(i,j)$ in the system equation of σ is the state that produces output value of a_σ at the place (i,j), while linear operator $h : X \to Y$ generates the output value $a_\sigma(i,j)$ at the place (i,j).

Remark 2: σ realizes a two-dimensional image a implies that σ is a faithful model for a.

Remark 3: Notice that a canonical commutative linear representation system $\sigma = ((X, F_\alpha, F_\beta), x^0, h)$ is a system with the most reduced space X among systems that have the behavior a_σ.

Example 9.2. 1) Let $K[z_\alpha, z_\beta]$ be a set of $K-$valued polynomials in two variables z_α, z_β. Let a linear operator z_α be $K[z_\alpha, z_\beta] \to K[z_\alpha, z_\beta]$; $\lambda \mapsto z_\alpha\lambda$, and let a linear operator z_β be $K[z_\alpha, z_\beta] \to K[z_\alpha, z_\beta]$; $\lambda \mapsto z_\beta\lambda$.
For any two-dimensional image a and the unit element $\mathbf{1} \in K[z_\alpha, z_\beta]$, $((K[z_\alpha, z_\beta], z_\alpha, z_\beta), \mathbf{1}, a)$ is a quasi-reachable commutative linear representation system which realizes a.
2) Let $F(N \times N, Y)$ be a set of any two-dimensional images.
For any two-dimensional image $a \in F(N \times N, Y)$, let
$S_\alpha a : N \times N \to Y; (i,j) \mapsto a(i+1, j)$ and
$S_\beta a : N \times N \to Y; (i,j) \mapsto a(i, j+1)$.
Then $S_\alpha, S_\beta \in L(F(N \times N, Y))$ and $S_\alpha S_\beta = S_\beta S_\alpha$ hold.

Let $(0,0) : F(N \times N, Y) \to Y; a \mapsto a(0,0)$ be a linear operator, and let a be any two-dimensional image. Then $((F(N \times N, Y), S_\alpha, S_\beta), a, (0,0))$ is a distinguishable commutative linear representation system which realizes a.

Theorem 9.3. *The following commutative linear representation systems are canonical realizations of any two-dimensional image $a \in F(N \times N, Y)$.*

1) $((K[z_\alpha, z_\beta]/_{=a}, \dot{z_\alpha}, \dot{z_\beta}), [\mathbf{1}], \dot{a})$,
where $K[z_\alpha, z_\beta]/_{=a}$ is a quotient space obtained by the following equivalence relation:
$$\sum_{i,j} \lambda_1(i,j) z_\alpha^i z_\beta^j = \sum_{i,j} \lambda_2(i,j) z_\alpha^i z_\beta^j \iff \sum_{i,j} \lambda_1(i,j) a(i,j) = \sum_{i,j} \lambda_2(i,j) a(i,j).$$
$\dot{z_\alpha}$ is given by a map $\dot{z_\alpha} : K[z_\alpha, z_\beta]/_{=a} \to K[z_\alpha, z_\beta]/_{=a}; [\lambda] \mapsto [z_\alpha\lambda]$,
$\dot{z_\beta}$ is given by a map $\dot{z_\beta} : K[z_\alpha, z_\beta]/_{=a} \to K[z_\alpha, z_\beta]/_{=a}; [\lambda] \mapsto [z_\beta\lambda]$ and
$\dot{a}$ is given by $\dot{a} : K[z_\alpha, z_\beta]/_{=a} \to Y$; $[\lambda] \mapsto \dot{a}([\lambda]) = \sum_{i,j} \lambda(i,j) a(i,j)$,
where $\lambda = \sum_{i,j} \lambda(i,j) z_\alpha^i z_\beta^j \in K[z_\alpha, z_\beta]$.

2) $((\ll S_\alpha^N S_\beta^N a \gg), S_\alpha S_\beta), a, (0,0))$,
where $\ll S_\alpha^N S_\beta^N a \gg$ *is the smallest linear space which contains* $S_\alpha^N S_\beta^N a := \{S_\alpha^i S_\beta^j a; (i,j) \in N \times N\}$.

Definition 9.4. Let $\sigma_1 = ((X_1, F_{\alpha_1} F_{\beta_1}), x_1^0, h_1)$ and $\sigma_2 = ((X_2, F_{\alpha_2} F_{\beta_2}), x_2^0, h_2)$ be commutative linear representation systems. Then a linear operator $T : X_1 \to X_2$ is said to be a commutative linear representation system morphism $T : \sigma_1 \to \sigma_2$ if T satisfies $TF_{\alpha_1} = F_{\alpha_2} T$, $TF_{\beta_1} = F_{\beta_2} T$, $Tx_1^0 = x_2^0$ and $h_1 = h_2 T$.
If $T : X_1 \to X_2$ is bijective, then $T : \sigma_1 \to \sigma_2$ is said to be an isomorphism.

Theorem 9.5. *Realization Theorem of Commutative Linear Representation Systems*
(1) Existence: For any two-dimensional image $a \in F(N \times N, Y)$, *there exist at least two canonical commutative linear representation systems which realize* a.
(2) Uniqueness: Let σ_1 *and* σ_2 *be any two canonical commutative linear representation systems that realize* $a \in F(N \times N, Y)$. *Then there exists an isomorphism* $T : \sigma_1 \to \sigma_2$.

We briefly introduce finite-dimensional commutative linear representation systems which can be treated by computer or non-linear circuits.

Commutative linear representation system $\sigma = ((X, F_\alpha, F_\beta), x^0, h)$ is called a finite (or n)-dimensional commutative linear representation system if the state space X is a finite (or n)-dimensional linear space.

Lemma 9.6. *For any image* $a \in F(N^2, Y)$, *the following three conditions are equivalent to each other.*
(1) a *has the behavior of a finite-dimensional canonical commutative linear representation system.*
(2) The quotient space $K[z_\alpha, z_\beta]/_{=a}$ *is finite-dimensional.*
(3) The linear space generated by $\{S_\alpha^i S_\beta^j a : i, j \in N\}$ *is finite-dimensional, where* $K[z_\alpha, z_\beta]/_{=a}$ *is a quotient space given by the following equivalence relations:* $a_1 = a_2 \Longleftrightarrow a_1(i,j) = a_2(i,j)$ *for any* $i, j \in N$.
Moreover $S_\alpha, S_\beta \in L(F(N^2, Y))$ *are given by* $S_\alpha a : N^2 \to Y$; $(i,j) \mapsto a(i+1, j)$ *and* $S_\beta a : N^2 \to Y; (i,j) \mapsto a(i, j+1)$.

9.2 Finite-Dimensional Commutative Linear Representation Systems

This section deals with the fundamental structures of finite-dimensional commutative linear representation systems based on the realization Theorem (9.5).

First, the conditions under which a finite-dimensional commutative linear representation system is canonical are given.

Second, the representation theorem for finite-dimensional canonical commutative linear representation systems is obtained. This involves demonstrating a standard systems are representatives in their equivalence class. The system is the Quasi-reachable Standard System.

Third, two criteria for the behavior of the finite-dimensional commutative linear representation systems are given. One is the rank condition of the infinite Hankel matrix, and the other is the application of Kleene's theorem from automata theory.

Finally, a procedure for obtaining the Quasi-reachable Standard System which realizes a given two-dimensional image is presented.

Corollary 9.7. *Let T be a commutative linear representation system morphism $T : \sigma_1 \to \sigma_2$. Then $a_{\sigma_1} = a_{\sigma_2}$ holds.*

There is a fact regarding finite-dimensional linear spaces that a n-dimensional linear space over the field K is isomorphic to K^n. Furthermore, $L(K^n, K^m)$ is isomorphic to $K^{m\times n}$. See Halmos [1958]. Therefore, without loss of generality, we can consider a n-dimensional commutative linear representation system as $\sigma = ((K^n, F_\alpha, F_\beta), x^0, h)$, where F_α, $F_\beta \in K^{n\times n}$, $x^0 \in K^n$ and $h \in K^{p\times n}$.

Theorem 9.8. *A commutative linear representation system $\sigma = ((K^n, F_\alpha, F_\beta), x^0, h)$ is canonical if and only if the following conditions 1) and 2) hold:*

1) $\text{rank}\,[x^0, F_\alpha x^0, F_\alpha^2 x^0, \cdots, F_\alpha^{n-1}x^0, F_\beta x^0, F_\alpha F_\beta x^0, F_\alpha^2 F_\beta x^0, \cdots, F_\alpha^{n-2}F_\beta x^0, F_\beta^2 x^0, F_\alpha F_\beta^2 x^0, F_\alpha^2 F_\beta^2 x^0, \cdots, F_\alpha^{n-3}F_\beta^2 x^0, F_\beta^3 x^0, F_\alpha F_\beta^3 x^0, F_\alpha^2 F_\beta^3 x^0, \cdots, F_\alpha^{n-4}F_\beta^3 x^0, \cdots, \cdots, F_\beta^{n-3}x^0, F_\alpha F_\beta^{n-3}x^0, F_\alpha^2 F_\beta^{n-3}x^0, F_\beta^{n-2}x^0, F_\alpha F_\beta^{n-2}x^0, F_\beta^{n-1}x^0] = n.$

2) $\text{rank}\,[h^T, (hF_\alpha)^T, (hF_\alpha^2)^T, \cdots, (hF_\alpha^{n-1})^T, (hF_\beta)^T, (hF_\alpha F_\beta)^T, (hF_\alpha^2 F_\beta)^T, \cdots, (hF_\alpha^{n-2}F_\beta)^T, \cdots, (hF_\beta^{n-3})^T, (hF_\alpha F_\beta^{n-3})^T, (hF_\alpha^2 F_\beta^{n-3})^T, (hF_\beta^{n-2})^T, (hF_\alpha F_\beta^{n-2})^T, (hF_\beta^{n-1})^T] = n.$

Definition 9.9. A canonical commutative linear representation system $\sigma_s = ((K^n, F_{\alpha_s}, F_{\beta_s}), \mathbf{e}_1, h_s)$ is said to be a Quasi-reachable Standard System with a vector index $\nu = (\nu_1, \nu_2, \cdots, \nu_k)$ if the following conditions hold:

1) An integer ν_j $(1 \le j \le k)$ satisfies $n = \sum_{j=1}^{k} \nu_j$ and $0 \le \nu_k \le \nu_{k-1} \le \cdots \le \nu_2 \le \nu_1$.

2) For any i, j $(1 \le j \le k,\ 1 \le i \le \nu_j)$,
$F_\beta^{j-1} F_\alpha^{i-1}\mathbf{e}_1 = \mathbf{e}_{\nu_1+\nu_2+\cdots+\nu_{j-1}+i}$,
$F_\beta^{j-1} F_\alpha^{\nu_j}\mathbf{e}_1 = \sum_{m=1}^{j}\sum_{l=1}^{\nu_m} c_{ml}^{j} F_\beta^{m-1}F_\alpha^{l-1}\mathbf{e}_1$, where $c_{ml}^{j} \in K$,
$F_\beta^{k}\mathbf{e}_1 = \sum_{m=1}^{k}\sum_{l=1}^{\nu_m} c_{ml}^{k+1} F_\beta^{m-1}F_\alpha^{l-1}\mathbf{e}_1$, where $c_{ml}^{k+1} \in K$,

where $\mathbf{e}_i = [0, \cdots, 0, \overset{i}{1}, 0, \cdots, 0]^T \in K^n$, and T denotes the transposition of matrices or vectors.

The $F_{\alpha s}$ and $F_{\beta s}$ of the Quasi-reachable Standard System with a vector index $\nu = (\nu_1, \nu_2, \cdots, \nu_k)$ are characterized by Fig. 9.1 and Fig. 9.2.

Each $\mathbf{c}^i$ in the figures is given by $\mathbf{c}^i := [c^i_{11}, \cdots, c^i_{1\nu_1}, \cdots, c^i_{i1}, \cdots, c^i_{i\nu_i}, \mathbf{0}]^T$ for $1 \le i \le k,\ 1 \le j \le i$, and $\mathbf{c}^{k+1} := [c^{k+1}_{11}, \cdots, c^{k+1}_{1\nu_1}, \cdots, c^{k+1}_{k1}, \cdots, c^{k+1}_{k\nu_k}]^T$.

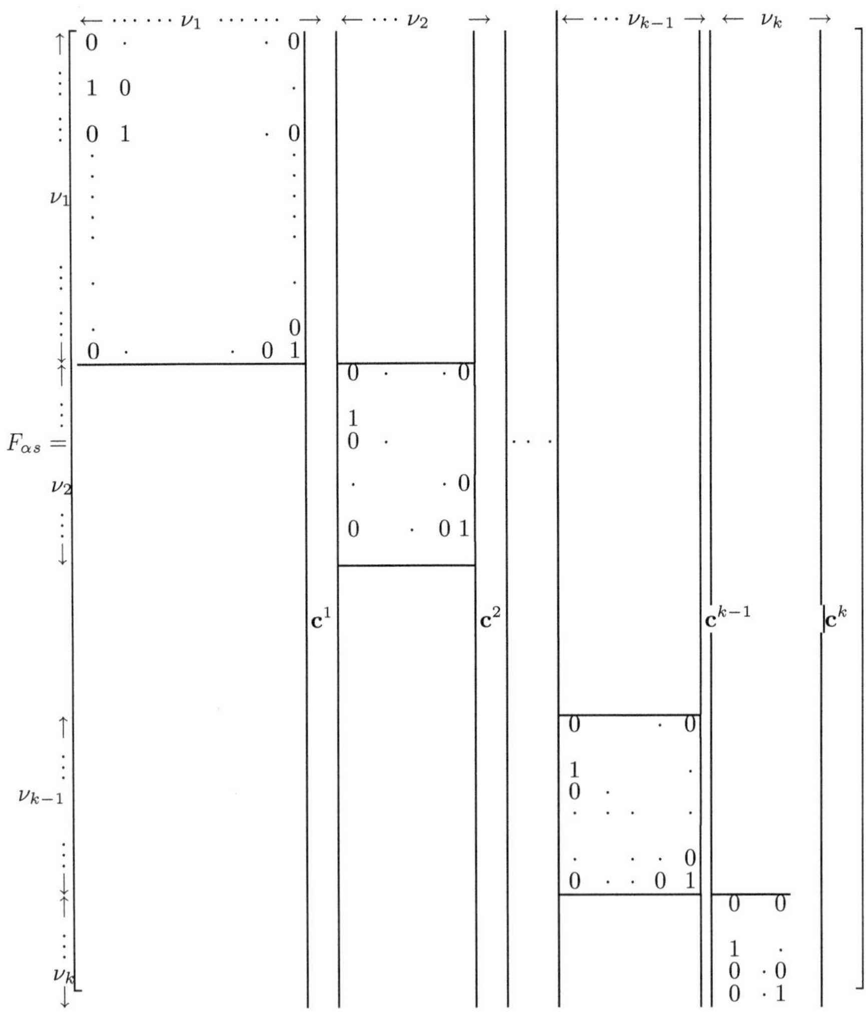

Fig. 9.1 $F_{\alpha s}$ of the Quasi-reachable Standard System σ_s defined in Definition (9.9)

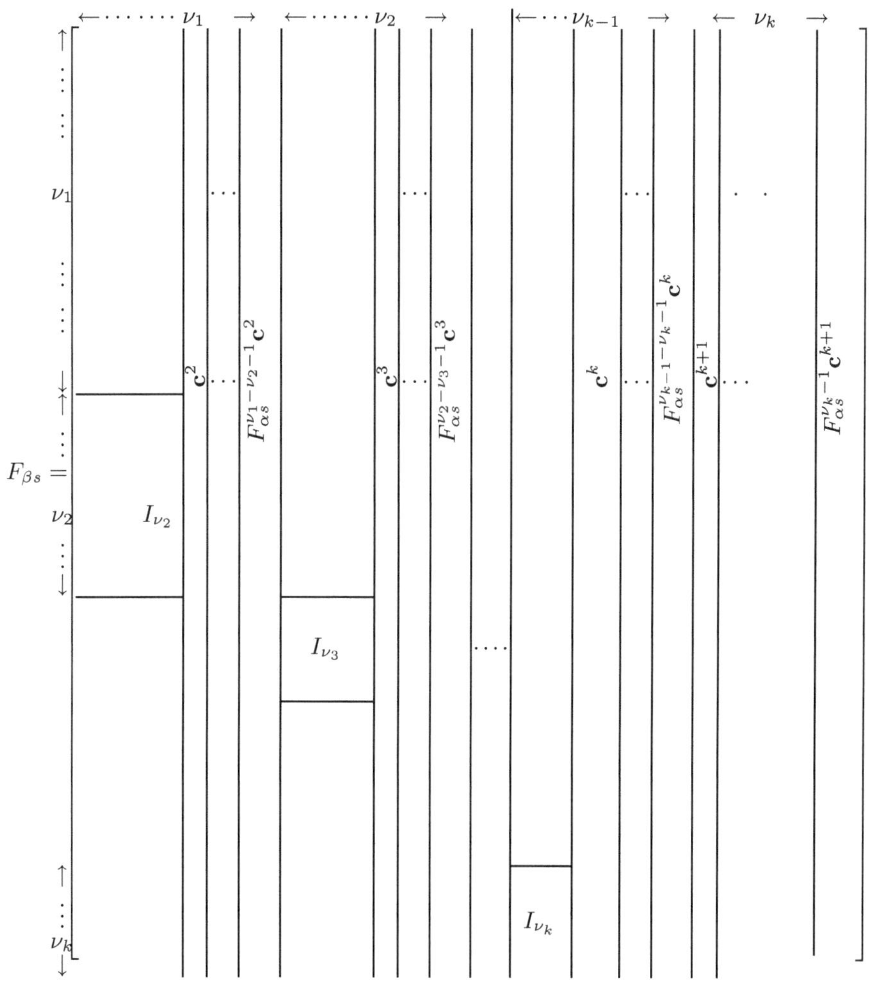

Fig. 9.2 $F_{\beta s}$ of the Quasi-reachable Standard System σ_s defined in Definition (9.9)

Theorem 9.10. *Representation Theorem for equivalence classes*
For any finite-dimensional canonical commutative linear representation system, there exists a uniquely determined isomorphic Quasi-reachable Standard System.

Definition 9.11. For any two-dimensional image $a \in F(N \times N, Y)$, the corresponding linear input/output map $A : (K[z_\alpha, z_\beta], z_\alpha, z_\beta) \to (F(N \times N, Y), S_\alpha, S_\beta)$ satisfies $A(z_\alpha^i z_\beta^j) = S_\alpha^i S_\beta^j a$ for $i,\ j \in N$.

Hence, A can be represented by the following infinite matrix H_a. This H_a is said to be a Hankel matrix of a:

$$H_a = \begin{array}{c} \\ \\ \\ (\tilde{i}, \tilde{j}) \end{array} \left(\begin{array}{ccc} & & (i,j) \\ & & \vdots \\ & & \vdots \\ & & \vdots \\ \cdots & \cdots & a(\tilde{i}+i, \tilde{j}+j) \\ & & \end{array} \right)$$

Note that the column vectors of the Hankel matrix H_a are represented by $S_\alpha^i S_\beta^j a$ for $i,\ j \in N$. Moreover, $a(\tilde{i}+i, \tilde{j}+j)$=$(0,0)S_\alpha^{\tilde{i}} S_\beta^{\tilde{j}} S_\alpha^i S_\beta^j a$ holds.
See Example (9.2) regarding it.

Theorem 9.12. *Theorem for existence criteria*
For a two-dimensional image $a \in F(N \times N, Y)$, the following conditions are equivalent:
1) The two-dimensional image $a \in F(N \times N, Y)$ is the behavior of the n-dimensional canonical commutative linear representation system.
2) There exist n linearly independent vectors and no more than n linearly independent vectors in a set $\{S_\alpha^i S_\beta^j a;\ i+j \le n-1 \text{ for } i,j \in N\}$.
3) The rank of the Hankel matrix H_a is n.

Let $K[z_\alpha, z_\beta]$ have the following operation $\times$.

$$\begin{aligned}
\times : & K[z_\alpha, z_\beta] \times K[z_\alpha, z_\beta] \to K[z_\alpha, z_\beta]; \\
& \left(\textstyle\sum_{i_1,j_1} \lambda_1(i_1,j_1) z_\alpha^{i_1} z_\beta^{j_1}, \sum_{i_2,j_2} \lambda_2(i_2,j_2) z_\alpha^{i_2} z_\beta^{j_2}\right) \\
& \quad \mapsto \textstyle\sum_{i=i_1+i_2, j=j_1+j_2} \lambda(i,j) z_\alpha^i z_\beta^j \\
& \quad = \left(\textstyle\sum_{i_1,j_1} \lambda_1(i_1,j_1) z_\alpha^{i_1} z_\beta^{j_1}\right) \times \left(\textstyle\sum_{i_2,j_2} \lambda(i_2,j_2) z_\alpha^{i_2} z_\beta^{j_2}\right).
\end{aligned}$$

Then $K[z_\alpha, z_\beta]$ is an algebra over K. Moreover $K[z_\alpha, z_\beta]$ is a free algebra over K.

$F(N \times N, K)$ can be considered to contain $K[z_\alpha, z_\beta]$, since $a \in F(N \times N, K)$ can be expressed as the formal power series $\bar{a} = \sum_{i,j} a(i,j) z_\alpha^i z_\beta^j$.

Any two-dimensional image $a \in F(N \times N, Y)$ is also expressed by the following formal power series, where $Y = K^p$: $\bar{a} = \sum_{i=0}^{\infty} \sum_{j=0}^{\infty} a(i,j) z_\alpha^{-i} z_\beta^{-j}$.

Theorem 9.13. *A two-dimensional image $a \in F(N \times N, K^p)$ is the behavior of a finite-dimensional commutative linear representation system if and only if the formal power series $\bar{a} = [\bar{a_1} \bar{a_2} \cdots \bar{a_p}]^T$ is expressed as follows:*

$$\bar{a_k} = \frac{z_\alpha z_\beta \left(\sum_{j=0}^m \sum_{i=0}^n \lambda_k(i,j) z_\alpha^i z_\beta^j \right)}{q_\alpha(z_\alpha) q_\beta(z_\beta)},$$

where k $(1 \le k \le p)$ is an integer, $q_\alpha(z_\alpha)$ and $q_\beta(z_\beta)$ are monic polynomials in z_α with order n, and a monic polynomial in z_β with order m, respectively. Furthermore, $\lambda_k(i,j) \in K$,

Theorem 9.14. *Theorem for a realization procedure*
Let a two-dimensional image $a \in F(N \times N, Y)$ satisfy the condition of Theorem (9.12). Then the Quasi-reachable Standard System $\sigma = ((K^n, F_{\alpha_s}, F_{\beta_s}), \mathbf{e}_1, h_s)$ which realizes the two-dimensional image a is obtained by the following procedure:

1) *Find an integer ν_1 and coefficients $\{c^1_{1l}; 1 \leq l \leq \nu_1\}$ such that the vectors $\{S^i_\alpha a; 1 \leq i \leq \nu_1 - 1\}$ of the set $\{S^i_\alpha a; i \leq n-1, i \in N\}$ are linearly independent and $S^{\nu_1}_\alpha a = \sum_{l=1}^{\nu_1} c^1_{1l} S^{l-1}_\alpha a$.*
2) *Find an integer ν_2 and coefficients $\{c^2_{ml}; 1 \leq l \leq \nu_m, 1 \leq m \leq 2\}$ such that the vectors $\{S^{j-1}_\beta S^{i-1}_\alpha a; 1 \leq i \leq \nu_j - 1, 1 \leq j \leq 2\}$ of the set $\{S^j_\beta S^i_\alpha a; i \leq n-1, j \leq n-2 \in N\}$ are linearly independent and $S_\beta S^{\nu_2}_\alpha a = \sum_{m=1}^{2} \sum_{l=1}^{\nu_m} c^2_{ml} S^{m-1}_\beta S^{l-1}_\alpha a$.*

⋮

k) *Find an integer ν_k and coefficients $\{c^k_{ml}; 1 \leq l \leq \nu_l, 1 \leq m \leq k\}$ such that the vectors $\{S^{j-1}_\beta S^{i-1}_\alpha a; 1 \leq i \leq \nu_j, 1 \leq j \leq k\}$ of the set $\{S^j_\beta S^i_\alpha a; i \leq n-1, j \leq k-1 \in N\}$ are linearly independent, $S^{k-1}_\beta S^{\nu_k}_\alpha a = \sum_{m=1}^{k} \sum_{l=1}^{\nu_m} c^k_{ml} S^{m-1}_\beta S^{l-1}_\alpha a$ and $S^k_\beta a = \sum_{m=1}^{k} \sum_{l=1}^{\nu_m} c^{k+1}_{ml} S^{m-1}_\beta S^{l-1}_\alpha a$.*

k+1) *Let the state space be K^n, and let the initial state be $\mathbf{e}_1$, where $n = \sum_{i=1}^{k} \nu_i$.*

k+2) *Let matrices F_{α_s} and F_{β_s} be those given in Fig. 9.1 and 9.2.*

k+3) *Let the output map h_s be*
$$h_s = [a(0,0), a(1,0), \cdots, a(\nu_1 - 1, 0), a(0,1), \cdots, a(\nu_2 - 1, 1), \cdots, a(0, k-1), \cdots, a(\nu_k - 1, k-1)].$$

9.3 Partial Realization Theory of Two-Dimensional Images

Here we consider a partial realization problem for two-dimensional images, namely, we will obtain a commutative linear representation system which describes given finite-sized two-dimensional images. Let $\underline{a}$ be an $(L+1) \times (M+1)$ sized two-dimensional image, that is, $\underline{a} \in F(\mathbf{L} \times \mathbf{M}, Y)$, where $L, M \in N$, $\mathbf{L} := \{0, 1, \cdots, L-1, L\}$ and $\mathbf{M} := \{0, 1, \cdots, M-1, M\}$. The $\underline{a}$ is said to be a finite-sized two-dimensional image. A finite-dimensional commutative linear representation system $\sigma = ((X, F_\alpha, F_\beta), x^0, h)$ is called a partial realization of $\underline{a}$ if $hF^i_\alpha F^j_\beta x^0 = \underline{a}(i,j)$ holds for any $(i,j) \in \mathbf{L} \times \mathbf{M}$.

A partial realization problem of commutative linear representation systems can be stated as follows:
< For any given $\underline{a} \in F(\mathbf{L} \times \mathbf{M}, Y)$, find a partial realization σ of $\underline{a}$ such

that the dimension of state space X of σ is minimum. This σ is said to be a minimal partial realization of $\underline{a}$. Moreover, show that the minimal partial realizations are unique modulo isomorphisms.>

Proposition 9.15. *For any given $\underline{a} \in F(\mathbf{L} \times \mathbf{M}, Y)$, there always exists a minimal partial realization of $\underline{a}$.*

Minimal partial realizations are, in general, not unique modulo isomorphisms. Therefore, a natural partial realization will be introduced, and it will be shown that natural partial realizations exist if and only if they are isomorphic.

Definition 9.16. For a commutative linear representation system $\sigma = ((X, F_\alpha, F_\beta), x^0, h)$ and some $l_1, m_1 \in N$, if $X = \ll \{F_\alpha^i F_\beta^j x^0; i \le l_1,\ j \le m_1\} \gg$, then σ is said to be (l_1, m_1)-quasi-reachable, where $\ll S \gg$ denotes the smallest linear space which contains a set S.

Let l_2, m_2 be some integer. If $hF_\alpha^i F_\beta^j x = 0$ implies $x = 0$ for any $i \le l_2$ and $j \le m_2$, then σ is said to be (l_2, m_2)-distinguishable.

For a given $\underline{a} \in F(\mathbf{L} \times \mathbf{M}, Y)$, if there exist l_1, m_1 and $l_2, m_2 \in N$ such that $l_1 + l_2 < L$, $m_1 + m_2 < M$ and σ of its partial realization is (l_1, m_1)-quasi-reachable and (l_2, m_2)-distinguishable, then σ is said to be a natural partial realization of $\underline{a}$.

For a partial finite-sized image $\underline{a} \in F(\mathbf{L} \times \mathbf{M}, Y)$, the following matrix, $H_{\underline{a}\,(l_1, m_1, L-l_1, M-m_1)}$, is said to be a finite-sized Hankel matrix of $\underline{a}$.

$$H_{\underline{a}\,(l_1,m_1,L-l_1,M-m_1)} = \begin{array}{c} \\ \\ \\ (l,m) \end{array} \left(\begin{array}{ccc} & & (i,j) \\ & & \vdots \\ & & \vdots \\ \cdots & \cdots & \underline{a}(i+l, j+m) \end{array} \right),$$

where $0 \le i \le l_1,\ 0 \le j \le m_1,\ 0 \le l \le L - l_1,\ 0 \le m \le M - m_1$.

Theorem 9.17. *Let $H_{\underline{a}\,(l_1,m_1,L-l_1,M-m_1)}$ be the finite Hankel matrix of $\underline{a} \in F(\mathbf{L} \times \mathbf{M}, Y)$. Then there exists a natural partial realization of $\underline{a}$ if and only if the following conditions hold:*

$$\begin{aligned} &\operatorname{rank} H_{\underline{a}\,(l_1,m_1,L-l_1,M-m_1)} \\ &= \operatorname{rank} H_{\underline{a}\,(l_1+1,m_1,L-l_1-1,M-m_1-1)} \\ &= \operatorname{rank} H_{\underline{a}\,(l_1,m_1+1,L-l_1,M-m_1-1)} \\ &= \operatorname{rank} H_{\underline{a}\,(l_1,m_1+1,L-l_1-1,M-m_1-1)} \\ &= \operatorname{rank} H_{\underline{a}\,(l_1,m_1,L-l_1-1,M-m_1-1)} \qquad \textit{for some } l_1 \in \mathbf{L}, m_1 \in \mathbf{M}. \end{aligned}$$

Theorem 9.18. *There exists a natural partial realization of a given partial finite-sized image $\underline{a} \in F(\mathbf{L} \times \mathbf{M}, Y)$ if and only if the minimal partial realizations of $\underline{a}$ are unique modulo isomorphisms.*

In order to discuss the partial realization problem for finite-sized two-dimensional images, define the following operators $\underline{S_\alpha}$ and $\underline{S_\beta}$ as:
$\underline{S_\alpha}: F(\mathbf{L}\times\mathbf{M},Y)\to F((\mathbf{L}-\mathbf{1})\times\mathbf{M},Y); \underline{a}\mapsto \underline{S_\alpha}\underline{a}\ [;(i,j)\mapsto \underline{a}(i+1,j)]$,
$\underline{S_\beta}: F(\mathbf{L}\times\mathbf{M},Y)\to F(\mathbf{L}\times(\mathbf{M}-\mathbf{1}),Y); \underline{a}\mapsto \underline{S_\beta}\underline{a}\ [;(i,j)\mapsto \underline{a}(i,j+1)]$.

Then the column vectors in $H_{\underline{a}\ (l_1,m_1,L-l_1,M-m_1)}$ are expressed as $\underline{S_\alpha^i}\underline{S_\beta^j}\underline{a}$ for $0\le i\le l_1$ and $0\le j\le m_1$.

Theorem 9.19. *Let a partial finite-sized image be $\underline{a}\in F(\mathbf{L}\times\mathbf{M},Y)$. There exists a natural partial realization of $\underline{a}$ if and only if the Quasi-reachable Standard System $\sigma_s = ((K^n,F_{\alpha s},F_{\beta s}),\mathbf{e}_1,h_s)$ which realizes $\underline{a}$ can be obtained by the following algorithm.*
Here, n is given by $n:=\sum_{i=1}^{k}\nu_i$.
1)
Check the independences of column vectors of the finite-sized Hankel matrices $H_{\underline{a}\ (0,0,L,M)}$, $H_{\underline{a}\ (1,0,L-1,M)}$, $H_{\underline{a}\ (2,0,L-2,M)},\cdots$, in turn.
Find the smallest integer ν_1 such that all column vectors $\{\underline{S_\alpha}^{i-1}\underline{a};1\le i\le\nu_1\}$ in $F((\mathbf{L}-\nu_1+\mathbf{1})\times\mathbf{M},Y)$ are linearly independent and column vectors $\{\underline{S_\alpha}^{i}\underline{a};0\le i\le\nu_1\}$ in $F((\mathbf{L}-\nu_1)\times\mathbf{M},Y)$ are linearly dependent.
Determine a set of coefficients $\{c^1_{1l};1\le l\le\nu_1\}$ such that $\underline{S_\alpha}^{\nu_1}\underline{a}=\sum_{l=1}^{\nu_1}c^1_{1l}\underline{S_\alpha}^{l-1}\underline{a}$ holds in the sense of $F((\mathbf{L}-\nu_1)\times\mathbf{M},Y)$.
2)
If column vectors $\{\underline{S_\alpha}^{i-1}\underline{a};1\le i\le\nu_1\}$ in $F((\mathbf{L}-\nu_1+\mathbf{1})\times(\mathbf{M}-\mathbf{1}),Y)$ in a finite-sized Hankel matrix $H_{\underline{a}\ (\nu_1-1,1,L-\nu_1+1,M-1)}$ are linearly dependent, then stop this algorithm.
Otherwise, find the smallest integer ν_2 such that column vectors
$\{\underline{S_\beta}^{m-1}\underline{S_\alpha}^{l-1}\underline{a}\in F((\mathbf{L}-\nu_1+\mathbf{1})\times(\mathbf{M}-\mathbf{1}),Y);1\le m\le 2,\ 1\le l\le\nu_m\}$
in a finite-sized Hankel matrix $H_{\underline{a}\ (\nu_1-1,1,L-\nu_1+1,M-1)}$ are linearly independent and column vectors
$\{\underline{S_\beta}^{m-1}\underline{S_\alpha}^{l-1}\underline{a},\ \underline{S_\beta}\underline{S_\alpha}^{\nu_2}\underline{a}\in F((\mathbf{L}-\nu_1)\times(\mathbf{M}-\mathbf{1}),Y);1\le m\le 2,\ 1\le l\le\nu_m\}$
in a finite-sized Hankel matrix $H_{\underline{a}\ (\nu_1,1,L-\nu_1,M-1)}$ are linearly dependent.
Determine a set of coefficients $\{c^2_{ml};1\le m\le 2,\ 1\le l\le\nu_m\}$ such that $\underline{S_\beta}\underline{S_\alpha}^{\nu_2}\underline{a}=\sum_{m=1}^{2}\sum_{l=1}^{\nu_m}c^2_{ml}\underline{S_\beta}^{m-1}\underline{S_\alpha}^{l-1}\underline{a}$ holds in the sense of $F((\mathbf{L}-\nu_1)\times(\mathbf{M}-\mathbf{1}),Y)$.

⋮

k)
If column vectors $\{\underline{S_\beta}^{m-1}\underline{S_\alpha}^{l-1}\underline{a}\in F((\mathbf{L}-\nu_1+\mathbf{1})\times(\mathbf{M}-\mathbf{k}+\mathbf{1}),Y);1\le m\le k,\ 1\le l\le\nu_m\}$ in a finite-sized Hankel matrix $H_{\underline{a}\ (\nu_1-1,k-1,L-\nu_1+1,M-k+1)}$ are linearly dependent, then stop this algorithm.
Otherwise find the smallest integer ν_k such that column vectors
$\{\underline{S_\beta}^{m-1}\underline{S_\alpha}^{l-1}\underline{a}\in F((\mathbf{L}-\nu_1+\mathbf{1})\times(\mathbf{M}-\mathbf{k}+\mathbf{1}),Y);1\le m\le k,\ 1\le l\le\nu_m\}$
in a finite-sized Hankel matrix $H_{\underline{a}\ (\nu_1-1,k-1,L-\nu_1+1,M-k+1)}$ are linearly inde-

pendent and column vectors
$\{\underline{S_\beta}^{m-1}\underline{S_\alpha}^{l-1}\underline{a},\ \underline{S_\beta}^{k-1}\underline{S_\alpha}^{\nu_k}\underline{a} \in F((\mathbf{L}-\nu_1)\times(\mathbf{M}-\mathbf{k}+\mathbf{1}),Y); 1 \le m \le k,\ 1 \le l \le \nu_m\}$ *in a finite-sized Hankel matrix* $H_{\underline{a}\ (\nu_1-1,k-1,L-\nu_1,M-k+1)}$ *are linearly dependent.*
Determine a set of coefficients $\{c^k_{ml}; 1 \le m \le k,\ 1 \le l \le \nu_m\}$ *such that* $\underline{S_\beta}^{k-1}\underline{S_\alpha}^{\nu_k}\underline{a} = \sum_{m=1}^{k}\sum_{l=l}^{\nu_m} c^k_{ml}\underline{S_\beta}^{m-1}\underline{S_\alpha}^{l-1}\underline{a}$ *holds in the sense of*
$F((\mathbf{L}-\nu_1)\times(\mathbf{M}-\mathbf{k}+\mathbf{1}),Y)$.

k+1)
If column vectors $\{\underline{S_\beta}^{m-1}\underline{S_\alpha}^{l-1}\underline{a} \in F((\mathbf{L}-\nu_1+\mathbf{1})\times(\mathbf{M}-\mathbf{k}),Y); 1 \le m \le k,\ 1 \le l \le \nu_m\}$ *in a finite-sized Hankel matrix* $H_{\underline{a}\ (\nu_1-1,k,L-\nu_1+1,M-k)}$ *are linearly dependent, then stop this algorithm.*
Otherwise find the smallest integer k such that column vectors
$\{\underline{S_\beta}^{k}\underline{a},\ \underline{S_\beta}^{m-1}\underline{S_\alpha}^{l-1}\underline{a} \in F((\mathbf{L}-\nu_1+\mathbf{1})\times(\mathbf{M}-\mathbf{k}),Y); 1 \le m \le k,\ 1 \le l \le \nu_m\}$ *in a finite-sized Hankel matrix* $H_{\underline{a}\ (\nu_1-1,k-1,L-\nu_1+1,M-k)}$ *are linearly dependent.*
Determine a set of coefficients $\{c^{k+1}_{ml}; 1 \le m \le k,\ 1 \le l \le \nu_m\}$ *such that* $\underline{S_\beta}^{k}\underline{a} = \sum_{m=1}^{k}\sum_{l=1}^{\nu_m} c^{k+1}_{ml}\underline{S_\beta}^{m-1}\underline{S_\alpha}^{l-1}\underline{a}$ *holds in the sense of*
$F((\mathbf{L}-\nu_1+\mathbf{1})\times(\mathbf{M}-\mathbf{k}),Y)$.

k+2)
For a set of obtained coefficients $\{c^j_{ml} \in K; 1 \le m \le k,\ 1 \le l \le \nu_m\}$ *for* $1 \le j \le k+1$, *set* $\mathbf{c}^j_i = [c^j_{i1}, c^j_{i2}, \cdots, c^j_{i\nu_i}]^T \in K^{\nu_i}$ $(1 \le i \le k)$, $\mathbf{c}^j = [\mathbf{c}^j_1, \mathbf{c}^j_2, \cdots, \mathbf{c}^j_k, 0, 0, \cdots, 0]^T$ *and* $\mathbf{c}^{k+1} = [\mathbf{c}^{k+1}_1, \mathbf{c}^{k+1}_2, \cdots, \mathbf{c}^{k+1}_k]^T$.

k+3)
Insert the vectors $\mathbf{c}^j$ $(1 \le j \le k+1)$ *obtained in the k+2) step into the* $F_{\alpha s}$ *and* $F_{\beta s}$ *of Fig. 9.1 and 9-2.*

k+4)
Let $h_s = [\underline{a}(0,0), \underline{a}(1,0), \cdots, \underline{a}(\nu_1-1,0), \underline{a}(0,1), \cdots, \underline{a}(\nu_2-1,1), \cdots, \underline{a}(0,k-1), \cdots, \underline{a}(\nu_k-1,k-1)]$.

9.4 Measurement Data with Approximate and Noisy Error

In this section, we will discuss the case where noises are added to two-dimensional axes.

For observed values $a(i,j) \in K^p$ of two-dimensional axes, a p-dimensional observed signal $\hat{a}(i,j) \in K^p$ and an additive noise $\bar{a}(i,j) \in K^p$ can be considered as the following equation: $a(i,j) = \hat{a}(i,j) + \bar{a}(i,j)$, where $i,\ j \in N$ and K is a finite field of residue class.

For a noisy case, the data processing problem can be stated roughly as follows:

For any given data $\{a(i,j) : i \le l, j \le m \text{ for some } l,\ m \in N\}$, find the signal $\{\hat{a}(i,j) : i \le l, j \le m\}$ which is the output of a dynamical system.

9.5 Analyses for Approximate and Noisy Data

In this section, we provide how to determine coffecients of a linear combination over a finite field.

Proposition 9.20. *Let K be a field, and let a matrix $\hat{\mathbf{X}} := \hat{\mathbf{H}}^{\mathbf{T}}\hat{\mathbf{H}}$ for $\hat{\mathbf{H}} := [\hat{\mathbf{x}}_1, \hat{\mathbf{x}}_2, \cdots, \hat{\mathbf{x}}_n] \in K^{L\times n}$ and $\hat{\mathbf{x}}_n = \alpha_1\hat{\mathbf{x}}_1 + \alpha_2\hat{\mathbf{x}}_2 + \cdots + \alpha_{n-1}\hat{\mathbf{x}}_{n-1}$.*
Let $\phi(x)$ be the determinant of a matrix $\hat{\mathbf{X}} - x * \mathbf{I}$.
And let Q be the matrix composed from the null vector $\{\phi_i(\hat{\mathbf{X}}) : 1 \le i \le m\}$, where $\phi(x) = \phi_1(x)\cdots\phi_m(x)x$, each $\phi_i(x)$ is a irreducible polynomial over a field K. Then $[\alpha_1, \alpha_2, \cdots, \alpha_{n-1}]^T = \mathbf{R}_{11}^{-1}\mathbf{R}_{12}$ holds for $\mathbf{Q}^{-1} := \begin{bmatrix} \mathbf{R}_{11} & \mathbf{R}_{12} \\ \mathbf{R}_{21} & \mathbf{R}_{22} \end{bmatrix}$, where $\mathbf{R}_{11} \in K^{(n-1)\times(n-1)}$, $\mathbf{R}_{12} \in K^{(n-1)\times 1}$, $\mathbf{R}_{21} \in K^{1\times(n-1)}$ and $\mathbf{R}_{22} \in K$. And where, $\mathbf{Q}^{-1}\hat{\mathbf{X}}\mathbf{Q}$ is given as follows:

$$\mathbf{Q}^{-1}\hat{\mathbf{X}}\mathbf{Q} = \begin{bmatrix} \chi & \chi & \cdots & \chi & 0 \\ \chi & \chi & \cdots & \chi & 0 \\ \vdots & \ddots & \ddots & \vdots & \vdots \\ \chi & \cdots & \cdots & \chi & 0 \\ 0 & 0 & \cdots & 0 & 0 \end{bmatrix}.$$

[proof] Since $\mathbf{Q}^{-1}\hat{\mathbf{X}} = \mathbf{Q}^{-1}\hat{\mathbf{H}}^T\hat{\mathbf{H}} = \mathbf{Q}^{-1}\hat{\mathbf{H}}^T[\hat{\mathbf{x}}_1, \hat{\mathbf{x}}_2, \cdots, \hat{\mathbf{x}}_n]$ $= \begin{bmatrix} \chi & \chi & \cdots & \chi & 0 \\ \chi & \chi & \cdots & \chi & 0 \\ \vdots & \ddots & \ddots & \vdots & \vdots \\ \chi & \cdots & \cdots & \chi & 0 \\ 0 & 0 & \cdots & 0 & 0 \end{bmatrix} \mathbf{Q}^{-1}$, $\mathbf{Q}^{-1}\hat{\mathbf{X}}\mathbf{e}_i = \begin{bmatrix} \chi & \chi & \cdots & \chi & 0 \\ \chi & \chi & \cdots & \chi & 0 \\ \vdots & \ddots & \ddots & \vdots & \vdots \\ \chi & \cdots & \cdots & \chi & 0 \\ 0 & 0 & \cdots & 0 & 0 \end{bmatrix} \mathbf{Q}^{-1}\mathbf{e}_i$ holds for $\mathbf{e}_i = [0, \cdots, 0, \overset{i}{1}, 0, \cdots, 0]^T$ and $0 \le i \le n$.

Hence $\mathbf{Q}^{-1}\hat{\mathbf{H}}^T\hat{\mathbf{x}}_i = \begin{bmatrix} \chi & \chi & \cdots & \chi & 0 \\ \chi & \chi & \cdots & \chi & 0 \\ \vdots & \ddots & \ddots & \vdots & \vdots \\ \chi & \cdots & \cdots & \chi & 0 \\ 0 & 0 & \cdots & 0 & 0 \end{bmatrix} \mathbf{Q}^{-1}\mathbf{e}_i$ is obtained.

By using $\hat{\mathbf{x}}_n = \alpha_1\hat{\mathbf{x}}_1 + \alpha_2\hat{\mathbf{x}}_2 + \cdots + \alpha_{n-1}\hat{\mathbf{x}}_{n-1}$, $\mathbf{0} = \mathbf{Q}^{-1}\hat{\mathbf{H}}^T(\alpha_1\hat{\mathbf{x}}_1 + \alpha_2\hat{\mathbf{x}}_2 + \cdots + \alpha_{n-1}\hat{\mathbf{x}}_{n-1} - \hat{\mathbf{x}}_n) = \begin{bmatrix} \chi & \chi & \cdots & \chi & 0 \\ \chi & \chi & \cdots & \chi & 0 \\ \vdots & \ddots & \ddots & \vdots & \vdots \\ \chi & \cdots & \cdots & \chi & 0 \\ 0 & 0 & \cdots & 0 & 0 \end{bmatrix} \mathbf{Q}^{-1}[\alpha_1, \alpha_2, \cdots, \alpha_{n-1}, -1]^T$ can be obtained. Hence, $[\mathbf{R}_{11}, \mathbf{R}_{12}][\alpha_1, \alpha_2, \cdots, \alpha_{n-1}, -1]^T = 0$ holds. Therefore, $[\alpha_1, \alpha_2, \cdots, \alpha_{n-1}]^T = \mathbf{R}_{11}^{-1}\mathbf{R}_{12}$.

9.6 Non-linear Integer Programming for Digital Images

In this section, we will discuss how the approximate and noisy realization problems are solved. We assume that values of digital two-dimensional images are the numbers that belong to a finite field modulo a prime number. We only introduce a sensuous metric for the nearness. As for it, we use the mean value for the absolute value of the difference in a range. For our purpose, we will propose the following method called non-linear integer programing for digital images:

Find a commutative linear representstion system σ with a vector index $\nu = (\nu_1, \nu_2, \cdots, \nu_k)$ which satisfies

$$\underset{\substack{-l_1 \leq e(i,j) \leq l_1 \\ l_1,\ l_2 \in N \\ 0 \leq i \leq 2*\nu_1 - 1}}{Minimize} \quad \left(\sum_{i=0}^{m}\sum_{j=0}^{n} |a(i,j) - a_\sigma(i,j)|\right)/((m+1)(n+1))$$

$$\text{subject to} \quad \begin{matrix} |\tilde{H}_1| = 0 \\ \vdots \\ |\tilde{H}_k| = 0 \\ |\tilde{H}_{k+1}| = 0, \end{matrix}$$

where $|\ \tilde{H}_i\ |$ abbreviates the determinant of a partial Hankel matrix $\tilde{H}_i((0,0), \cdots, (0, \nu_1 - 1), ((1,0), \cdots, (1, \nu_2 - 1), \cdots, (i-1, 0), \cdots, (i-1, \nu_i))$ which is a non-linear function in integers $e[l, m]$ for $0 \leq l,\ m \leq k$ and $1 \leq i \leq k$. Especially, $|\ \tilde{H}_{k+1}\ |$ abbreviates the determinant of a partial Hankel matrix $\tilde{H}_{k+1}((0,0), \cdots, (0, \nu_1 - 1), ((1,0), \cdots, (1, \nu_2 - 1), \cdots, (i-1, 0), \cdots, (i-1, \nu_i - 1), (0, k))$. And $\tilde{H}_i((0,0), \cdots, (0, \nu_1 - 1), (1,0), \cdots, (1, \nu_2 - 1), \cdots, (i-1, 0), \cdots, (i-1, \nu_i))$ is given by the following:

$\tilde{H}_i((0,0), \cdots, (0, \nu_1 - 1), (1,0), \cdots, (1, \nu_2 - 1), \cdots, (i-1, 0), \cdots, (i-1, \nu_i))=$
$H_i((0,0), \cdots, (0, \nu_1 - 1), (1,0), \cdots, (1, \nu_2 - 1), \cdots, (i-1, 0), \cdots, (i-1, \nu_i))-$
$E_i((0,0), \cdots, (0, \nu_1 - 1), (1,0), \cdots, (1, \nu_2 - 1), \cdots, (i-1, 0), \cdots, (i-1, \nu_i))$,
where
$H_i((0,0), \cdots, (0, \nu_1 - 1), (1,0), \cdots, (1, \nu_2 - 1), \cdots, (i-1, 0), \cdots, (i-1, \nu_i))=$

$$\begin{bmatrix} a(0,0) & \cdot & a(0,\nu_1-1) & a(1,0) & \cdot & a(1,\nu_2-1) & \cdot & a(i-1,\nu_i) \\ \cdot & \cdot & \cdot & \cdot & & \cdot & & \cdot \\ a(0,\nu_1-1) & \cdot & a(0,2*\nu_1-2) & a(1,\nu_1-1) & \cdot & a(1,\nu_1+\nu_2-2) & \cdot & a(i-1,\nu_1+\nu_i-2) \\ a(1,0) & \cdot & a(1,\nu_1-1) & a(2,\nu_1-1) & \cdot & a(2,\nu_2-1) & \cdot & a(i,\nu_i) \\ \cdot & \cdot & \cdot & \cdot & \cdot & \cdot & \cdot & \cdot \\ a(1,\nu_2-1) & \cdot & a(1,\nu_1+\nu_2-2) & a(1,2*\nu_2-2) & \cdot & a(2,2*\nu_2-2) & \cdot & a(i,\nu_2+\nu_i-2) \\ \cdot & \cdot & \cdot & \cdot & \cdot & \cdot & \cdot & \cdot \\ a(i-1,\nu_i) & \cdot & \cdot & a(i,\nu_i+\nu_1-1) & \cdot & \cdot & \cdot & a(2*i-2,2*\nu_i) \end{bmatrix}$$

and
$E_i((0,0), \cdots, (0, \nu_1 - 1), (1,0), \cdots, (1, \nu_2 - 1), \cdots, (i-1, 0), \cdots, (i-1, \nu_i))=$

$$\begin{bmatrix} e(0,0) & \cdot & e(0,\nu_1-1) & e(1,0) & \cdot & e(1,\nu_2-1) & \cdot & e(i-1,\nu_i) \\ \cdot & \cdot & \cdot & \cdot & \cdot & \cdot & \cdot & \cdot \\ e(0,\nu_1-1) & \cdot & e(0,2*\nu_1-2) & e(1,\nu_1-1) & \cdot & e(1,\nu_1+\nu_2-2) & \cdot & e(i-1,\nu_1+\nu_i-2) \\ e(1,0) & \cdot & e(1,\nu_1-1) & e(2,\nu_1-1) & \cdot & e(2,\nu_2-1) & \cdot & e(i,\nu_i) \\ \cdot & \cdot & \cdot & \cdot & \cdot & \cdot & \cdot & \cdot \\ e(1,\nu_2-1) & \cdot & e(1,\nu_1+\nu_2-2) & e(1,2*\nu_2-2) & \cdot & e(2,2*\nu_2-2) & \cdot & e(i,\nu_2+\nu_i-2) \\ \cdot & \cdot & \cdot & \cdot & \cdot & \cdot & \cdot & \cdot \\ e(i-1,\nu_i) & \cdot & \cdot & e(i,\nu_i+\nu_1-1) & \cdot & \cdot & \cdot & e(2*i-2,2*\nu_i) \end{bmatrix}.$$

9.7 Algebraically Approximate Realization of Two-Dimensional Images

In this section, we discuss algebraically approximate realization problems of two-dimensional images.

For a finite field modulo a prime number, we cannot introduce the topology. Hence, we adopt that the approximate degree between signal and obtained signal error is judged by the difference of the smallness and largeness of the number.

Roughly speaking, the algebraically approximate realization of two-dimensional images can be stated as follows:

< For any given partial two-dimensional image, find, using only algebraic calculations, a commutative linear representation system which has a lower dimension than the given minimal state space of a commutative linear representation system and approximates the given data. >
In order to make our discussion simple only in this section, we assume that the set Y of output's value is the set $\boldsymbol{K}$ of $N/241N$ which is the quotient field modulo the prime number 241, namely 1-output. And the numbers 0 and 240 denote black and white respectively.

In order to obtain a solution, the non-linear integer programing for digital images discusssed in section 9.6 is used.

Theorem 9.21. *Algebraic algorithm for approximate realization*
Let an image a be a considered object which is a given commutative linear representation system. Then an approximate realization $\sigma = ((\boldsymbol{K}^n, F_{\alpha s}, F_{\beta s}), x^0, h_s)$ *of a is given by the following algorithm:*

1) Based on a relatively no change located near the zero of the characteristic polynomial from $H_{((0,0),(0,1),\cdots,(0,\nu_1-1))}$ *into* $H_{((0,0),(0,1),\cdots,(0,\nu_1-1),(0,\nu_1))}$*, determine the number* ν_1 *in a vector index* $\nu = (\nu_1)$*, namely, determine the value* ν_1 *in a vector index* $\nu = (\nu_1)$ *of an approximate realization* σ*.*

2) Based on a relatively no change located near the zero of the characteristic polynomial from $H_{((0,0),(0,1),\cdots,(0,\nu_1-1),(1,0),\cdots,(1,\nu_2-1)}$ *into* $H_{((0,0),(0,1),\cdots,(0,\nu_1-1),(1,0),\cdots,(1,\nu_2-1),(1,\nu_2))}$*, determine the number* ν_2 *in a vector index* $\nu = (\nu_1, \nu_2)$*.*

$\vdots$

k) Based on a relatively no change located near the zero of the characteristic polynomial from $H_{((0,0),(0,1),\cdots,(0,\nu_1-1),\cdots,(k-1,0),\cdots,(k-1,\nu_k-1)}$ *into*

$H_{((0,0),(0,1),\cdots,(0,\nu_1-1),\cdots,(k-1,0),\cdots,(k-1,\nu_k-1),(k-1,\nu_k))}$ and $H_{((0,0),(0,1),\cdots,(0,\nu_1-1),\cdots,(k-1,0),\cdots,(k-1,\nu_k-1),(k,0))}$, determine the number ν_k in a vector index $\nu = (\nu_1, \nu_2, \cdots, \nu_k)$.

k+1) The non-linear integer programing for digital images is used as follows: Using $|\tilde{H}_i| = 0$, determine vectors $\mathbf{c}^i$ for $1 \leq i \leq k+1$.

k+2) Determine $F_{\alpha s}$, $F_{\beta s}$, h_s and x_s^0 on the basis of Definition (9.9), Fig. 9.1 and Fig. 9.2 which has the least mean value for the absolute value of the difference in the range l_1, where the range l_1 is listed in Section 9.6.

Remark : For $|\tilde{H}_i| = 0$, refer to Section 9.6 Non-linear integer programing for digital images.

Example 9.22. Let a signal be a digital image generated by the following 3-dimensional commutative linear representation system $\sigma = ((K^3, F_\alpha, F_\beta), x^0, h)$ with a vector index $\nu = (3)$,

where $F_\alpha = \begin{bmatrix} 0 & 0 & 34 \\ 1 & 0 & 16 \\ 0 & 1 & 59 \end{bmatrix}$, $F_\beta = \begin{bmatrix} 104 & 195 & 111 \\ 76 & 54 & 148 \\ 27 & 223 & 197 \end{bmatrix}$, $x^0 = [1,0,0]^T$, $h = [12,7,3]$.

Then the approximate realization problem is solved by the following algorithm:

characteristic polynomial for	values of variable and polynomial					
$H_{((0,0),(0,1))}$	0	16	225	231	240	
	228	3	1	237	3	
$H_{((0,0),(0,1),(0,2))}$	0	8	10	12	228	
	4	9	235	238	13	
$H_{((0,0),(0,1),(0,2),(0,3))}$	0	1	3	223	234	239
	0	228	3	240	11	237
$H_{((0,0),(0,1),(1,0))}$	2	19	228	238		
	15	6	1	11		

Numbers in the upper stand (or the lower stand) denote the values of variable in the characteristic polynomial (or the polynomial values corresponding to the value in the upper stand).

1) As we can find a relatively no change located near zero of the characteristic polynomial from $H_{((0,0),(0,1))}$ into $H_{((0,0),(0,1),(0,2))}$, we determine the number ν_1 of dimension which is 2.
2) As we can find a relatively no change located near zero of the characteristic polynomial from $H_{((0,0),(0,1))}$ to $H_{((0,0),(0,1),(1,0))}$, we determine the number ν_2 of dimension which is 0.

Then the non-linear integer programing for digital images produces a commutative linear representative system $\sigma_s = (K^2, F_{\alpha s}, F_{\beta s}, x_s^0, h_s)$ with a vector index $\nu = (2)$ which has the least mean value 73.9 for the absolute value of the difference in the range 8.

Therefore, the quasi-reachable standard system $\sigma_s = (K^2, F_{\alpha s}, F_{\beta s}, x_s^0, h_s)$ with a vector index $\nu = (2)$ can be obtained as follows:

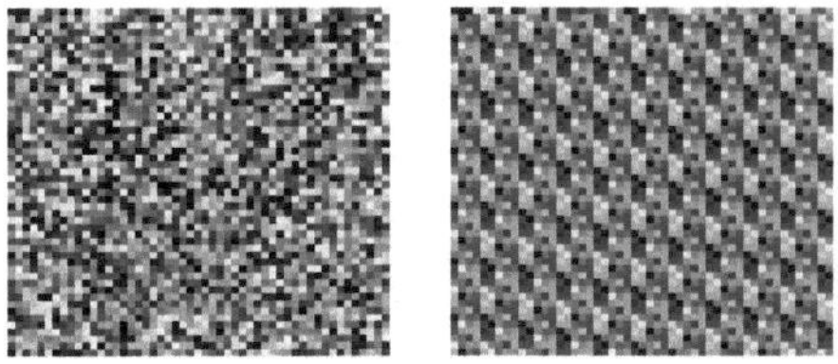

Fig. 9.3 In Example (9.22), the left is a 50×51 sized image of the original image a_σ for a given 3-dimensional commutative linear representation system σ. The right is a 50×51 sized image of an approximate system for the given σ.

$$F_{\alpha s} = \begin{bmatrix} 0 & 166 \\ 1 & 3 \end{bmatrix},\ F_{\beta s} = \begin{bmatrix} 156 & 115 \\ 127 & 55 \end{bmatrix},\ x^0 = [1,\ 0]^T, h = [10,\ 10].$$

Based on Fig. 9.3, it is felt that the image generated by the approximated system shows some patterns characterized by the original image.

Example 9.23. Let a signal be a digital image generated by the following 3-dimensional commutative linear representation system
$\sigma = ((K^3, F_\alpha, F_\beta), x^0, h)$ with a vector index $\nu = (3)$,

where $F_\alpha = \begin{bmatrix} 0 & 0 & 58 \\ 1 & 0 & 47 \\ 0 & 1 & 21 \end{bmatrix},\ F_\beta = \begin{bmatrix} 53 & 66 & 118 \\ 28 & 40 & 145 \\ 51 & 135 & 224 \end{bmatrix},\ x^0 = [1, 0, 0]^T, h = [12, 7, 3].$

Then the approximate realization problem is solved by the following algorithm:

characteristic polynomial for	values of variable and polynomial				
$H_{((0,0),(0,1))}$	0	16	225	231	240
	228	3	1	237	3
$H_{((0,0),(0,1),(0,2))}$	2	13	236	238	239
	235	235	0	228	230
$H_{((0,0),(0,1),(0,2),(0,3))}$	0	12	23		
	0	233	8		
$H_{((0,0),(0,1),(1,0))}$	4	5	12	236	
	6	235	7	1	

Numbers in the upper stand (or the lower stand) denote the values of variable in the characteristic polynomial (or the polynomial values corresponding to the value in the upper stand).

1) As we can find a relatively no change located near zero of the characteristic polynomial from $H_{((0,0),(0,1))}$ into $H_{((0,0),(0,1),(0,2))}$, we determine the number ν_1 of dimension which is 2.
2) As we can find a relatively no change located near zero of the characteristic polynomial from $H_{((0,0),(0,1))}$ to $H_{((0,0),(0,1),(1,0))}$, we determine the number ν_2 of dimension which is 0.

Then the non-linear integer programing for digital images produces a commutative linear representative system $\sigma_s = (K^2, F_{\alpha s}, F_{\beta s}, x_s^0, h_s)$ with a vector

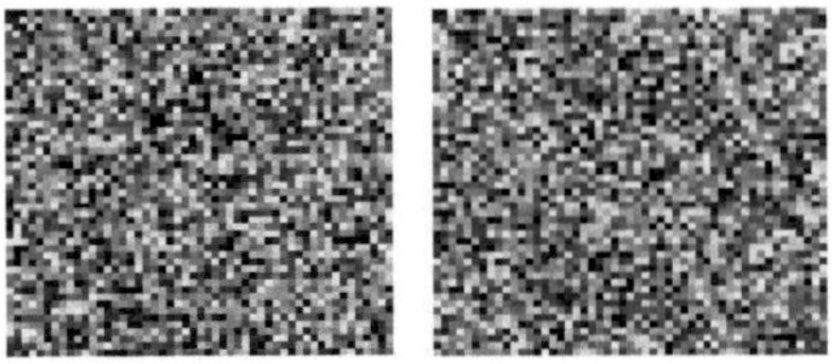

Fig. 9.4 In Example (9.23), the left is a 50×51 sized image of the original image a_σ for a given 3-dimensional commutative linear representation system σ. The right is a 50×51 sized image of an approximate system for the given σ.

index $\nu = (2)$ which has the least mean value 75.8 for the absolute value of the difference in the range 8.

Therefore, the quasi-reachable standard system $\sigma_s = (K^2, F_{\alpha s}, F_{\beta s}, x_s^0, h_s)$ with a vector index $\nu = (2)$ can be obtained as follows:

$$F_{\alpha s} = \begin{bmatrix} 0 & 118 \\ 1 & 87 \end{bmatrix}, F_{\beta s} = \begin{bmatrix} 134 & 1 \\ 96 & 51 \end{bmatrix}, x^0 = [1,\ 0]^T, h = [11,\ 10].$$

Based on Fig. 9.4, it is felt that the image generated by the approximated system shows patterns characterized by the original image.

Example 9.24. Let a signal be a digital image generated by the following 3-dimensional commutative linear representation system $\sigma = ((K^3, F_\alpha, F_\beta), x^0, h)$ with a vector index $\nu = (3)$,

$$\text{where } F_\alpha = \begin{bmatrix} 0 & 0 & 58 \\ 1 & 0 & 13 \\ 0 & 1 & 21 \end{bmatrix}, F_\beta = \begin{bmatrix} 103 & 21 & 235 \\ 38 & 62 & 136 \\ 71 & 83 & 118 \end{bmatrix}, x^0 = [1, 0, 0]^T, h = [12, 7, 3].$$

Then the approximate realization problem is solved by the following algorithm:

characteristic polynomial for	values of variable and polynomial					
$H_{((0,0),(0,1))}$	0	16	225	231	240	
	228	3	1	237	3	
$H_{((0,0),(0,1),(0,2))}$	0	11	230	233	234	238
	12	18	0	239	238	233
$H_{((0,0),(0,1),(0,2),(0,3))}$	0	13	229			
	0	1	9			
7 $H_{((0,0),(0,1),(1,0))}$	0	15	223	226		
	1	230	14	13		

Numbers in the upper stand (or the lower stand) denote the values of variable in the characteristic polynomial (or the polynomial values corresponding to the value in the upper stand).

1) As we can find a relatively no change located near zero of the characteristic polynomial from $H_{((0,0),(0,1))}$ into $H_{((0,0),(0,1),(0,2))}$, we determine the number ν_1 of dimension which is 2.

2) As we can find a relatively no change located near zero of the characteristic polynomial from $H_{((0,0),(0,1))}$ to $H_{((0,0),(0,1),(1,0))}$, we determine the number ν_2 of dimension which is 0.

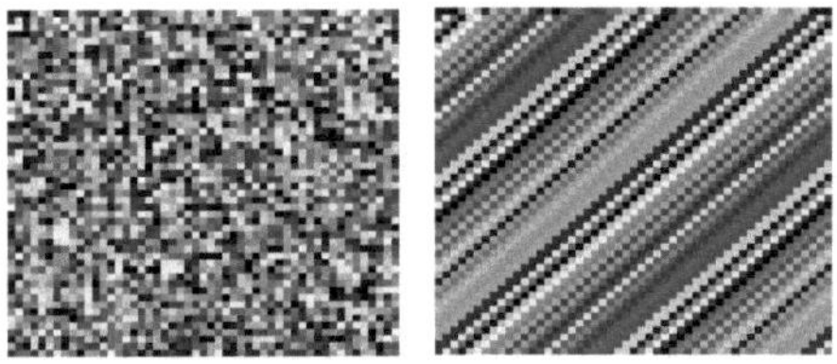

Fig. 9.5 In Example (9.24), the left is a 50×51 sized image of the original image a_σ for a given 3-dimensional commutative linear representation system σ. The right is a 50×51 sized image of an approximate system for the given σ.

Then the non-linear integer programing for digital images produces a commutative linear representative system $\sigma_s = (K^2, F_{\alpha s}, F_{\beta s}, x_s^0, h_s)$ with a vector index $\nu = (2)$ which has the least mean value 74.1 for the absolute value of the difference in the range 9.

Hence, the quasi-reachable standard system $\sigma_s = (K^2, F_{\alpha s}, F_{\beta s}, x_s^0, h_s)$ with a vector index $\nu = (2)$ can be obtained as follows:

$$F_{\alpha s} = \begin{bmatrix} 0 & 230 \\ 1 & 83 \end{bmatrix},\ F_{\beta s} = \begin{bmatrix} 102 & 10 \\ 21 & 158 \end{bmatrix},\ x^0 = [1,\ 0]^T,\ h = [17,\ 11].$$

Based on Fig. 9.5, it is felt that the image generated by the approximated system shows a pattern characterized by the original image.

Example 9.25. Let a signal be a digital image generated by the following 4-dimensional commutative linear representation system
$\sigma = ((K^4, F_\alpha, F_\beta), x^0, h)$ with a vector index $\nu = (4)$,

$$\text{where } F_\alpha = \begin{bmatrix} 0 & 0 & 0 & 43 \\ 1 & 0 & 0 & 12 \\ 0 & 1 & 0 & 24 \\ 0 & 0 & 1 & 43 \end{bmatrix},\ F_\beta = \begin{bmatrix} 32 & 145 & 11 & 225 \\ 31 & 235 & 64 & 85 \\ 29 & 196 & 73 & 212 \\ 37 & 174 & 207 & 57 \end{bmatrix},$$

$x^0 = [1,\ 0,\ 0,\ 0]^T$, $h = [12,\ 9,\ 15,\ 3]$.
Then the approximate realization problem is solved by the following algorithm:

characteristic polynomial for	values of variable and polynomial					
$H_{((0,0),(0,1))}$	4	5	226	231		
	7	23	6	228		
$H_{((0,0),(0,1),(0,2))}$	4	14	16	229	233	240
	1	2	229	10	10	237
$H_{((0,0),(0,1),(0,2),(0,3))}$	15	239				
	232	230				
$H_{((0,0),(0,1),(0,2),(0,3),(0,4))}$	0	3	17	223	235	240
	0	237	6	226	2	224
$H_{((0,0),(0,1),(0,2),(1,0))}$	2	14	229	230	234	
	223	1	2	4	4	

Numbers in the upper stand (or the lower stand) denote the values of variable in the characteristic polynomial (or the polynomial values corresponding to the value in the upper stand).

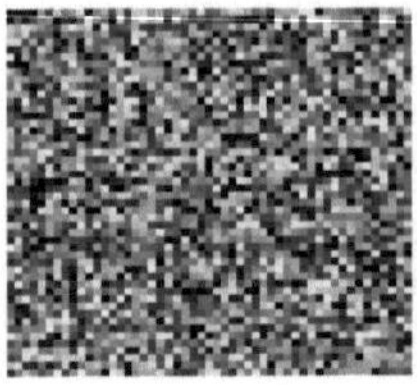 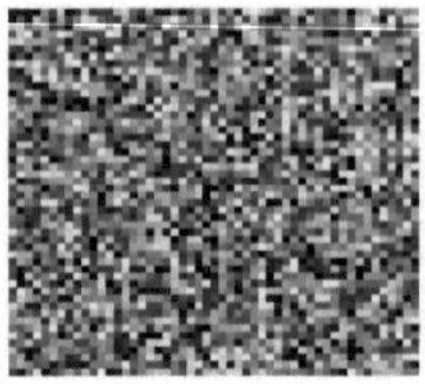

Fig. 9.6 In Example (9.25), the left is a 50×51 sized image of the original image a_σ for a given 3-dimensional commutative linear representation system σ. The right is a 50×51 sized image of an approximate system for the given σ.

1) As we can find a relatively no change located near zero of the characteristic polynomial from $H_{((0,0),(0,1))}$ into $H_{((0,0),(0,1),(0,2))}$, we determine the number ν_1 of dimension which is 3.
2) As we can find a relatively no change located near zero of the characteristic polynomial from $H_{((0,0),(0,1))}$ to $H_{((0,0),(0,1),(1,0))}$, we determine the number ν_2 of dimension which is 0.

Then the non-linear integer programing for digital images produces a commutative linear representative system $\sigma_s = (K^3, F_{\alpha s}, F_{\beta s}, x_s^0, h_s)$ with a vector index $\nu = (3)$ which has the least mean value 76.0 for the absolute value of the difference in the range 3.

Hence, the quasi-reachable standard system $\sigma_s = (K^3, F_{\alpha s}, F_{\beta s}, x_s^0, h_s)$ with a vector index $\nu = (3)$ can be obtained as follows:

$$F_{\alpha s} = \begin{bmatrix} 0 & 0 & 214 \\ 1 & 0 & 31 \\ 0 & 1 & 122 \end{bmatrix},\ F_{\beta s} = \begin{bmatrix} 169 & 146 & 41 \\ 78 & 46 & 90 \\ 66 & 177 & 191 \end{bmatrix},\ x^0 = [1,0,0]^T,\ h = [10,12,15].$$

Based on Fig. 9.6, it is felt that the image generated by the approximated system shows some patterns characterized by the original image.

Example 9.26. Let a signal be a digital image generated by the following 4-dimensional commutative linear representation system
$\sigma = ((K^4, F_\alpha, F_\beta), x^0, h)$ with a vector index $\nu = (4)$,

$$\text{where } F_\alpha = \begin{bmatrix} 0 & 0 & 0 & 44 \\ 1 & 0 & 0 & 23 \\ 0 & 1 & 0 & 54 \\ 0 & 0 & 1 & 21 \end{bmatrix},\ F_\beta = \begin{bmatrix} 13 & 178 & 64 & 144 \\ 61 & 117 & 80 & 205 \\ 37 & 148 & 86 & 191 \\ 15 & 111 & 69 & 89 \end{bmatrix},$$

$x^0 = [1,0,0,0]^T$, $h = [22,9,11,9]$.

Then the approximate realization problem is solved by the following algorithm:

characteristic polynomial for	values of variable and polynomial					
$H_{((0,0),(0,1))}$	6	233	239			
	240	7	231			
$H_{((0,0),(0,1),(0,2))}$	3	228	230			
	228	1	238			
$H_{((0,0),(0,1),(0,2),(0,3))}$	0	13	230	232		
	227	239	225	237		
$H_{((0,0),(0,1),(0,2),(0,3),(0,4))}$	0	16	17	21	225	
	0	232	228	232	8	
$H_{((0,0),(0,1),(0,2),(1,0))}$	0	21	213	215	229	236
	233	10	236	0	228	0

Numbers in the upper stand (or the lower stand) denote the values of variable in the characteristic polynomial (or the polynomial values corresponding to the value in the upper stand).

Fig. 9.7 In Example (9.26), the left is a 50×51 sized image of the original image a_σ for a given 3-dimensional commutative linear representation system σ. The right is a 50×51 sized image of an approximate system for the given σ.

1) As we can find a relatively no change located near zero of the characteristic polynomial from $H_{((0,0),(0,1))}$ into $H_{((0,0),(0,1),(0,2))}$, we determine the number ν_1 of dimension which is 2.
2) As we can find a relatively no change located near zero of the characteristic polynomial from $H_{((0,0),(0,1))}$ to $H_{((0,0),(0,1),(1,0))}$, we determine the number ν_2 of dimension which is 0.

Then the non-linear integer programing for digital images produces a commutative linear representative system $\sigma_s = (K^3, F_{\alpha s}, F_{\beta s}, x_s^0, h_s)$ with a vector index $\nu = (3)$ which has the least mean value 74.8 for the absolute value of the difference in the range 4.

Therefore, the quasi-reachable standard system $\sigma_s = (K^3, F_{\alpha s}, F_{\beta s}, x_s^0, h_s)$ with a vector index $\nu = (3)$ can be obtained as follows:

$$F_{\alpha s} = \begin{bmatrix} 0 & 0 & 30 \\ 1 & 0 & 135 \\ 0 & 1 & 158 \end{bmatrix}, \; F_{\beta s} = \begin{bmatrix} 85 & 77 & 181 \\ 203 & 70 & 48 \\ 107 & 239 & 236 \end{bmatrix}, \; x^0 = [1,0,0]^T, \; h = [25,9,12].$$

Based on Fig. 9.7, it is felt that the image generated by the approximated system shows some patterns characterized by the original image.

Example 9.27. Let a signal be a digital image generated by the following 5-dimensional commutative linear representation system $\sigma = ((K^5, F_\alpha, F_\beta), x^0, h)$ with a vector index $\nu = (5)$,

$$\text{where } F_\alpha = \begin{bmatrix} 0 & 0 & 0 & 0 & 4 \\ 1 & 0 & 0 & 0 & 6 \\ 0 & 1 & 0 & 0 & 5 \\ 0 & 0 & 1 & 0 & 11 \\ 0 & 0 & 0 & 1 & 13 \end{bmatrix}, F_\beta = \begin{bmatrix} 27 & 48 & 234 & 11 & 193 \\ 32 & 99 & 158 & 130 & 180 \\ 14 & 92 & 30 & 232 & 70 \\ 23 & 146 & 133 & 0 & 100 \\ 12 & 179 & 63 & 229 & 85 \end{bmatrix},$$

$x^0 = [1,\ 0,\ 0,\ 0,\ 0]^T, h = [32,\ 13,\ 17,\ 9,\ 25]$.

Then the approximate realization problem is solved by the following algorithm:

characteristic polynomial for	values of variable and polynomial					
$H_{((0,0),(0,1))}$	3	231	239			
	237	1	236			
$H_{((0,0),(0,1),(0,2))}$	8	232	236			
	232	23	236			
$H_{((0,0),(0,1),(0,2),(0,3))}$	0	4	8			
	8	237	229			
$H_{((0,0),(0,1),(0,2),(0,3),(0,4))}$	4	9	17	224	239	
	239	10	1	10	7	
$H_{((0,0),(0,1),(0,2),(0,3),(0,4),(0,5))}$	0	1	6	11	218	240
	0	4	230	0	1	237
$H_{((0,0),(0,1),(0,2),(0,3),(1,0))}$	4	8	12	21		
	228	9	232	239		

Numbers in the upper stand (or the lower stand) denote the values of variable in the characteristic polynomial (or the polynomial values corresponding to the value in the upper stand).

1) As we can find a relatively no change located near zero of the characteristic polynomial from $H_{((0,0),(0,1),(0,2))}$ into $H_{((0,0),(0,1),(0,2),(0,3))}$, we determine the number ν_1 of dimension which is 4.
2) As we can find a relatively no change located near zero of the characteristic polynomial from $H_{((0,0),(0,1),(0,2),(0,3))}$ to $H_{((0,0),(0,1),(0,2),(0,3),(1,0))}$, we determine the number ν_2 of dimension which is 0.

Then the non-linear integer programing for digital images produces a commutative linear representative system $\sigma_s = (K^4, F_{\alpha s}, F_{\beta s}, x_s^0, h_s)$ with a vector index $\nu = (4)$ which has the least mean value 74.3 for the absolute value of the difference in the range 3.

Therefore, the quasi-reachable standard system $\sigma_s = (K^4, F_{\alpha s}, F_{\beta s}, x_s^0, h_s)$ with a vector index $\nu = (4)$ can be obtained as follows:

$$F_{\alpha s} = \begin{bmatrix} 0 & 0 & 0 & 122 \\ 1 & 0 & 0 & 235 \\ 0 & 1 & 0 & 117 \\ 0 & 0 & 1 & 159 \end{bmatrix}, F_{\beta s} = \begin{bmatrix} 15 & 87 & 206 & 127 \\ 222 & 149 & 227 & 180 \\ 234 & 19 & 70 & 11 \\ 58 & 57 & 165 & 36 \end{bmatrix},$$

$x^0 = [1,\ 0,\ 0,\ 0]^T, h = [29,\ 16,\ 19,\ 10]$.

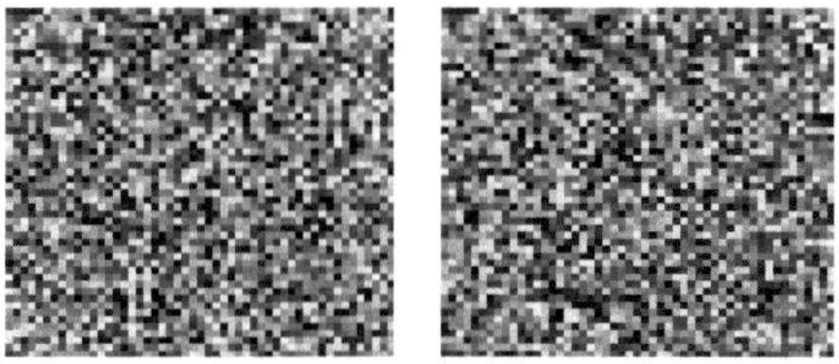

Fig. 9.8 In Example (9.27), the left is a 50×51 sized image of the original image a_σ for a given 3-dimensional commutative linear representation system σ. The right is a 50×51 sized image of an approximate system for the given σ.

Based on Fig. 9.8, it is felt that the image generated by the approximated system shows some patterns characterized by the original image.

9.8 Algebraically Noisy Realization of Two-Dimensional Images

In this section, we discuss the algebraically noisy realization problem of two-dimensional images.

For noise $\{\bar{a}(i,j) : i,\ j \in N\}$ added to a behavior $\hat{a}$ of the unknown commutative linear representation system σ, we will obtain the observed data $\{\hat{a}(i,j) + \bar{a}(i,j) : i,\ j \in N\}$.

For any given $\{\hat{a}(i,j) + \bar{a}(i,j) : i,\ j \in N\}$, σ which satisfies $a_\sigma(i,j) \approx \hat{a}(i,j) : i,\ j \in N$ is called a noisy realization of a.

Roughly speaking, we can propose the following algebraically noisy realization problem:

For any given $\{\hat{a}(i,j) + \bar{a}(i,j) : i,\ j \in N\}$, find, a commutative linear representation system σ which satisfies $a_\sigma(i,j) \approx \hat{a}(i,j)$ for any $i,\ j \in N$.

In order to make our discussion simple in this section, we assume that the set Y of output is the set $\boldsymbol{K}$ of $N/241N$ which is the quotient field modulo the prime number 241, namely 1-output. And the numbers 0 and 240 denote black and white respectively.

Let the observed object be a commutative linear representation system and noise be added to output. Then we will obtain the data $\{a(i,j) = \hat{a}(i,j) + \bar{a}(i,j) : 0 \le i \le \underline{M}, 0 \le j \le \underline{N}\}$ for some integers $\underline{M}, \underline{N} \in N$, where $\hat{a}(i,j)$ is the exact signal which comes from the observed commutative linear representation system and $\bar{a}(i,j)$ is the noise added at the place of observation.

Problem 9.28. Problem statement of an algebraically noisy realization for commutative linear representation systems.

Let $H_{\underline{a}\ (p,\bar{p})}$ be the measured finite-sized Hankel matrix. Then find, using only algebraic calculations, the cleaned-up Hankel matrix $\hat{H}_{\underline{a}\ (p,\bar{p})}$ such that $H_{\underline{a}\ (p,\bar{p})} = \hat{H}_{\underline{a}\ (p,\bar{p})} + \bar{H}_{\underline{a}\ (p,\bar{p})}$ holds.

Namely, find, using only algebraic calculations, a minimal dimensional commutative linear representation system $\sigma = ((\boldsymbol{K}^n, F_\alpha, F_\beta), x^0, h))$ which realizes $\hat{H}_{\underline{a}\ (p,\bar{p})}$.

In order to solve the noisy realization problem, the non-linear integer programing for digital images discussed in Section 9.6 is used.

Theorem 9.29. *Algebraic algorithm for noisy realization*
Let a partial image $\underline{a}$ be a considered object which is a commutative linear representation system. Then a noisy realization $\sigma = ((\boldsymbol{K}^n, F_{\alpha s}, F_{\beta s}), x^0, h_s)$ of $\underline{a}$ is given by the following algorithm:

1) Based on a marked change located near the zero of the characteristis polynomial from $H_{((0,0),(0,1),\cdots,(0,\nu_1-1))}$ into $H_{((0,0),(0,1),\cdots,(0,\nu_1-1),(0,\nu_1))}$, determine the number ν_1 in a vector index $\nu = (\nu_1)$, namely, determine the value ν_1 in a vector index $\nu = (\nu_1)$ of a noisy realization σ.

2) Based on a marked change located near the zero of the characteristis polynomial from $H_{((0,0),(0,1),\cdots,(0,\nu_1-1),(1,0),\cdots,(1,\nu_2-1)}$ into $H_{((0,0),(0,1),\cdots,(0,\nu_1-1),(1,0),\cdots,(1,\nu_2-1),(1,\nu_2))}$, determine the number ν_2 in a vector index $\nu = (\nu_1, \nu_2)$.

$\vdots$

k) Based on a marked change located near the zero of the characteristis polynomial from $H_{((0,0),(0,1),\cdots,(0,\nu_1-1),\cdots,(k-1,0),\cdots,(k-1,\nu_k-1)}$ into $H_{((0,0),(0,1),\cdots,(0,\nu_1-1),\cdots,(k-1,0),\cdots,(k-1,\nu_k-1),(k-1,\nu_k))}$ and $H_{((0,0),(0,1),\cdots,(0,\nu_1-1),\cdots,(k-1,0),\cdots,(k-1,\nu_k-1),(k,0))}$, determine the number ν_k in a vector index $\nu = (\nu_1, \nu_2, \cdots, \nu_k)$.

k+1) The non-linear integer programing for digital images is used as follows: Using $\mid \tilde{H}_i \mid= 0$, determine vectors $\mathbf{c}^i$ for $1 \leq i \leq k+1$.

k+2) Determine $F_{\alpha s}$, $F_{\beta s}$, h_s and x_s^0 on the basis of Definition (9.9), Fig. 9.1 and Fig. 9.2 which has the least mean value for the absolute value of the difference in the range l_1, where the range l_1 is listed in Section 9.6.

Remark : For $\mid \tilde{H}_i \mid= 0$, refer Section 9.6 Non-linear integer programing for digital images .

Example 9.30. Let a signal be a digital image generated by the following 2-dimensional commutative linear representation system
$\sigma = ((K^2, F_\alpha, F_\beta), x^0, h)$ with a vector index $\nu = (2)$, where $F_\alpha = \begin{bmatrix} 0 & 34 \\ 1 & 16 \end{bmatrix}$,
$F_\beta = \begin{bmatrix} 104 & 174 \\ 76 & 115 \end{bmatrix}$, $x^0 = [1,\ 0]^T$, $h = [12,\ 7]$.
Let added noises be given as shown in Fig. 9.9.

Then the noisy realization problem is solved by the following algorithm:

characteristic polynomial for	values of variable and polynomial			
$H_{((0,0))}$	12			
	0			
$H_{((0,0),(0,1))}$	9			
	3			
$H_{((0,0),(0,1),(0,2))}$	8	12		
	4	2		
$H_{((0,0),(0,1),(0,2),(0,3))}$	11	40	206	
	13	1	14	
$H_{((0,0),(0,1),(0,2),(0,3),(0,4))}$	13	14	18	222
	4	231	5	4
$H_{((0,0),(0,1),(1,0))}$	10	233		
	240	240		
$H_{((0,0),(0,1),(1,0),(1,1))}$	10	226	233	
	14	3	231	

Numbers in the upper stand (or the lower stand) denote the values of variable in the characteristic polynomial (or the polynomial values corresponding to the value in the upper stand).

1) As we can find the marked change located near zero of the characteristic polynomial from $H_{((0,0),(0,1))}$ into $H_{((0,0),(0,1),(0,2))}$, we determine the number ν_1 of dimension which is 2.
2) As we can find the marked change located near zero of the characteristic polynomial from $H_{((0,0),(0,1))}$ to $H_{((0,0),(0,1),(1,0))}$, we determine the number ν_2 of dimension which is 0.

Then the non-linear integer programing for digital images produces a commutative linear representative system $\sigma_s = (K^2, F_{\alpha s}, F_{\beta s}, x_s^0, h_s)$ with a vector index $\nu = (2)$ which has the least mean value 3.87 for the absolute value of the difference in the range 5, and can be obtained as follows:

$$F_{\alpha s} = \begin{bmatrix} 0 & 34 \\ 1 & 16 \end{bmatrix}, F_{\beta s} = \begin{bmatrix} 104 & 174 \\ 76 & 115 \end{bmatrix}, x^0 = [1,\ 0]^T, h = [12,\ 7].$$

The obtained commutative linear representation system is completely reconstructed.

Fig. 9.9 In Example (9.30), the left is a 50×51 sized image of the original image a_σ for a given 2-dimensional commutative linear representation system σ, the middle is added noises. The right is a 50×51 sized image which displays the original image and added noise.

For reference, we list the following commutative linear representative system $\sigma_{s1} = (K^2, F_{\alpha 1}, F_{\beta 1}, x_1^0, h_1)$ with a vector index $\nu = (2)$ which has the second least mean value 75.5 for the absolute value of the difference in the range 5.

$$F_{\alpha 1} = \begin{bmatrix} 0 & 8 \\ 1 & 153 \end{bmatrix}, F_{\beta 1} = \begin{bmatrix} 22 & 148 \\ 139 & 81 \end{bmatrix}, x_1^0 = [1,\ 0]^T,\ h_1 = [13,\ 9].$$

In this example, the noise is completely removed from a given noisy signal by using non-linear integer programming for digital images. This example also shows that the commutative linear representation system obtained by our method has the second least mean value which is large. Therefore, our method implies that the obtained noisy realization is robust over our finite field.

Example 9.31. Let a signal be a digital image generated by the following 2-dimensional commutative linear representation system $\sigma = ((K^2, F_\alpha, F_\beta), x^0, h)$ with a vector index $\nu = (2)$, where $F_\alpha = \begin{bmatrix} 0 & 75 \\ 1 & 156 \end{bmatrix}$,

$$F_\beta = \begin{bmatrix} 77 & 63 \\ 184 & 102 \end{bmatrix}, x^0 = [1,\ 0]^T, h = [12,\ 7].$$

Let added noises be given as shown in Fig. 9.10.

Then the noisy realization problem is solved by the following algorithm:

characteristic polynomial for	values of variable and polynomial			
$H_{((0,0))}$	13			
	0			
$H_{((0,0),(0,1))}$	12			
	0			
$H_{((0,0),(0,1),(0,2))}$	9	232		
	8	0		
$H_{((0,0),(0,1),(0,2),(0,3))}$	22	219	233	
	7	5	8	
$H_{((0,0),(0,1),(0,2),(0,3),(0,4))}$	5	13	19	235
	2	2	2	1
$H_{((0,0),(0,1),(1,0))}$	0	17		
	5	6		
$H_{((0,0),(0,1),(1,0),(1,1))}$	3	232	239	
	8	235	5	

Numbers in the upper stand (or the lower stand) denote the values of variable in the characteristic polynomial (or the polynomial values corresponding to the value in the upper stand).

1) As we can find the marked change located near zero of the characteristic polynomial from $H_{((0,0),(0,1))}$ into $H_{((0,0),(0,1),(0,2))}$, we determine the number ν_1 of dimension which is 2.

2) As we can find the marked change located near zero of the characteristic polynomial from $H_{((0,0),(0,1))}$ to $H_{((0,0),(0,1),(1,0))}$, we determine the number ν_2 of dimension which is 0.

Then the non-loinear integer programing for digital images produces a commutative linear representative system $\sigma_s = (K^2, F_{\alpha s}, F_{\beta s}, x_s^0, h_s)$ with a vector index $\nu = (2)$ which has the least mean value 5.45 for the absolute value of the difference in the range 8, and can be obtained as follows:

$$F_{\alpha s} = \begin{bmatrix} 0 & 75 \\ 1 & 156 \end{bmatrix},\ F_{\beta s} = \begin{bmatrix} 77 & 63 \\ 184 & 102 \end{bmatrix},\ x_s^0 = [1,\ 0]^T,\ h_s = [12,\ 7].$$

The obtained commutative linear representation system is completely reconstructed.

Fig. 9.10 In Example (9.31), the left is a 50×51 sized image of the original image a_σ for a given 2-dimensional commutative linear representation system σ, the middle is added noises. The right is a 50×51 sized image which displays the original image and added noise.

For reference, we list the following commutative linear representative system $\sigma_s = (K^2, F_{\alpha 1}, F_{\beta 1}, x_1^0, h_1)$ with a vector index $\nu = (2)$ which has the second least mean value 74.1 for the absolute value of the difference in the range 8.

$$F_{\alpha 1} = \begin{bmatrix} 0 & 72 \\ 1 & 151 \end{bmatrix},\ F_{\beta 1} = \begin{bmatrix} 114 & 208 \\ 110 & 95 \end{bmatrix},\ x_1^0 = [1,\ 0]^T,\ h_1 = [8,\ 3].$$

In this example, the noise is completely removed from a given noisy signal by using non-linear integer programming for digital images. This example also shows that the commutative linear representation system obtained by our method has the second least mean value which is large. Therefore, our method implies that the obtained noisy realization is robust over our finite field.

Example 9.32. Let a signal be a digital image generated by the following 3-dimensional commutative linear representation system
$\sigma = ((K^3, F_\alpha, F_\beta), x^0, h)$ with a vector index $\nu = (3)$,

where $F_\alpha = \begin{bmatrix} 0 & 0 & 73 \\ 1 & 0 & 21 \\ 0 & 1 & 23 \end{bmatrix}$, $F_\beta = \begin{bmatrix} 52 & 226 & 128 \\ 56 & 84 & 210 \\ 13 & 114 & 55 \end{bmatrix}$,

$x^0 = [1,\ 0,\ 0]^T$, $h = [15,\ 9,\ 6]$.
Let added noises be given as shown in Fig. 9.11.

Then the noisy realization problem is solved by the following algorithm:

characteristic polynomial for	values of variable and polynomial				
$H_{((0,0))}$	16				
	0				
$H_{((0,0),(0,1))}$	231	238	239		
	2	12	228		
$H_{((0,0),(0,1),(0,2))}$	227	237			
	0	235			
$H_{((0,0),(0,1),(0,2),(0,3))}$	3	236	240		
	229	4	3		
$H_{((0,0),(0,1),(0,2),(0,3),(0,4))}$	1	8	12	235	236
	12	4	235	12	10
$H_{((0,0),(0,1),(0,2),(1,0))}$	15	227	235		
	238	13	228		

Numbers in the upper stand (or the lower stand) denote the values of variable in the characteristic polynomial (or the polynomial values corresponding to the value in the upper stand).

1) As we can find the marked change located near zero of the characteristic polynomial from $H_{((0,0))}$ into $H_{((0,0),(0,1))}$, we determine the number ν_1 of dimension which is 2.
2) As we can find the marked change located near zero of the characteristic polynomial from $H_{((0,0),(1,0))}$ into $H_{((0,0),(1,0),(1,1))}$ and $H_{((0,0),(1,0),(2,0))}$, we determine the number ν_2 of dimension which is 0.

Then the non-linear integer programing for digital images produces a commutative linear representative system $\sigma_s = (K^3, F_{\alpha s}, F_{\beta s}, x_s^0, h_s)$ with a vector index $\nu = (3)$ which has the least mean value 3.62 for the absolute value of the difference in the range 3, and can be obtained as follows:

$$F_{\alpha s} = \begin{bmatrix} 0 & 0 & 73 \\ 1 & 0 & 21 \\ 0 & 1 & 23 \end{bmatrix}, \; F_{\beta s} = \begin{bmatrix} 52 & 226 & 128 \\ 56 & 84 & 210 \\ 13 & 114 & 55 \end{bmatrix}, \; x_s^0 = [1,\ 0]^T,\ h_s = [12,\ 9,\ 6].$$

The obtained commutative linear representation system is completely reconstructed.

Fig. 9.11 In Example (9.32), the left is a 50 × 51 sized image of the original image a_σ for a given 3-dimensional commutative linear representation system σ, the middle is added noises. The right is a 50 × 51 sized image which displays the original image and added noise.

For reference, we list the following quasi-reachable standard system $\sigma_s = (K^3, F_{\alpha 1}, F_{\beta 1}, x_1^0, h_1)$ with a vector index $\nu = (3)$ which has the second least mean value 75.6 for the absolute value of the difference in the range 3.

$$F_{\alpha 1} = \begin{bmatrix} 0 & 0 & 152 \\ 1 & 0 & 28 \\ 0 & 1 & 186 \end{bmatrix}, F_{\beta 1} = \begin{bmatrix} 198 & 38 & 4 \\ 55 & 205 & 229 \\ 181 & 222 & 45 \end{bmatrix}, x_1^0 = [1,\ 0]^T, h_1 = [13,\ 12,\ 3].$$

In this example, the noise is completely removed from a given noisy signal by using non-linear integer programming for digital images. This example also shows that the commutative linear representation system obtained by our method has the second least mean value which is large. Therefore, our method implies that the obtained noisy realization is robust over our finite field.

Example 9.33. Let a signal be a digital image generated by the following 4-dimensional commutative linear representation system
$\sigma = ((K^4, F_\alpha, F_\beta), x^0, h)$ with a vector index $\nu = (4)$,

$$\text{where } F_\alpha = \begin{bmatrix} 0 & 0 & 0 & 7 \\ 1 & 0 & 0 & 8 \\ 0 & 1 & 0 & 1 \\ 0 & 0 & 1 & 9 \end{bmatrix}, F_\beta = \begin{bmatrix} 34 & 70 & 197 & 90 \\ 25 & 114 & 123 & 231 \\ 7 & 35 & 211 & 67 \\ 10 & 97 & 185 & 189 \end{bmatrix},$$

$x^0 = [1, 0, 0, 0]^T$, $h = [12, 7, 3, 4]$.

Let added noises be given as shown in Fig. 9.12.

Then the noisy realization problem is solved by the following algorithm:

characteristic polynomial for	values of variable and polynomial					
$H_{((0,0),(0,1),(0,2))}$	3	13	15	228		
	15	230	2	238		
$H_{((0,0),(0,1),(0,2),(0,3))}$	22	218	225			
	7	2	237			
$H_{((0,0),(0,1),(0,2),(0,3),(0,4))}$	7	240				
	4	21				
$H_{((0,0),(0,1),(0,2),(0,3),(0,4),(0,5))}$	5	9	17	232	237	
	20	236	237	12	15	
$H_{((0,0),(0,1),(0,2),(0,3),(1,0))}$	9	14	15	228	231	233
	0	221	16	8	8	236

Numbers in the upper stand (or the lower stand) denote the values of variable in the characteristic polynomial (or the polynomial values corresponding to the value in the upper stand).

1) As we can find the marked change located near zero of the characteristic polynomial from $H_{((0,0))}$ into $H_{((0,0),(0,1))}$, we determine the number ν_1 of dimension which is 2.

2) As we can find the marked change located near zero of the characteris-

tic polynomial from $H_{((0,0),(1,0))}$ into $H_{((0,0),(1,0),(1,1))}$ and $H_{((0,0),(1,0),(2,0))}$, we determine the number ν_2 of dimension which is 0.

Then the non-linear integer programing for digital images produces a commutative linear representative system $\sigma_s = (K^4, F_{\alpha s}, F_{\beta s}, x_s^0, h_s)$ with a vector index $\nu = (4)$ which has the least mean value 3.53 for the absolute value of the difference in the range 3, and can be obtained as follows:

$$F_{\alpha s} = \begin{bmatrix} 0 & 0 & 0 & 7 \\ 1 & 0 & 0 & 8 \\ 0 & 1 & 0 & 1 \\ 0 & 0 & 1 & 9 \end{bmatrix}, \ F_{\beta s} = \begin{bmatrix} 34 & 70 & 197 & 90 \\ 25 & 114 & 123 & 231 \\ 7 & 35 & 211 & 67 \\ 10 & 97 & 185 & 189 \end{bmatrix},$$

$x_s^0 = [1, 0, 0, 0]^T$, $h_s = [12, 7, 3, 4]$.

The obtained commutative linear representation system is completely reconstructed.

Fig. 9.12 In Example (9.33), the left is a 50×51 sized image of the original image a_σ for a given 4-dimensional commutative linear representation system σ, the middle is added noises. The right is a 50×51 sized image which displays the original image and added noise.

For reference, we list the following quasi-reachable standard system $\sigma_s = (K^4, F_{\alpha 1}, F_{\beta 1}, x_1^0, h_1)$ with a vector index $\nu = (3)$ which has the second least mean value 75.4 for the absolute value of the difference in the range 3.

$$F_{\alpha 1} = \begin{bmatrix} 0 & 0 & 0 & 162 \\ 1 & 0 & 0 & 75 \\ 0 & 1 & 0 & 0 \\ 0 & 0 & 1 & 27 \end{bmatrix}, \ F_{\beta 1} = \begin{bmatrix} 206 & 217 & 154 & 38 \\ 43 & 61 & 83 & 2 \\ 240 & 43 & 61 & 83 \\ 223 & 236 & 149 & 228 \end{bmatrix},$$

$x_1^0 = [1, 0, 0, 0]^T$, $h_1 = [9, 9, 9, 8]$.

In this example, the noise is completely removed from a given noisy signal by using non-linear integer programming for digital images. This example also shows that the commutative linear representation system obtained by our method has the second least mean value which is large. Therefore, our method implies that the obtained noisy realization is robust over our finite field.

Example 9.34. Let a signal be a digital image generated by the following 5-dimensional commutative linear representation system

$\sigma = ((K^5, F_\alpha, F_\beta), x^0, h)$ with a vector index $\nu = (5)$,

where $F_\alpha = \begin{bmatrix} 0 & 0 & 0 & 0 & 7 \\ 1 & 0 & 0 & 0 & 8 \\ 0 & 1 & 0 & 0 & 1 \\ 0 & 0 & 1 & 0 & 9 \\ 0 & 0 & 0 & 1 & 21 \end{bmatrix}$, $F_\beta = \begin{bmatrix} 34 & 35 & 82 & 158 & 169 \\ 25 & 74 & 232 & 56 & 179 \\ 7 & 30 & 189 & 48 & 149 \\ 10 & 52 & 101 & 220 & 162 \\ 5 & 115 & 57 & 93 & 4 \end{bmatrix}$,

$x^0 = [1, 0, 0, 0, 0]^T$, $h = [12, 7, 3, 4, 7]$.

Let added noises be given as shown in Fig. 9.13.

Then the noisy realization problem is solved by the following algorithm:

characteristic polynomial for	values of variable and polynomial					
$H_{((0,0),(0,1),(0,2))}$	0	3	6	12	229	234
	235	15	11	3	8	8
$H_{((0,0),(0,1),(0,2),(0,3))}$	0	6	240			
	20	231	3			
$H_{((0,0),(0,1),(0,2),(0,3),(0,4))}$	18	229				
	237	5				
$H_{((0,0),(0,1),(0,2),(0,3),(0,4),(0,5))}$	9	14	229	231	240	
	1	3	237	2	9	
$H_{((0,0),(0,1),(0,2),(0,3),(0,4),(1,0))}$	10	20	230	237	238	
	230	1	238	16	235	

Numbers in the upper stand (or the lower stand) denote the values of variable in the characteristic polynomial (or the polynomial values corresponding to the value in the upper stand).

1) As we can find the marked change located near zero of the characteristic polynomial from $H_{((0,0))}$ into $H_{((0,0),(0,1))}$, we determine the number ν_1 of dimension which is 2.

2) As we can find the marked change located near zero of the characteristic polynomial from $H_{((0,0),(1,0))}$ into $H_{((0,0),(1,0),(1,1))}$ and $H_{((0,0),(1,0),(2,0))}$, we determine the number ν_2 of dimension which is 0.

Then the non-linear integer programing for digital images produces a commutative linear representative system $\sigma_s = (K^5, F_{\alpha s}, F_{\beta s}, x_s^0, h_s)$ with a vector index $\nu = (5)$ which has the least mean value 2.63 for the absolute value of the difference in the range 3, and can be obtained as follows:

$F_{\alpha s} = \begin{bmatrix} 0 & 0 & 0 & 0 & 7 \\ 1 & 0 & 0 & 0 & 8 \\ 0 & 1 & 0 & 0 & 1 \\ 0 & 0 & 1 & 0 & 9 \\ 0 & 0 & 0 & 1 & 21 \end{bmatrix}$, $F_{\beta s} = \begin{bmatrix} 34 & 35 & 82 & 158 & 169 \\ 25 & 74 & 232 & 56 & 179 \\ 7 & 30 & 189 & 48 & 149 \\ 10 & 52 & 101 & 220 & 162 \\ 5 & 115 & 57 & 93 & 4 \end{bmatrix}$,

$x_s^0 = [1, 0, 0, 0]^T$, $h_s = [12, 7, 3, 4, 7]$.

The obtained commutative linear representation system is completely reconstructed.

For reference, we list the following quasi-reachable standard system $\sigma_s = (K^5, F_{\alpha 1}, F_{\beta 1}, x_1^0, h_1)$ with a vector index $\nu = (5)$ which has the second least mean value 74.3 for the absolute value of the difference in the range 3.

Fig. 9.13 In Example (9.34), the left is a 50×51 sized image of the original image a_σ for a given 5-dimensional commutative linear representation system σ, the middle is added noises. The right is a 50×51 sized image which displays the original image and added noise.

$$F_{\alpha 1} = \begin{bmatrix} 0 & 0 & 0 & 0 & 208 \\ 1 & 0 & 0 & 0 & 222 \\ 0 & 1 & 0 & 0 & 17 \\ 0 & 0 & 1 & 0 & 132 \\ 0 & 0 & 0 & 1 & 136 \end{bmatrix}, \; F_{\beta 1} = \begin{bmatrix} 225 & 56 & 238 & 19 & 170 \\ 90 & 155 & 120 & 154 & 168 \\ 65 & 178 & 47 & 81 & 8 \\ 41 & 82 & 190 & 212 & 124 \\ 232 & 22 & 182 & 119 & 8 \end{bmatrix},$$

$x_1^0 = [1,0,0,0,0]^T$, $h_1 = [12,8,9,9,9]$.

In this example, the noise is completely removed from a given noisy signal by using non-linear integer programming for digital images. This example also shows that the commutative linear representation system obtained by our method has the second least mean value which is large. Therefore, our method implies that the obtained noisy realization is robust over our finite field.

9.9 Historical Notes and Concluding Remarks

Based on the fundamental facts about relations between two-dimensional images and commutative linear representation systems established in [Hasegawa and Suzuki, 2006], we discussed algebraically approximate and noisy realization of two dimensional images by using the state space method. In particular, we treated the problems in monochrome images whose values belong to the finite field, namely, the quotient field modulo a prime number. For the two-dimensional images whose values belong to the real number field, we did not discuss the algebraically approximate and noisy realization problem because we can discuss them in the same manner as in preceding chapters.

For our approximate and noisy realizatio problems, we proposed a new method called non-linear integer programming for digital images by introducing an intuitive norm which is the mean value for the absolute value of the difference in the range l_1. Since two-dimensional digital image processing based on the state space method has been well established, we can discuss an approximate problem for digital images which could not be theoretically treated.

In order to show that our non-linear integer programming for digital image is effective, we concretely solved some examples. However, because of our inexpert numerical solving method within a small range, we could not obtain good results for approximate realization. Since our numerical solving was performed for all combinations, huge amounts of time were wasted. Speaking bravely, it is necessary that a new algorithm to solve non-linear integer programming for digital images with a wide range be produced quickly.

For our noisy realization of two-dimensional images, in order to show that our method is effective, we treated some examples, which brought about good results. However, our numerical solving method needs huge amounts of time. We hope that a new and rapid computation for non-linear integer programming with a wide range will appear in the near future.

References

Akaike, H.: A New Look at Statistical Model Identification. IEEE Trans. on Automatic Control AC-19, 716–723 (1974)

Burger, W., Burge, M.J.: Digital Image Processing. Springer, Heidelberg (2008)

Bourbaki, N.: Elements de Mathematique. In: Algebre, Herman, Paris (1958)

Elements of Mathematics, Theory of Sets, Hermann, Paris (1968) (English Translation)

Elements of Mathematics, Algebra Part 1, Hermann, Paris (1974) (English Translation)

Chevalley, C.: Fundamental concept of algebra. Academic Press, London (1956)

D'alessandro, D., Isidori, A., Ruberti, A.: Realization and structure theory of bilinear dynamical systems. SIAM J. Contr. 12, 517–534 (1974)

Davison, E.J.: A method for simplifying dynamic systems. IEEE Trans. on Automatic Control AC-11, 93–101 (1966)

Eckart, C., Young, G.: The approximation of one matrix by another of lower rank. Psychometrika 1, 211–218 (1936)

Fornasini, E., Marchesini, G.: State-space realization theory of two-dimensional filters. IEEE Trans. on Automatic Control AC-21, 484–492 (1976)

Gantmacher, F.R.: The theory of matrices, vol. 2. Chelsea, New York (1959)

Glover, K.: All optimal Hankel-norm approximation of linear multivariable systems and their L^{∞}-error bounds. Int. J. Control 39(6), 1115–1193 (1984)

Haber, R., Keviczky, L.: Nonlinear system identification input-output modeling approach, vol. 1. Kluwer Academic Publishers, Dordrecht (1999)

Halmos, P.R.: Finite dimensional vector spaces. D. Van. Nos. Com. (1958)

Hasegawa, Y.: Approximate and Noisy Realization of Discrete-Time Dynamical Systems. LNCIS, vol. 376. Springer, Heidelberg (2008)

Hasegawa, Y.: Approximate and Noisy Realizations of Non-Linear Systems. In: International Conference Dedicated to the Centennial Anniversary of L. S. Pontryagin (2008a)

Hasegawa, Y., Hamada, K., Matsuo, T.: A Realization Algorithm for Discrete-Time Linear Systems. In: Problems in Modern Applied Mathematics, pp. 192–197. World Scientifics, Singapore (2000)

Hasegawa, Y., Matsuo, T.: Realization Theory of discrete-time linear representation systems. SICE Transaction 15(3), 298–305 (1979a) (in Japanese)

On the discrete time finite-dimensional linear representation systems. SICE Transaction 15(4), 443–450 (1979b) (in Japanese)

Realization theory of discrete-time Pseudo-Linear Systems. SICE Transaction 28(2), 199–207 (1992) (in Japanese)

Realization theory of Discrete-Time finite-dimensional Pseudo-Linear systems. SICE Transaction 29(9), 1071–1080 (1993) (in Japanese)

Realization theory of discrete-time Almost-Linear Systems. SICE Transaction 30(2), 150–157 (1994a) (in Japanese)

Realization theory of Discrete-Time finite-dimensional Almost-Linear systems. SICE Transaction 30(11), 1325–1333 (1994b) (in Japanese)

Partial Realization & Real-Time, Partial Realization Theory of Discrete-Time Pseudo-Linear System. SICE Transaction 31(4), 471–480 (1995) (in Japanese)

Modeling for Hysteresis Characteristic by Affine Dynamical Systems. In: International Conference on Power Electronics, Drivers and Energy Systems, New-Delhi, pp. 1006–1011 (1996a)

Real-Time, Partial Realization Theory of Discrete-Time Non-Linear Systems. SICE Transaction 32(3), 345–354 (1996b) (in Japanese)
A relation between Discrete-Time Almost-Linear Systems and " So-called " Linear Systems. SICE Transaction 32(5), 782–784 (1996c) (in Japanese)
Partial realization and real-time partial realization theory of discrete-time almost linear systems. SICE Transaction 32(5), 653–662 (1996d)

Hasegawa, Y., Matsuo, T., Hirano, T.: On the Partial Realization Problem of discrete-time linear representation systems. SICE Transaction 18(2), 152–159 (1982) (in Japanese)

Hasegawa, Y., Niinomi, S., Matsuo, T.: Realization Theory of Continuous-Time Pseudo-Linear Systems. In: IEEE International Symposium on Circuits and Systems, USA, vol. 3, pp. 186–189 (1996)
Realization Theory of Continuous-Time Finite Dimensional Pseudo-Linear Systems. In: Recent Advances in Circuits and Systems, pp. 109–114. World Scientific, Singapore (1998)

Hasegawa, Y., Suzuki, T.: Realization Theory and Design of Digital Images. LNCIS, vol. 342. Springer, Heidelberg (2006)

Horn, R.A., Johnson, C.R.: Matrix Analysis. Cambridge University Press, Cambridge (1985); Topics in Matrix Analysis. Cambridge University Press, Cambridge (1991)

Isidori, A.: Direct construction of minimal bilinear realization from nonlinear input-output maps. IEEE Trans. on Autom. Contr. AC-18, 626–631 (1973)

Kaasshocek, M.A., van Schuppen, J.H., Ran, A.C.M. (eds.): Realization and modeling in system theory, Proc. of the Int. Symp. MTNS-1989, 1, Progress in systems and Control Theory. Birkhauser, Basel (1990)

Kalman, R.E.: On the general theory of control systems. In: Proc. 1st IFAC Congress, Moscow. Butterworths, London (1960)
Mathematical description of linear dynamical systems. SIAM J. Contr. 1, 152–192 (1963)
Algebraic structure of linear dynamical systems. I, The module of Σ. Proc. Nat. Acad. Sci. (USA) 54, 1503–1508 (1965)
Algebraic aspects of the theory of dynamical systems. In: Hale, J.K., Lasale, J.P. (eds.) Differential Equations and Dynamical systems, pp. 133–143. Academic Press, New York (1967); Lectures on controllability and observability, note for a course held at C.I.M.E., Bologna, Italy. Cremonese, Roma;
A theory for the identification of linear systems. In: Brezis, H., Giariet, P.G. (eds.) Proceedings, Colloque Lions (1989)
Nine Lectures on Identification in Swiss Federal Institute of Technology, Zurich, Switzerland (1997)

Kalman, R.E., Falb, P.L., Arbib, M.A.: Topics in mathematical system theory. McGraw-Hill, New York (1969)

Kung, S.Y., Lin, D.W.: Optimal Hankel-Norm Model Reductions: Multivariable Systems. IEEE Transaction Automatic Control AC-26(4) (August 1981)

Matsuo, T.: System theory of linear continuous-time systems using generalized function theory. Res. Rept. of Auto. Contr. Lab. 15 (1968) (in Japanese)
Mathematical theory of linear continuous-time systems. Res. Rept. of Auto. Contr. Lab. 16, 11–17 (1969)
On the realization theory linear infinite dimensional systems. Res. Rept. No. SC-75-9, The Institute of Electrical Engineer of Japan (1975) (in Japanese)
Foundations of mathematical system theory. Journal of SICE 16, 648–654 (1977)
On-line partial realization of a discrete-time non-linear black-box by linear repre-

sentation systems. In: Proc. of 18th Annual Allerton Con. on Com. Contr. and Computing. Monticello, Illinois (1980)

Realization theory of continuous-time dynamical systems. LNCIS, vol. 32. Springer, Heidelberg (1981)

Matsuo, T., Hasegawa, Y.: On linear representation systems and affine dynamical systems. Res. Rept. No. SC-77-27, The institute of Electrical Engineers of Japan (1977) (in Japanese)

Realization theory of discrete-time systems. SICE Transaction 152, 178–185 (1979)

Two-Dimensional Arrays and Finite-Dimensional Commutative Linear Representation Systems. Electronics and Communications in Japan 64-A, 5, 11–19 (1981)

Partial Realization & Real-Time, Partial Realization Theory of Discrete-Time Almost-Linear System. SICE Transaction 32(5), 653–662 (1996) (in Japanese)

Realization Theory of Discrete-Time Dynamical Systems. LNCIS, vol. 296. Springer, Heidelberg (2003)

Matsuo, T., Hasegawa, Y., Niinomi, S., Togo, M.: Foundations on the Realization Theory of Two-Dimensional Arrays. Electronics and Communications in Japan 64-A, 5, 1–10 (1981)

Matsuo, T., Hasegawa, Y., Okada, Y.: The partial Realization Theory of Finite Size Two-Dimensional Arrays. IECE Trans. 64-A, 10, 811–818 (1981) (in Japanese)

Matsuo, T., Niinomi, S.: The realization theory of continuous-time affine dynamical systems. SICE Trans. 17(1), 56–63 (1981) (in Japanese)

Mohler, R.R.: Nonlinear Systems, Applications to Bilinear Control, vol. II. Prentice-Hall, Englewood Cliffs (1991)

Niinomi, S., Hasegawa, Y.: Approximately Partial Realizations of Affine Dynamical Systems. In: International Conference Dedicated to the Centennial Anniversary of L. S. Pontryagin (2008)

Niinomi, S., Matsuo, T.: Relation between discrete-time linear representation systems and affine dynamical systems. SICE Trans. 17(3), 350–357 (1981) (in Japanese)

Netravali, A., Limb, J.O.: Picture Coding: A Review. Proceeding of IEEE 68(3), 366–406 (1980)

Pareigis, B.: Category and functors. Academic Press, New York (1970)

Scherzer, O., Grasmair, M., Grossauer, H., Haltmeier, M., Lenzen, F.: Variational Methods in Imaging. Springer, Heidelberg (2009)

Smale, S.: Differential dynamical systems. Bull. Amer. Soc. 73, 747–817 (1967)

Index

MIX
Papier aus verantwortungsvollen Quellen
Paper from responsible sources
FSC® C105338

If you have any concerns about our products, you can contact us on
ProductSafety@springernature.com

In case Publisher is established outside the EU, the EU authorized representative is:
Springer Nature Customer Service Center GmbH
Europaplatz 3, 69115 Heidelberg, Germany

Printed by Libri Plureos GmbH
in Hamburg, Germany